MODELLING AND MANAGING AIRPORT PERFORMANCE

Aerospace Series List

Modelling and Managing Airport Performance	Zografos	July 2013
Advanced Aircraft Design: Conceptual Design, Analysis and Optimization of Subsonic Civil Airplanes	Torenbeek	June 2013
Design and Analysis of Composite Structures: With applications to aerospace Structures, Second Edition	Kassapoglou	April 2013
Aircraft Systems Integration of Air-Launched Weapons	Rigby	April 2013
Design and Development of Aircraft Systems, 2nd Edition	Moir and Seabridge	November 2012
Understanding Aerodynamics: Arguing from the Real Physics	McLean	November 2012
Aircraft Design: A Systems Engineering Approach	Sadraey	October 2012
Introduction to UAV Systems 4e	Fahlstrom and Gleason	August 2012
Theory of Lift: Introductory Computational Aerodynamics with MATLAB and Octave	McBain	August 2012
Sense and Avoid in UAS: Research and Applications	Angelov	April 2012
Morphing Aerospace Vehicles and Structures	Valasek	April 2012
Gas Turbine Propulsion Systems	MacIsaac and Langton	July 2011
Basic Helicopter Aerodynamics, 3rd Edition	Seddon and Newman	July 2011
Advanced Control of Aircraft, Spacecraft and Rockets	Tewari	July 2011
Cooperative Path Planning of Unmanned Aerial Vehicles	Tsourdos et al.	November 2010
Principles of Flight for Pilots	Swatton	October 2010
Air Travel and Health: A Systems Perspective	Seabridge et al.	September 2010
Unmanned Aircraft Systems: UAVS Design, Development and Deployment	Austin	April 2010
Introduction to Antenna Placement & Installations	Macnamara	April 2010
Principles of Flight Simulation	Allerton	October 2009
Aircraft Fuel Systems	Langton et al.	May 2009
The Global Airline Industry	Belobaba	April 2009
Computational Modelling and Simulation of Aircraft and the Environment: Volume 1 – Platform Kinematics and Synthetic Environment	Diston	April 2009
Handbook of Space Technology	Ley, Wittmann Hallmann	April 2009
Aircraft Performance Theory and Practice for Pilots	Swatton	August 2008
Aircraft Systems, 3rd Edition	Moir & Seabridge	March 2008
Introduction to Aircraft Aeroelasticity And Loads	Wright & Cooper	December 2007
Stability and Control of Aircraft Systems	Langton	September 2006
Military Avionics Systems	Moir & Seabridge	February 2006
Design and Development of Aircraft Systems	Moir & Seabridge	June 2004
Aircraft Loading and Structural Layout	Howe	May 2004
Aircraft Display Systems	Jukes	December 2003
Civil Avionics Systems	Moir & Seabridge	December 2002

MODELLING AND MANAGING AIRPORT PERFORMANCE

Editors

Konstantinos G. Zografos

Department of Management Science,
Lancaster University Management School,
Lancaster University, UK

Giovanni Andreatta

University of Padova, Italy

Amedeo R. Odoni

Massachusetts Institute of Technology, USA

Library of Congress Cataloguing-in-Publication Data
Modelling and managing airport performance / [edited by] Konstantinos G. Zografos, Giovanni Andreatta,
Amedeo R. Odoni.
   1 online resource.
   Includes bibliographical references and index.
   Description based on print version record and CIP data provided by publisher; resource not viewed.
   ISBN 978-1-118-53547-9 (ePub) – ISBN 978-1-118-53585-1 (ePDF) – ISBN 978-1-118-53586-8
(MobiPocket) – ISBN 978-0-470-97418-6 (cloth)   1. Airports–Management.   2. Airports–Management–
Evaluation.   3. Airports–Management–Simulation methods.   I. Zografos, Kostas G.   II. Andreatta, G.
(Giovanni)   III. Odoni, Amedeo R.
   TL725.3.M2
   387.7'360684–dc23

2013015971

A catalogue record for this book is available from the British Library.

ISBN: 978-0-470-97418-6

Set in 10/12pt Times by SPi Publisher Services, Pondicherry, India
Printed and bound in Malaysia by Vivar Printing Sdn Bhd

1   2013

About the Editors

Konstantinos G. Zografos is Chair Professor at the Department of Management Science of Lancaster University Management School (LUMS). He served as Professor and for four years as Chairman at the Department of Management Science and Technology of the Athens University of Economics and Business (AUEB), where he founded and directed the Transportation Systems and Logistics (TRANSLOG) Laboratory. His professional expertise, research and teaching interests include applications of Operations Research and Information Systems in Transportation and Logistics with particular emphasis on: the optimization of transportation and logistical decisions, supply chain performance assessment, airport planning and operations, and project management. He has published more than 60 papers in refereed academic journals and edited volumes. He has served as a member of the editorial board of 5 academic journals and has acted as a reviewer for numerous academic journals, including major journals in the fields of transportation, logistics and management science. He has served as member of several Transportation Research Board (TRB) committees in the areas of airfield and airspace capacity and delay, airport terminals and ground access, freight transportation planning and logistics, hazardous materials transportation, and transportation network modeling. In addition, he has been a member of European Commission Committees for the development and management of European R&D Framework Programmes in the area of Transport. Currently, he is a member of the Scientific Committee of the Single European Sky ATM Research (SESAR) Joint Undertaking. He has received teaching and research awards including the Edelman Laureate Honorary Medal in 2008 for Achievement in Operations Research and the Management Sciences, awarded by the Institute of Operations Research and the Management Sciences (INFORMS), and the 2005 President's Medal Award by the British Operational Research Society. Professor Zografos has been involved as a principal investigator in more than 60 R&D projects funded by national and international organizations and companies in USA, Europe and Greece and he has acted as consultant to projects funded by governmental agencies, companies and international organizations including the European Commission, United Nations Economic Commission for Europe (UNECE) and EUROCONTROL.

Giovanni Andreatta is Full Professor of Operations Research at the School of Sciences of the University of Padova (Italy). He received a Laurea degree in mathematics from the same university and a Masters in Computer Science from the University of Bologna (Italy). He completed his academic formation at MIT (Cambridge, MA, USA) as Visiting Scholar for a couple of years (August 1979–July 1981). He has been teaching several courses, mostly within the domain of Operations Research, in Italy: University of Padova (School of Sciences, of Engineering, of Economy), University of Venice and Bicocca University in Milan. He has been teaching for a short period at MIT (Cambridge, USA) and in other foreign universities: University of Luanda (Angola), University of Mogadishu (Somaly), HIBA in Damascus (Syria). He has been actively engaged in research, both in Italy and abroad, in the field of operations research, with particular interest in the applications to air transportation problems and has been the Scientific Leader of the Padova team for several national and European research programs (TAPE, OPAL, THENA, SPADE, SPADE-2, AAS). He is an Associate Editor of the *Journal of Aerospace Operations* and a member of national (AIRO, UMI, etc.) and international scientific associations (ATCA, INFORMS, etc.). He has served as a member of the Scientific Program Committee of national (such as AIRO) and international conferences (e.g. Eurocontrol-FAA ATM Seminar, ICRAT and RIVF).

Amedeo R. Odoni is Professor of Aeronautics and Astronautics and Professor of Civil and Environmental Engineering at MIT. He specializes in the application of operations research methodologies to airport planning and operations, as well as to air traffic management. Among other positions, he has served as Co-Director of MIT's Operations Research Center (1985–1991), of the FAA's National Center of Excellence in Aviation Operations Research – NEXTOR (1996–2002), of the Global Airline Industry Center at MIT (1999–2009), and of the Future Urban Mobility research project of SMART, the Singapore-MIT Alliance for Research and Technology (2009–present). Dr Odoni is the author, co-author, or co-editor of nine books and more than 100 professional publications. He served as editor-in-chief of *Transportation Science* from 1985–1991, and is a current or past member of the editorial boards of many professional journals. He is an elected member of the US National Academy of Engineering, a Fellow of the Institute for Operations Research and Management Science (INFORMS) and the recipient of many awards for his teaching and research. He has supervised more than 50 PhD students, several of whom have won major prizes for their research and dissertations, and has been a member of the PhD thesis committees of more than 100 other students. He has served as consultant to national and international organizations, and to many of the busiest airports in the world on projects related to practically every aspect of airport planning, design and operations, as well as to air traffic management in terminal airspace.

Contents

List of Contributors

Giovanni Andreatta
Department of Mathematics, University of Padova, Italy

Michael O. Ball
Robert H. Smith School of Business and Institute for Systems Research, University
of Maryland, USA

Henk A.P. Blom
Aerospace Engineering Department, Delft University of Technology & Air Transport Safety
Institute, National Aerospace Laboratory NLR, The Netherlands

David K. Chin
Federal Aviation Administration, Office of Performance Analysis, USA

Anderson Ribeiro Correia
Department of Air Transportation, Aeronautics Institute of Technology, Brazil

Michel J.A. van Eenige
Environment and Policy Support Department, National Aerospace Laboratory (NLR),
The Netherlands

Philippe Enaud
Performance Review Unit, EUROCONTROL, Belgium

John Gulding
Federal Aviation Administration, Office of Performance Analysis, USA

Mark Hansen
National Center of Excellence for Aviation Operations Research, Department of Civil and
Environmental Engineering, University of California, Berkeley, USA

Holger P. Hegendoerfer
Performance Review Unit, EUROCONTROL, Belgium

Dennis Klingebiel
Physics Institute IIIA, RWTH Aachen University, Germany

David A. Knorr
Office of International Affairs, Federal Aviation Administration, USA

Daniel Kösters
Frankfurt Airport, FRA Vorfeldkontrolle GmbH, Fraport AG, Germany

Michael A. Madas
Transportation Systems and Logistics Laboratory (TRANSLOG), Department of
Management Science and Technology, Athens University of Economics and
Business, Greece

Richard F. Marchi
Senior Advisor, Policy and Regulatory Affairs, Airports Council International, North
America, USA

Alius J. Meilus
Federal Aviation Administration, Office of Performance Analysis, USA

Daniel Murphy
Federal Aviation Administration, Office of Performance Analysis, USA

Johannes Reichmuth
RWTH Aachen Institute of Transport Science VIA, and DLR Institute of Air Transport and
Airport Research, German Aerospace Center (DLR), RWTH Aachen University, Germany

Alfred Roelen
Air Transport Safety Institute, National Aerospace Laboratory (NLR), The Netherlands

Marc Rose
MCR LLC, USA

Megan S. Ryerson
Department of Civil and Environmental Engineering, University of Tennessee, USA

Prem Swaroop
Robert H. Smith School of Business and Institute for Systems Research, University of Maryland, USA

Prabhakar Thyagarajan
Federal Aviation Administration, Office of Performance Analysis, USA

S.C. Wirasinghe
Department of Civil Engineering, Schulich School of Engineering, University of Calgary, Canada

Konstantinos G. Zografos
Department of Management Science, Lancaster University Management School, Lancaster University, UK

Bo Zou
Civil and Materials Engineering, University of Illinois at Chicago, USA

Series Preface

The Wiley *Aerospace Series* recognizes that the aerospace industry is multi-faceted and multi-disciplinary, involving a wide range of professionals and stakeholders that include not only aeronautical engineers but manufacturers, operators, and policy-makers, in addition to academics and students. The goal of this series is to provide practical and topical information to the diverse set of people in this innovative and dynamic industry.

A systems approach to aerospace engineering is essential to a full understanding of the industry, from the conception, design, testing, and production through to the operation of aerospace vehicles. In this systems perspective, relevant topics include air transportation, airline industry performance and infrastructure operations.

In *Modelling and Managing Airport Performance*, the editors have compiled an impressive set of articles devoted to an important part of the aviation system – air transportation infrastructure. Specifically, this book provides a comprehensive treatment of methodologies for analyzing, forecasting and improving the performance of airports and air traffic flows. Given the limited capability to expand existing airport infrastructure in most countries, the use of advanced airport performance modelling is critical to understanding and ameliorating the impacts of the growing demand for air travel on airport congestion and delays.

The contributors to this book are global experts on air traffic and airport management, from well-known academics to internationally recognized professionals in consultancy, government agencies and airport authorities. Models for analyzing airport capacities, measuring and forecasting air traffic delays are described, and the impacts of airport operations on airline costs and environmental impacts are explored. The implications for improving airport safety and a discussion of air traffic demand management strategies and their implementation round out this collection of articles, which is a welcome addition to the *Aerospace Series*.

Peter Belobaba, Jonathan Cooper and
Allan Seabridge

Acknowledgements

We wish to acknowledge Mr Tom Carter, Project Editor at John Wiley & Sons, Ltd, who provided critical assistance in managing the complex process of coordinating and stream-lining the contributions of a diverse set of authors. The preparation of this book has been supported by the SPADE-2 Integrated R&D Project (Supporting Platform for Airport Decision-making and Efficiency analysis: Phase 2, 2006–2009) partly financed by the European Commission: Directorate General for Energy and Transport (DG TREN) within the Sixth Framework Programme. We would like to thank the SPADE-2 Project Officers, Mr Cesare Bernabei, Mrs Elizabeth Martin and Mr Hoang Vu Duc, for providing useful feedback throughout the project's duration. The preparation of the book has also been supported by the General Secretariat of Research and Technology of Greece under contract NC-1375-01 and by the Research Center of the Athens University of Economics and Business (AUEB-RC) through project EP-1809-01.

List of Abbreviations

AAR	Airport Acceptance Rate
AAS	Amsterdam Airport Schiphol
A-CDM	Airport Collaborative Decision-Making
ACI	Airports Council International
ACI Europe	Airport Council International Europe
ACM	Airport Capacity Management
AEDT	Aviation Environmental Design Tool
AFP	Airspace Flow Program
AHP	Analytical Hierarchy Process
AIA	Athens International Airport
AIR-21	Wendall H. Ford Aviation Investment and Reform Act
AIXM	Aeronautical Information Exchange Model
AMAN	Arrival Manager
AMS	Amsterdam-Schiphol Airport
ANS	Air Navigation Service
ANSP	Air Navigation Service Providers
APT	Airport Passenger Terminal
ARC	Airport Research Center
ARMT	Aviation Environmental Portfolio Management
ASA	Atlantic Southeast Airways
ASDE	Airport Surface Detection Equipment
ASDE-X	Airport Surface Detection Equipment, Model X
ASMA	Arrival Sequencing and Metering Area
A-SMGCS	Advanced Surface Movement Guidance and Control System
ASPM	Aviation System Performance Metrics (published by FAA)
ATA	Air Transport Assocation
ATC	Air Traffic Control
ATCSCC	Air Traffic Control System Command Centre
ATFM	Air Traffic Flow Management
ATH	Athens Airport
ATM	Air Traffic Management

ATO	Air Traffic Organization
BAA	British Airport Authority
BRU	Brussels Airport
CAA	Civil Aviation Authority
CAEP	Committee on Aviation Evironmental Protection
CANSO	Civil Air Navigation Service Organization
CDA	Continuous Descent Arrivals
CDG	Charles DeGaulle Airport (Paris)
CDM	Collaborative Decision Making
CENEL	California's Day-Night Sound Level measurement
CFMU	Central Flow Management Unit
CH_4	Methane
CO	Carbon Monoxide
CO_2	Carbon Dioxide
CO_2e	Carbon Dioxide Equivalents
CODA	Central Office for Delay Analysis
CPH	Copenhagen Airport
CTA	Controlled Time of Arrival
DbA	Decibal, A-weighted
DC	District of Columbia
DFS	Deutsche Flugsicherung GmbH (German ANSP)
DLTA	Difference from Long-Term Average
DNL	Day-Night Sound Level
DSS	Decision Support System
DUS	Düsseldorf Airport
ECU	Effective Curb Utilization
EDCT	Estimated Departure Clearance Times
EDMS	Emissions and Dispersion Modeling System
ERA	Environmental Protection Agency
ESRA	Eurocontrol Statistical Reference Area
EU	European Union
EU-27	27 member European Union
EU-ETS	European Union Emissions Trading Scheme
EUR	Europe
EUROCONTROL	European Organisation for the Safety of Air Navigation
EWR	Newark Airport (New York)
FAA	(U.S.) Federal Aviation Administration
FAA AEE	Federal Aviation Administration Office of Environment and Energy
FAQ	Frequently Asked Question
FMS	Flight Management Systems
FRA	Frankfurt Airport
GAO	Government Accountability Office
GDP	Ground Delay Program
GHG	Greenhouse Gas
GS	Ground Stop
GSE	Ground Service Equipment
GSRT	Hellenic General Secretariat of Research and Technology
GUI	Graphical User Interface
HAP	Hazardous Air Pollutants
HC	Hydrocarbons

HCAA	Hellenic Civil Aviation Authority
HDR	High Density Rule
IATA	International Air Transport Association
ICAO	International Civil Aviation Organization
IFR	Instrument Flight Rules
IMC	Instrument Meteorological Conditions (poor weather)
ISA	Innovation for Sustainable Aviation
JFK	John F. Kennedy Airport (New York)
KPA	Key Performance Area
LA_{eq}	United Kingdom's Day-Night Sound Level measurement
L_{day}	Level of sound integrated over a day
L_{den}	European Union's Day-Night Sound Level measurement
L_{eq}	Level of sound integrated over an hour
LGA	La Guardia Airport (New York)
LHR	London Heathrow Airport
L_{max}	Peak Sound Level
LMI	Logistics Management Institute
LOS	Level of Service
LTO	Landing-takeoff
MDI	Minimum Departure Interval
MIT	Miles in Trail
MOVES	Motor Vehicle Emission Simulator
MTOW	Maximum Take-Off Weight
MUC	Munich Airport
MVC	Model-View-Controller
N_2O	Nitrous Oxide
NAAQS	National Air Quality Standards
NAS	(U.S.) National Airspace System
NEXTGEN	Next Generation Air Traffic Control System (FAA)
NEXTOR	Center for Excellence in Operations Research
NLA	New Large Aircraft
NLR	National Aerospace Laboratory
NM	Nautical mile (1.852 km)
NO_x	Nitrogen Oxides
OAG	Official Airline Guide
OEP35	35 U.S. airports forming the "Operational Evolution Partnership"
OPAL	Optimisation Platform for Airports including Landside
PANYNJ	The Port Authority of New York and New Jersey
PAS	Proportionate Allocation Scheme
PASSUR	Passive Surveillance system
PM	Particular Matter
PRC	EUROCONTROL Performance Review Commission
PRU	EUROCONTROL Performance Review Unit
R&D	Research and Development
RAS	Reward-based Allocation Scheme
RMS	Root Mean Square
RNAV	Area Navigation capabilities
SAGE	System for assessing Aviations Global Emissions
SEL	Sound level of an event integrated over its duration
SES	Single European Sky (EU)

SESAR	Single European Sky ATM Research (Eurocontrol)
SID	Standard Instrument Departure
SIPs	State Implementation Plans
SO_x	Sulfur Oxides
SPADE	Supporting Platform for Airport Decision-making and Efficiency Analysis
STAR	Standard Terminal Arrival Route
SWMM	Storm Water Management Model
TAPE	Total Airport Performance and Evaluation
TBM	Time Based Metering
TM	TradeMark
TMA	Terminal manoeuvring area
TRB	Transportation Research Board
UC	Use Case
US	United States of America
VCE	Venice Airport
VMC	Visual Meteorological Conditions (good weather)
VOC	Volatile Organize Gasses
ZRH	Zurich Airport

Introduction

The increasing demand for air transport in conjunction with technical, physical and political constraints on providing capacity has resulted in a serious mismatch between demand and capacity. According to Eurocontrol, the planned capacity at the 138 Eurocontrol Statistical Reference Area (ESRA) airports is expected to increase by 41% in total by 2030, while the corresponding demand is foreseen to exceed airport capacity by as many as 2.3 million flights (or 11% of demand) in the most-likely growth forecast scenario for 2030 (Eurocontrol, 2008). Similarly, the FAA expects a quick resumption of US traffic growth, with traffic reaching 2007 levels by 2013, and growing by an additional 32% by 2025 (FAA, 2011).

The anticipated traffic volumes have to be accommodated by a system of airports with limited capacity, which in many cases has already been exceeded. Airports, as the terminal nodes of the air transport network, are the locations where delays generated and propagated throughout the network become most evident. At the same time, airports are also the most important 'triggers' of delay events, as a result of their often-reduced capacity due to poor weather or other problems. Direct consequences of airport congestion and delays include large external costs, poor level of service to the travelling public, inefficiency in airport operations, and negative impacts on the quality of the surrounding environment and the safety of the entire air transport system. Even during the current economic crisis, unconstrained demand (i.e. demand in the absence of slot controls) at several of the busiest European airports would have exceeded capacity for most of the day or, in a few cases, throughout the day. The percentage of departures delayed reached 37% (36% for arrivals), with an average delay per delayed flight for departures reaching 28 min (29 min for arrivals) in 2011 (Eurocontrol, 2012). The economic costs of these delays, operational inefficiencies and bottlenecks have been staggering. Ball et al. (2010) have estimated that the total economic impact of air transportation delays on the US economy amounted to $28.9 billion in 2007. Unavoidably, there has been increasing political pressure for improvements in airport performance through better and sustainable management of existing airport resources. But in order to improve performance, one should first be able to assess it. This has stimulated vigorous research efforts aimed at modelling all aspects of airport operations and evaluating quantitatively their impacts on delays and congestion, safety, the environment and the economy at large.

The assessment of airport performance is a complex task that requires a thorough understanding of the numerous aspects of airport operations and processes. By definition, a large variety of performance measures (e.g. capacity, delays, level of service, safety, security, emissions, noise, economic costs and benefits) should be considered along with their interdependencies and trade-offs. The airport decision making process is further complicated by the diversity of entities processed (passengers, baggage, cargo and aircraft) and the range of strategic, tactical, and operational considerations that need to be addressed throughout the airport, from ground access to the terminal airspace. Most importantly, these decisions should account for the often-conflicting needs and interests of the multiple stakeholders involved (civil aviation authorities, airlines, airport operators, passengers and shippers, airport neighbours, other government agencies). In such a multifaceted and complex environment, airport decision makers and planners must be supported by advanced airport modelling capabilities complemented by policies and strategies aimed at minimizing congestion and the externalities of airport operations.

The objective of this book is to provide an integrated view of state-of-the-art research on the performance of airport systems. Multiple facets of performance, such as capacity, delays, noise, emissions and safety, are considered. Furthermore, some of the chapters aim to go beyond modelling of operations, by shedding light on ways in which airport performance can be improved through policies for managing capacity allocation and congestion, as well as mitigating the externalities of noise, emissions, and safety/risks. Taken together, the chapters that are included herein have been selected with a view to (1) covering both landside and airside elements of the airport, (2) considering a broad spectrum of airport performance measures and (3) coupling reviews of modelling capabilities with the development of concepts and strategies for managing and improving airport performance.

The book consists of 10 chapters that are conceptually organized into three thematic sections. Section I (Chapters 1 and 2) focuses on the modelling and assessment of airport performance both on landside and airside. Section II (Chapters 3–7) deals with the quantification, measurement and forecast of the costs of airport delays, while further elaborating on the assessment of additional impacts (externalities) of airport operations. Finally, Section III (Chapters 8–10) covers topics related to the management of airport congestion and the efficiency of the scheduling and capacity allocation process on both sides of the North Atlantic.

In the first chapter of the book, Correia and Wirasinghe develop a methodology for evaluating the operational performance and quality of service offered by airport passenger terminals. They provide an overview of existing level of service (LOS) standards for airport passenger terminals along with their use for planning and operational management purposes. Furthermore, they recognize that LOS standards vary substantially with local passenger characteristics, and therefore employ a methodology for deriving quantitative standards by use of passenger surveys and observations. They initially use the psychometric scaling technique in order to develop quantitative LOS standards for individual components of the terminal. Then, they use the Analytical Hierarchy Process (AHP) in order to derive the importance weights that passengers assign to individual components and attributes of an airport passenger terminal. The application of the proposed methodology is demonstrated for São Paulo/Guarulhos International Airport, the busiest airport in Latin America. It is concluded that the proposed AHP methodology is appropriate for LOS modelling and can be further used to obtain a global LOS measure of an airport passenger terminal as a function of the LOS of its constituent components.

Chapter 2 by Zografos, Andreatta, van Eenige and Madas presents the development of an integrated Decision Support System (DSS) for total airport performance assessment. The presented DSS seamlessly integrates a variety of analytical models (e.g. MACAD, SLAM, INM, TRIPAC) and simulation tools (e.g. TAAM, RAMS Plus, SIMMOD, CAST) to capture the interdependencies among various performance measures (e.g. capacity, delay, noise, safety) and enable trade-off analyses at various levels of detail (e.g. strategic, tactical/operational) for the entire airport complex (both airside and landside) simultaneously. The chapter presents the architecture and operational concept of the DSS and demonstrates its capabilities through two pilot studies at Athens International Airport and Amsterdam Airport Schiphol. Based on the system's demonstration and validation process, the authors conclude that it fulfils the requirements of potential end users and produces useful results for airport decision making with an adequate approximation of reality. The proposed DSS offers some important benefits to the user/airport decision maker since it adopts a problem-oriented rather than tool-driven approach which shields the user from the technical complexity of the tools, thus enabling him/her to focus on the real decision making issues to be addressed.

In Chapter 3 (Section II), a joint FAA/Eurocontrol team of researchers (Gulding, Knorr, Rose, Enaud and Hegendoerfer) use an analysis of extensive data to compare and discuss airport-related flight delays and other measures of operational efficiency at the 34 busiest airports of the Unites States and of Europe. These airports handle more than two thirds of all passengers in each region every year. The authors initially define several different metrics and then use them to draw comparisons between US and European airports, placing their emphasis on air traffic delays that can be influenced by Air Traffic Management (ATM) actions. The chapter adopts two alternative reference points when it comes to measuring delays. One reference is the airline schedules: delays are measured relative to scheduled times, that is, they provide estimates of punctuality. This type of delay is of special interest to air travellers, but may not capture the full extent of delays because airlines tend to 'pad' their schedules in order to improve their on-time performance. For this reason, the second reference is the unimpeded travel times between each origin-destination pair: delays relative to unimpeded travel times provide a more complete picture of performance and are particularly useful in identifying operational bottlenecks and inefficiencies in the ATM system that may be hidden by the schedule padding practices of the airlines. For this second type of analysis, delays are attributed to specific flight segments (pre-take-off, airborne, post-landing) and compared with unimpeded times for each segment. The results may be very informative and valuable to Air Navigation Service Providers (ANSP), the operators of ATM systems. Furthermore, delay comparisons between US and European airports can be the basis for the identification of best practices and opportunities for focused improvements. As an example of the insights provided, the chapter concludes that reductions in taxiway delays in the US may be achievable through the use of European queue management practices, while Europe might benefit from US best practices for absorbing arrival delays through collaboration with the airlines in implementing air traffic flow management interventions.

Delays are also the main subject of Chapter 4 by Chin, Meilus, Murphy and Thyagarajan. This chapter presents a new initiative/programme led by the FAA's Air Traffic Organization (ATO) that forecasts delays at major (Core 30) US airports 6–12 months in advance. The starting points for these forecasts are the airlines' published schedules, as well as historical operational data. To increase the reliability of the exercise, airline schedules are supplemented by a short-term, economics-based forecast of demand for each one of the thirty airports, based on the FAA's Terminal Area Forecast (TAF) methodology. Demand forecasts, in the form of planned daily schedules of flights, are then combined with estimates of airport capacities to project delays six months into the future. Two alternative approaches are used for this purpose. The first is highly 'macroscopic' and based on a simple model of airport capacity (Annual Service Volume) along with parameterized relationships between annual operations and average delays for individual airports. The second approach uses a far more detailed network model to compute system-wide delays for each day of operations. The authors describe in detail both approaches as well as the extensive work that has been done in order to validate each by using historical data. At this point, the FAA is in a position to predict with a high level of confidence airport-specific and nationwide delays in the near future. In turn, the accurate projection of delays sufficiently far in advance enables the FAA not only to implement short-term mitigation measures, but also to provide early signals to policy makers regarding expected airport congestion. Affected groups (e.g. passengers) may also use this information to adjust their travel plans accordingly.

A comprehensive assessment of airport operational performance necessitates the quantification of the cost of delays. This topic is discussed in detail in Chapter 5 by Hansen and Zou. The chapter attempts to estimate the cost of air traffic delays to airlines in the United States. The authors take into account both the cost of delays relative to schedule and the cost of the additional time that airlines build into their schedules in anticipation of delay (schedule padding). They review critically and compare two fundamentally different approaches for estimating the impact of imperfect operational performance on airline costs. The first – the 'cost factor' method – is a 'bottom-up' approach that applies unit costs to different categories of delay (e.g. according to flight phase, aircraft type, delay duration, primary vs reactionary delay, etc.). The second – the 'total cost' method – is a 'top-down' approach based on aggregate costs. It posits the existence of a relationship between the total costs of an airline and the amount of delay (relative to schedule, as well as due to schedule padding) incurred by the flights of the airline. To derive this latter relationship, the authors develop an econometric model that investigates the statistical relationship between the various aspects of the operational performance of an airline and its costs. This econometric approach avoids some of the oversimplified assumptions and shortcomings of competing approaches. It is also rigorous, methodologically sophisticated, and probably more accurate. The econometric model is not, however, free of limitations with the major one being its intensive data requirements and its applicability at only a relatively aggregate level. The selection of the appropriate delay cost estimation approach should consider explicitly the scope/context of analysis and required level of detail. The authors argue that the existing estimates of the cost of air traffic delays in the US are somewhat re-assuring in that they are of roughly similar magnitude, despite the fundamental methodological differences between the various approaches utilized.

The next chapter, written by Hansen, Ryerson and Marchi, shifts the focus to externalities of airport operations other than delay-related measures. Chapter 6 discusses four key undesirable byproducts of airport and aviation operations: noise, water run-off, air pollutants, and Greenhouse Gas (GHG) emissions. It reviews these pollutants and their impact on airports and the surrounding environment. Furthermore, it discusses relevant mitigation policies and presents current methods/models for analysing and mitigating the GHG and noise impacts of airports, as well as the policy challenges faced in controlling these impacts. The chapter concludes with the development of an integrated modelling framework which combines environmental impact models with environmental policy impact models. Environmental impact models assess the level of emissions and resulting ecological and welfare impact from a given aviation system, while environmental policy impact models estimate how the system will evolve in response to a given environmental policy.

The analysis of the impacts of airport operations (Section II) concludes with Chapter 7 by Roelen and Blom dealing with aspects of safety performance. The chapter makes use of historical data of worldwide accidents of scheduled commercial flights by fixed wing aircraft with a maximum take-off weight of more than 5700 kg over the period 1990–2008 in order to analyse how safety performance has evolved for the ground segment of a flight relative to the airborne segment. It provides a systematic comparison of the evolution of accident rates over this period for Take-off, Landing and Ground Operations versus other accident categories such as Airborne and Weather. The comparative analysis reveals that the accident rate for Take-off, Landing and Ground Operations is not really improving, a fact that is in contrast with the overall improvements in the safety statistics of commercial aviation (decrease of safety risk per flight) during the period. A more detailed analysis of the accidents related to the ground segment of a flight at the level of the main accident categories (e.g. runway excursion/ incursion, abnormal runway contact, ground collision) shows that the non-decreasing accident rate applies to each of these main accident categories. Overall, the authors claim that the decrease of safety risk per flight during the last two decades has been caused predominantly from a reduction in airborne and aircraft related accidents rather than a reduction in accidents during take-off, landing and ground operations. The chapter concludes with a discussion of current as well as recommended safety initiatives (e.g. Flight Safety Foundation, Commercial Aviation Safety Team, FAA/Eurocontrol Action Plan 15) and driving mechanisms (e.g. technological developments, regulation, safety culture) addressing both the airborne and ground segments of a flight.

Section III starts with Chapter 8 by Klingebiel, Kösters and Reichmuth that deals with the efficiency of the slot scheduling and coordination process. The chapter analyses the airport's scheduling performance based on the 'scheduled delays' criterion expressing the difference between requested and allocated slot times during initial slot coordination. The authors introduce a deterministic modelling approach aiming to calculate scheduled delays depending on the level of slot utilization. The model uses declared capacity restrictions and initial slot requests as input parameters to provide a first estimate of the scheduled delays as a result of conflict resolution within the initial slot coordination. The impact on airport scheduling performance of both varying declared capacity restrictions and different slot demand and utilization patterns are demonstrated and further analysed with the use of the model. The validation of the model at five German coordinated airports produced results of sufficient accuracy for certain levels of slot utilization and led to recommendations for further research and calibration efforts towards incorporating crucial coordination parameters (e.g. priority classes per slot request, grandfather rights). The authors suggest that future research be focused on the estimation of minimum declared capacity values being capable of complying with desired scheduling performance indicators and the benchmarking of actual coordination results against ideal, optimization-based slot allocation outcomes.

The subsequent chapter by Madas and Zografos (Chapter 9) covers a wide range of airport demand management approaches and strategies aiming at mitigating airport congestion through the allocation of scarce capacity (slots) at European Union's (EU) airports. It provides initial evidence on the need for a new slot allocation regime (e.g. slot misuse, allocation inefficiencies, declared capacity considerations, barriers to new entrants, pricing effectiveness) and provides some quantitative insights into the potential impacts of a new congestion management regime. In particular, the authors demonstrate that the aggregate impact is significant in both operational and economic terms. By quantifying delays and their associated costs and comparing them with the existing landing fee scheme applied at European airports, they also show that the actual pricing system of scarce airport infrastructure is quite far from perfect, thus rendering some (mainly the largest and busiest) European airports extremely under-priced. Furthermore, the chapter presents a strategic policy framework aiming to provide guidance for the implementation of a new congestion-based pricing regime at different types of EU airports. The proposed framework addresses directly the airport congestion problem by means of varying congestion fees. It is simple and inexpensive to implement since it is directly compatible with the IATA schedule coordination approach currently in use worldwide, but can be also easily customized with local airport needs. The chapter concludes with a discussion of recent efforts and future research directions towards improving or complementing, rather than substituting, the existing slot allocation practice by means of optimization models aimed at controlling strategically the distribution of traffic at the airport level.

The concluding chapter (Chapter 10) of Section III by Ball, Hansen, Swaroop and Zou provides the US perspective and experiences with airport congestion management. This chapter provides a critical review of recent attempts in the United States to implement a slot control system that includes the use of auctions for allocating some of the slots. It also discusses institutional and political considerations that led to the existing congestion management regime in the United States. Furthermore, the chapter gives an overview of models that provide economic justification of slot controls and discusses a number of design issues/considerations such as the definition of appropriate slot levels, access to small communities, (re)distribution of slot revenues, and slot ownership. The authors also address the trade-off between queuing delay and schedule delay as a key determinant of the need for slot controls and the optimal slot control levels. They estimate the slopes of the two curves (queuing and schedule delay) for the largest 35 US airports using two alternative models that attempt to capture the reaction of airlines to reductions in available slots. Despite the substantial reluctance to implement slot controls in the United States, the analysis supports the hypothesis that many US airports are overscheduled and provides strong justification for setting slot limits to a number that is often less than the peak airport capacity.

References

Ball, M.O., Barnhart, C., Dresner, M., et al. (2010) *Total Delay Impact Study: A Comprehensive Assessment of the Costs and Impacts of Flight Delay in the United States.* NEXTOR Technical Report, October 2010.

Eurocontrol (2008) *Long-Term Forecast: IFR Flight Movements 2008–2030.* Forecast prepared as part of the Challenges of Growth 2008 project, Brussels, Belgium.

Eurocontrol Central Office for Delay Analysis (CODA) (2012) *CODA Digest: Delays to Air Transport in Europe (Annual 2011).* Report prepared by Eurocontrol's Central Office for Delay Analysis (CODA), issue published on 14 March, 2012, Brussels, Belgium.

Federal Aviation Administration (2011). *FAA Aerospace Forecast: Fiscal Years 2011–2031.* [Online] Available from: http://www.faa.gov/about/office_org/headquarters_offices/apl/aviation_forecasts/aerospace_forecasts/2011-2031/media/2011%20Forecast%20Doc.pdf [Accessed 9 December, 2012].

1

Modeling Airport Landside Performance

Anderson Ribeiro Correia[A] and S. C. Wirasinghe[B]

[A] *Department of Air Transportation, Aeronautics Institute of Technology, Brazil*
[B] *Schulich School of Engineering, University of Calgary, Canada*

1.1 Motivation for Level of Service Modeling

The motivation for developing level of service (LOS) measures is twofold. First, given that one of the goals of airport planning is to improve, or at least maintain, the level of service experienced by the airport user, it is necessary to be able to measure LOS in order to know whether this goal is being achieved. Second, airport passenger terminal improvements rarely are without expense. To know whether a particular expenditure is justified it is necessary to be able to measure the change in LOS resulting from it (Gosling, 1988).

Establishing measures to evaluate operational performance of the airport landside and quality of service is one of the major problems facing the airlines and airport operators. Humphreys and Francis (2000) affirm that LOS evaluation in American airports have been undertaken at individual airports, with no standard method or reporting system on a national scale. Research is also needed in developing countries, mainly to generate references for planning airport infrastructure. In this regard, Fernandes and Pacheco (2002) stress that the lack of studies, for example in Brazil, to enable parameters reflecting local conditions to be estimated means that estimates made on the basis of conditions at foreign airports are used without proper evaluation. According to them, the issues of domestic traffic, in particular, deserve special attention in terms of Brazilian specifics.

Airport landside LOS and capacity have been topics of research interest over the past two decades or so. More recently, owing to the critical nature of airport LOS issues, a number of studies have been initiated on the identification of the landside problem in general, and on

Modelling and Managing Airport Performance, First Edition. Edited by Konstantinos G. Zografos, Giovanni Andreatta and Amedeo R. Odoni.
© 2013 John Wiley & Sons, Ltd. Published 2013 by John Wiley & Sons, Ltd.

capacity and service measures in particular. Despite all the airport LOS studies developed in the last two decades, the subject is in a rudimentary state of development in comparison with the status of LOS analysis in highway engineering. In 1986 the FAA responded to concerns of inadequate understanding of landside capacity constraints by commissioning the Transportation Research Board to conduct a study of ways to measure airport landside capacity. This study (TRB, 1987) recognized that the capacity of any given landside facility cannot be evaluated without defining acceptable LOS standards, but that there is currently little agreement on how to do this. Lemer (1988) reviewed the study's principal findings and recommendations. He concluded that the effort represented a valuable first step toward definitive guidelines for capacity assessment, but that much remained to be done. Thus, the development of appropriate ways to measure airport landside LOS is a critical research need.

The measurement of LOS of airport passenger terminals is also an important issue considering the recent trends for airport privatization and the need for regulation of privatized facilities. There are concerns that efforts to regulate the prices charged by airports can result in under-investment and decline of service standards. In Australia for example, newly privatized airports would be subjected to price regulation in the form of price-caps on aeronautical charges; there is a concern about their effect on the incentive faced by the enterprise to downgrade quality (Forsyth, 1997). This makes it important to monitor not only the cost-efficiency and cost-effectiveness, but also the service effectiveness of airports (Hooper and Hensher, 1997). Gillen and Lall (1997) agree and state that while airports should be asked to adhere to private financial standards, they must also be judged in the context of their overall goals.

This chapter is composed of three main sections. The first section (*Airport Landside Components*) provides an overview of current LOS standards developed by researchers or industry organizations. Additionally, this section provides basic recommendations for the use of such standards at planning and management levels. The second section (*Methodology for Deriving Quantitative Standards for Individual Components*) presents a method, based on the psychometric scaling technique, that any airport authority or organization could apply to develop standards. Finally, the third section (*Degree of Importance of Landside Components and Attributes*) presents a method, based on the Analytical Hierarchy Process, aimed at obtaining the importance that passengers attribute to individual components and attributes. These measures could be used to obtain an overall terminal LOS evaluation.

1.2   Relationship between Measures of Capacity and Level of Service

LOS and capacity measures are intrinsically correlated. Generally, the efficiency of an airport landside component is evaluated by comparing its capacity values with standard measures of the LOS to passengers. In this concern, the IATA (1995), in its airport development reference manual, has proposed LOS standards in order to evaluate the capacity of airport landside components. Some of these LOS standards will be presented in the next sections. Another initiative to evaluate the relationship between LOS and capacity was developed by Brunetta et al. (1999); in that paper, the authors proposed a model, called SLAM (Simple Landside Aggregate Model), in order to measure capacity and to identify reference values for LOS standards.

Since LOS measures, as space standards, directly affect the capacity of any airport landside component, airport managers and planners should also give special attention to dwell time in these components. Among the several alternatives to improve the efficiency of airport operations and thus decreasing dwell time, include (de Neufville and Odoni, 2002): (1) using electronic check-in kiosks, (2) accelerating board of aircraft, and (3) speeding up passport control processing times.

1.3 Airport Landside Components

In this section, we present the level of service standards and general recommendations for the main domestic airport passenger terminal components: emplaning curbside, check-in counter, security screening, departure lounges and baggage claim.

1.3.1 Emplaning Curbside

1.3.1.1 Level of Service Standards

The function of the curbside is to provide an interface between ground access and the airport passenger terminal. It is the first airport component that most passengers pass through for their departing trip. For this reason, the LOS of this facility can influence the first impression passengers have regarding the whole airport. In addition to that, because the main function of a curbside is to transfer passengers from the ground transportation system to the terminal building, the entire ground/air linkage will be unbalanced if this area does not operate properly.

For many years, approaches to LOS and capacity problems at the curbside have dealt with vehicles rather than people (Siddiqui, 1994). Most of the previous studies dealt primarily with the length of curbside areas, not considering passengers' perceptions of other attributes that might influence the LOS evaluation of these facilities, including minimizing walking distances, reducing level changes, and providing space availability for circulation, weather protection, better visual information, lighting and aesthetics.

The traditional approach for curbside LOS evaluation is assessing the curbside utilization as proposed by Mandle et al. (1982), which adopted definitions of LOS for airport curbside planning and design, on the range from A to E, as follows:

- Level A: No traffic queues, no double parking.
- Level B: Effective curb utilization equal to 1.1 times actual curb frontage.
- Level C: Effective curb utilization equal to 1.3 times actual curb frontage.
- Level D: Effective curb utilization equal to 1.7 times actual curb frontage.
- Level E: Operational breakdowns, effective curb utilization equal to 2.0 times actual curb frontage.

The effective curb utilization is defined as the effective length of the area occupied by vehicles as the actual curb length may differ from effective curb length, due to double or triple parking or undesirable loading/unloading areas. In this viewpoint, the effective length of curb is directly related to the LOS provided at the curb (Figure 1.1).

According to the suggested LOS standards provided in Figure 1.1, the curbside will have an excellent (LOS A) evaluation if the utilization of the curbside is lower than 100% (ECU ≤ 1). On the other hand, it will provide a good level of service (LOS B – indicated for design of new airports) if the ECU is close to 1.1. This indicates that it is not necessary to provide space at the curbside for all vehicles, because during short peak periods, some double parking could be acceptable. However, it is necessary to verify that this double line of parked vehicles will not provide

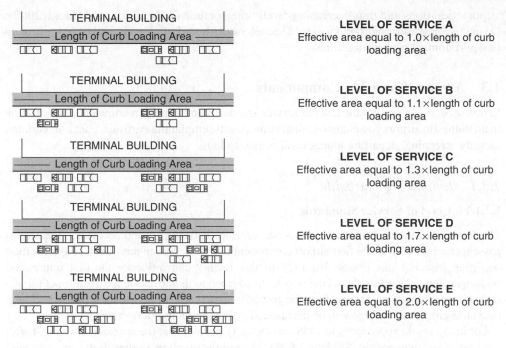

Figure 1.1   Airport curbside levels of service (Mandle et al., 1982)

additional disturbance to the road traffic, especially if there are not enough lanes. If that is the case, a more conservative provision of space for all vehicles (100% – ECU=1.0) would be more appropriate.

1.3.1.2   General Recommendations

Two things can be done to reduce the ECU, (1) increasing the length of the curbside or (2) decreasing the virtual length of cars. The first action is usually adopted at the planning level of a terminal. The length of the curbside required for unloading passengers and baggage is determined by the type and volume of ground vehicle traffic anticipated in the peak period of the design day. The curbside geometry is usually conformed to the geometry of the terminal; any further improvements to this area might not be practical. On the other hand, the virtual (total) length of vehicles can be decreased by reducing the demand on the curbside or reducing the interval that vehicles stay at the curbside. The virtual (total) length of vehicles can also be decreased by reducing their size (e.g., segregating the buses in a special remote area).

Several airports provide free short-term parking. In most of the cases, this time is usually enough to drop off or pick up a passenger. Consequently, the curbside is rarely operated over its capacity. Adding new transit alternatives to the airport can alleviate the demand for the use of cars at the curbside. Some airports are also moving in this direction to alleviate the air pollution in its vicinity. However, the transit alternatives are usually very expensive and just a few airports have enough traffic to afford them.

The most economical way to improve the efficiency of the curbside is by reducing the time vehicles stay at the curbside. For instance, if it were possible to reduce the waiting time at the curbside by 50%, this would result in a 100% increase of the ECU. One way to reduce the time vehicles spend at the curbside is to provide a good orientation system so that drivers will spend a minimum of time looking for space at the curbside.

Enforcement has been applied at many airports to make sure cars do not wait too long or Double Park at the curbside. There is also an anecdotal example from Seattle Airport; the management had a tow truck parked at the curbside on a permanent basis. That procedure motivated the curbside users to stay a shorter time than usual, in the face of the practical possibility of having the car towed away (Correia, 2009).

1.3.2 Check-in Counter

1.3.2.1 Level of Service Standards

The check-in counter is the most studied airport passenger terminal component. It is usually the first processing component in the terminal, through which passengers pass during the enplaning trip. In this facility, passengers can get their seats assigned, baggage checked, and receive a boarding pass which includes the gate number. However, it is noted that passengers can bypass the check-in at many modern airports if they have no checked baggage and have printed their own boarding pass at home.

The level of service provided at the check-in counter reflects both the airport and airline images. In addition to that, because it is one of the first components in the passenger's pathway, it can cause delay to other activities and flights. Not only can a poor level of service cause operational problems to airlines and airport administration, it can add to passenger stress, when they are trying to get to the airplane as soon as possible.

One of the first approaches to evaluate check-in LOS, developed by Mumayiz (1985), defined three LOS according to passenger perception of delay. The levels for check-in subsystems for scheduled long-haul flights, for example, are defined as:

- Level A (good): T < 15 min;
- Level B (tolerable): 15 min < T < 25 min;
- Level C (bad): T > 25 min.

where T is the time spent at check-in (including waiting).

Table 1.1 represents LOS standards obtained from a study of user perceptions at two airports (São Paulo International Airport and Calgary International Airport). The range of waiting time

Table 1.1 Suggested check-in counter LOS standards (Correia and Wirasinghe, 2007)

LOS	Waiting Time (min)	
	São Paulo	*Calgary*
A	<1	<7
B	1–17	7–18
C	17–34	18–26
D	34–58	26–34
E	>58	>34

Table 1.2 Suggested check-in counter LOS
standards (Correia and Wirasinghe, 2007)

LOS	Processing Time (min)	
	São Paulo	*Calgary*
A	<1	<5
B	1–14	5–17
C	14–20	17–19
D	20–25	19–20
E	>25	>20

Table 1.3 Proposed check-in counter LOS standards

Level of Service Standards (m²/Occupant)					
A	B	C	D	E	F
1.8	1.6	1.4	1.2	1.0	Breakdown

Source: IATA (1995)

values from LOS A to LOS E at São Paulo is greater than the range at Calgary. That might be due to the fact that in São Paulo there is more variability of flight types than in Calgary. Most of the flights from Calgary have destinations in North America with a few to Europe, as opposed to flights from São Paulo that have domestic destinations, as well as inter-continental flights to Europe, Africa, Asia, North America and Australia. It is also clear that passengers in South America may have very different perceptions about time spent at check-in in comparison with North American passengers. These differences might indicate that LOS standards might differ depending on the flight type (international/intercontinental or domestic/transborder) and the country.

Using the same approach, Table 1.2 presents suggested LOS standards for processing times at check-in counters.

Besides considering waiting and processing times, another approach to evaluate check-in LOS is considering space standards, which is the approach proposed by the International Air Transport Association. It provides LOS standards for space available at check-in lines, as follows (Table 1.3).

Past research on LOS for check-in facilities has been concentrated mainly on service times and space available for passengers. A survey with members of the Airports Council International (ACI, 2000) showed that this practice has been widely used. Among the airports surveyed, the main objective criteria used for check-in facilities are the check-in waiting time/queue and check-in transaction time. Nonetheless, Martel and Seneviratne (1990) indicate that, for instance, aesthetics is an important attribute of check-in facilities. An objective measurement of the aesthetics variable is difficult to undertake. Obviously any passenger could provide his/her perception about the terminal aesthetics, expressing it by a linguistic variable as good or excellent; however, it is difficult to propose any performance variable relative to aesthetics, which could be correlated to the aesthetics LOS passengers ratings. Anyway, it is important that airport planners and managers consider subjective aspects when implementing or managing check-in facilities.

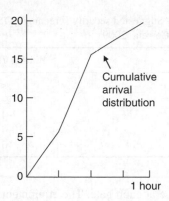

Figure 1.2 Cumulative arrival distribution (TRB, 1987)

1.3.2.2 General Recommendations

An airport will offer a good level of service at the check-in counter when the service is fast and reasonable space is available for passengers. These two characteristics are intrinsically dependent on the number and type of check-in counter desks and automated checking machines and upon the peak demand.

At the planning level, the number of check-in counter desks required is determined as a function of the peak demand and the waiting time acceptable to the passengers. The demand rate at the check-in counter is clearly not uniform.

According to Figure 1.2, during the first 15 min, 6 passengers arrive at a counter at fairly uniform intervals of 2.5 min. The arrival rate then increases, so that by the end of the first half-hour, 10 more people have arrived, for a total of 16. All peak-hour passengers, a total of 20 passengers, have arrived by the end of 55 min. No passenger arrivals are expected during the final 5 min of the peak hour. As it can be noticed, the arrival of passengers is not uniform. This characteristic influences the design and management of check-in counter facilities. In addition to these cumulative arrival distribution curves, there has been a tendency to apply micro-simulation models to the check-in planning, design and management.

In some markets, a considerable number of passengers may be pre-ticketed, and a higher percentage of automated check-in positions may be warranted either within the terminal building or at the curb front. Self-service kiosks are effective at reducing check-in lines; they can also be equipped with a combination of biometrics to perform identity checks for automated "fast-track" immigration control, self-service check-in or even at security checkpoints (de Barros, 2001). For instance, the adoption of common-use self-service kiosks at Sydney and Melbourne Airports reduced the processing time for check-in to less than one minute (Correia, 2009). Shorter processing times result in shorter waiting times, which mean less people in the queues, reducing the need for space provision. The adoption of common use self-service and/or remote kiosks should be adopted in airport passenger terminals that are faced with long check-in lines.

1.3.3 Security Screening

All airline passengers are required to pass through a security screening to ensure in-flight safety. This procedure is required prior to departure both for passengers and for hand luggage before passengers enter the departure lounge. In some airports it is done at the entrance to

Table 1.4   Suggested security screening LOS standards (Correia, 2009)

LOS	Waiting Time (min)
A	<2
B	2–7
C	7–10
D	10–12
E	>12

a concourse. In others, it is done at each gate. The equipment used for this process is X-ray screening and magnetometers, among others. Full body scanners are also rapidly coming into use at international airports for the purpose of secondary screening.

Several factors influence the level of service of passenger security areas, including the number of channels, space availability, type of equipment and the courtesy extended by staff. One of the main attributes used to evaluate the level of service of these areas is the waiting time for each passenger. LOS standards for waiting times at the security screening are demonstrated in Table 1.4, based on a user survey at São Paulo International Airport (Correia, 2009). Accordingly, passengers will rate the security screening component as good if the waiting time falls between 2–7 min.

It is important to notice that longer waiting times could reduce the level of passenger satisfaction as well as cause departing and connecting passengers to miss flights. In this case, the capacity of these facilities should be increased by providing more resources (more equipment, employees, and channels).

1.3.4   Departure Lounge

1.3.4.1   Level of Service Standards

The main purpose of the departure lounge is to assemble passengers that are waiting to board a flight. People accompanying passengers are not allowed into this area. It is also usual to have separate groups of lounges to separate international from domestic passengers. There are airports where airlines have exclusive departure lounges; in other cases they have to share.

Usually the airlines stipulate the time before departure when passengers are supposed to arrive at the departure lounge. Of course there are some passengers who arrive later, but they take the risk of missing the flight. There are others who prefer to enter the lounge in advance just so sit and relax before the flight. As we can see, the airline procedures and human behavior are very important factors when planning the space and facilities to be available in the departure lounge. Wirasinghe and Shehata (1988) have proposed a procedure to plan single gate as well as multi gate departure lounges, while de Barros and Wirasinghe (1998) have analyzed departure lounges for the new large aircraft (NLA).

The LOS evaluation of departure lounges is very particular and subjective, because we have to deal with parameters such as comfort, convenience and aesthetics. In the processing components, there are not many things the airport and airlines can do to improve the quality the passenger experience while he/she waits. Conversely, in holding components, many factors can influence the passenger perception of LOS, for example the availability of seats, concessions in the area and availability of TV. A survey conducted by Seneviratne and Martel (1991) showed that the availability of seats is the most significant performance indicator of waiting areas in airport passenger terminals. The other relevant factors are:

- good seating arrangements;
- space available for circulation;
- lighting;
- comfort;
- proximity of concessions and amenities;
- aesthetics.

The number of seats is an important attribute for departure lounges. In this concern, Wirasinghe and Shehata (1988) defined and obtained an equation for the optimal number of seats (N_o), in a lounge with a security check, as that which minimizes the sum of the cost of the lounge (proportional to its area), the cost of seats (including the extra space required) and a penalty for compulsory standing time spent by passengers:

$$No = Q - r(1 - r/b)(\gamma_c / \gamma_p) \qquad (1.1)$$

where,
Q=maximum passenger accumulation in the lounge, based on cumulative arrival and departure curves;
r=rate of output of passengers from the security check to the lounge (passengers per hour);
b=rate of output of passengers from the lounge to the aircraft after boarding has commenced (passengers per hour);
γ_p=cost of the penalty for each unit of compulsory standing time per passenger prior to boarding ($ per hour per passenger); and

$$\gamma_c = \alpha \, \gamma_L (m_1 - m_2) + \gamma_s \qquad (1.2)$$

where,
α=one plus the fractional increase in lounge area to allow for passenger circulation and airline activities;
γ_L=cost of the lounge (discounted construction cost plus maintenance cost) in $ per square meter per aircraft departure;
m_1, m_2=lounge area (in square meters) allocated per passenger per seat, and per passenger per standing space, respectively; and
γ_s=cost in $ of a seat (chair) per aircraft departure.

Using the aforementioned concept, Seneviratne and Martel (1994) defined a seating availability index for the evaluation of LOS in departure lounges, as follows:

$$PI_{as} = \frac{N_a}{N_o} \qquad (1.3)$$

Table 1.5 Proposed hold room LOS standards

Level of Service Standards (m²/Occupant)					
A	B	C	D	E	F
1.4	1.2	1.0	0.8	0.6	Breakdown

Source: IATA (1995)

where,

N_a = number of available seats in area considered at a given time;

N_o = optimal number of seats;

PI_{as} = performance index for availability of seats.

Thus, LOS in relation to availability of seats was defined as:

- Level A: $PI_{as} \geq 1.0$.
- Level B: $0.9 > PI_{as} > 0.7$.
- Level C: $0.6 > PI_{as} > 0.4$.
- Level D: $0.3 > PI_{as} > 0.2$.
- Level E: $0.2 > PI_{as} > 0.1$.
- Level F: $PI_{as} < 0.1$.

Besides the various attributes that have been previously presented, the current approach to evaluate departure lounge LOS is proposed by the International Air Transport Association, as shown in Table 1.5.

1.3.4.2 General Recommendations

The sizing of the departure lounge must be planned as a function of the number of passengers during the peak period of the design day and the crowding level acceptable by the passengers. The number of passengers in the lounge depends on the number of flights during the peak period, flight passenger load, and the time in advance that passengers arrive in the lounge. This can be determined with the help of cumulative arrival curves and micro-simulation models.

Provision of shared use lounges is recommended (de Neufville and Odoni, 2002) to cope with the fluctuating demand of passengers. By using this procedure, the overall needs for space are greatly reduced. It is not recommended to aggregate international and domestic passenger in the same room; however, it is possible to implement reversible rooms that can be used by domestic passengers or international passengers. Such rooms exist for example at São Paulo/Guarulhos International Airport.

Seating should be provided in the lounges; however this need is inversely related to the number of amenities inside the lounge (food areas, shops). If there are no such facilities, passengers have no option but to sit and wait for the departing flight.

1.3.5 Baggage Claim

There is not too much information available regarding baggage claim level of service, since the emphasis of most studies is concentrated on departing passengers.

In air transport it is customary to separate passengers from their baggage during the line haul portion of the trip. This adds substantially to the complexity of handling the air trip and seriously

Table 1.6 Demand and operating factors influencing service level and capacity of the baggage claim area

Factor	Description
Equipment configuration and claim area	Type, layout, feed mechanism, and rate of baggage display; space available for waiting passengers; relation of wait area to display frontage; access to and amount of feed belt available
Staffing practices	Availability of porters (sometimes called "sky caps") and inspection of baggage at exit; rate of baggage loading/unloading from cart to feed belt
Baggage load	Numbers of bags per passengers, fraction of passengers with baggage, time of baggage arrival from aircraft
Passenger characteristics	Rate of arrival from gate, ability to handle luggage, use of carts, number of visitors

Source: TRB (1987)

complicates the design of passenger terminals, since it is essential that the separation and reuniting of passenger and baggage be carried out with maximum efficiency and at an extremely high level of reliability (Ashford and Wright, 1992). That separation imposes a challenge to airport operators, since unloading of passengers is usually done faster than unloading of baggage.

The baggage claim lobby should be located so that checked baggage may be returned to terminating passengers in reasonable proximity to the terminal deplaning curb. Most airports have installed mechanical delivery and electronic display equipment. The number of claim devices required is determined by the number and type of aircraft that will arrive during the peak hour, the time distribution of these arrivals, the number of terminating passengers, the amount of baggage checked on these flights, and the mechanism used to transport baggage from aircraft to the claim area (Horonjeff and McKelvey, 1994). The distance from baggage claim lobby to terminal curb has been for many years one of the main quality indicators of baggage claim facilities. That is a very important issue, since arriving passengers are tired, and transporting bags will be a cumbersome task for them. Not only must the distance be short, but also enough orientation must be provided, since many passengers may not be familiar with the terminal.

The inbound baggage area is composed of the following spaces (Hart, 1985):

1. Baggage off-load area, consisting of a cart drive aligned with the section of the claim device assigned for off-loading with 3 ft (0.9 m) of space between the cart and the device for handling off-loading, which can be direct feed or remote feed. Because this area is not open to passengers, it does not directly affect the LOS.
2. The section of the claim device located in the public area.
3. Claim area around the device.
4. Public milling and waiting area.
5. Concessions, mainly car rental counters, service counters, courtesy phones, public telephones.

The principal demand and operating factors influencing service level and capacity in the baggage claim are summarized in Table 1.6.

Table 1.7   Suggested LOS standards: processing time, Calgary

LOS	Processing Time at Baggage Claim (min)
A	<1
B	1–14
C	14–20
D	20–26
E	>26

Source: Correia and Wirasinghe (2010)

Research on baggage claim LOS is a critical need. According to a study developed by the transportation research board (TRB, 1987), additional data are needed on characteristics of bags, passengers, and equipment, as well as on airline and airport procedures. Because of the importance of the baggage claim to the passenger's overall perception of an airport and an airline, research into what levels of delay passengers may tolerate and under what conditions is also needed. These data would be valuable as well in mathematical modeling of baggage handling operations, a necessary tool for exploring consequences of new larger aircraft and changes in flight schedules (de Barros and Wirasinghe, 2004).

Martel and Seneviratne (1990) suggest the following factors as they might influence quality of service of baggage claim facilities:

- Processing time;
- service variability range;
- area size;
- pedestrian density;
- claim frontage;
- care of handling;
- aids to handicapped;
- proximity to curb.

Although this list provides a summary of factors to be considered when evaluating baggage claim LOS, some of them are very subjective making their measurement a difficult task. For instance, care of handling is expected to be very important for passengers, but its measurement is difficult. The same can be said about aids to handicapped travelers.

de Neufville and Odoni (2000) suggest the most important factor for baggage claim areas is claim frontage. According to them baggage claim areas must first provide enough claim presentation length, that is, length along the conveyor belt or race tracks, for people to identify and pick up their bags. The IATA standards recommend about 70 m for wide body, and about 40 m for narrow-body aircraft being served at the same time. This standard implies about 0.3 m of claim presentation per passenger. The FAA alternatively defines the length required in terms of the number of aircraft arriving in the peak 20 min, and assumes that passengers check 1.3 bags per person. Either standard leads to approximately the same results. However, these standards should be modified according to local realities such as the average number of bags checked, and the possibility that passengers are passing previously through another queue such as immigration clearance. In a study conducted at the Calgary International Airport, Correia and Wirasinghe (2010) proposed the following standards for baggage claim facilities (see Table 1.7).

According to the suggested LOS standards, it is recommended that a processing time that is lower than 14 min be targeted to obtain a good (B) level of service evaluation. Processing times longer than 20 minutes will provide a poor (D) or unacceptable (E) level of service evaluation.

1.4   Methodology for Deriving Quantitative Standards for Individual Components

1.4.1   Introduction

The previous sections presented LOS standards proposed during the last decades. Airport planners and managers could employ them during the phases of planning, design and operations. However, considering that airport and customer characteristics differ from one another, eventually airport authorities should be able to implement their own standards. This section presents a methodology that is able to derive quantitative LOS standards, based on user perceptions, for a given airport or group of airports. The methodology presented herein is based directly on the psychometric scaling technique developed by Bock and Jones (1968) and further applied by Muller (1987), Ndoh and Ashford (1993) and Correia (2009).

Psychometrics and psychological scaling theory have given extensive consideration to the behavior of subjects, sampled from a specific population, in choosing among alternatives (Bock and Jones, 1968). These ideas can be applied to passenger level of service evaluation of an airport terminal by considering passengers as subject to the experience of being processed at the terminal during the transition between their access and egress mode (whether by ground or air), and then being asked to choose a rating for the quality of that experience (Muller and Gosling, 1991). Most of the studies on this subject are developed from the work of Thurstone (1959). He introduced the fundamental concept of a sensory continuum, which remains an essential part of current psychological theory.

There are many methods available based on psychometric scaling theory. We could divide them into two categories. There are the methods where judges assess a stimulus directly in terms of other objects in which category is included the constant, paired comparisons and rank order methods. The other category, successive-categories judgments, however, depends upon passenger evaluations of the stimulus as a function of rating categories. For the purpose of measuring terminal LOS, it is supposed that the passenger will experience a stimulus only once during his/her trip experience, which is being measured; in this case constant, paired comparisons and rank order methods are not useful for measuring performance variables LOS of different terminal components. Considering this, the successive categories method will be employed, since it is the most suitable for measuring airport passenger terminal LOS. The method has been mathematically developed by Bock and Jones (1968), as presented next.

1.4.2   The Method of Successive Categories

The methodology for obtaining LOS quantitative values will be illustrated with a practical example consisting of a survey applied to 119 passengers at the check-in counter at São Paulo/Guarulhos International Airport. They were asked to rate the experience at the check-in into five ordered level of service categories. In general these categories will be defined by k, which

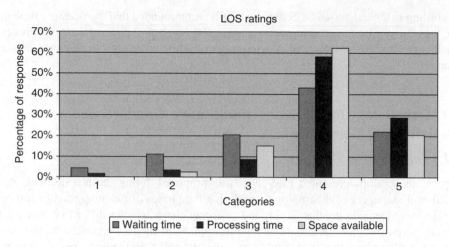

Figure 1.3 Distribution of responses across categories

Table 1.8 Distribution of responses as a function of check-in waiting time, São Paulo/Guarulhos
International Airport

Group	WT Range (min)	Average WT (min)	(1) Unacceptable	(2) Poor	(3) Fair	(4) Good	(5) Excellent	Total
1	WT=0	0.0	0	0	1	8	7	16
2	WT=1	1.0	0	0	1	4	4	9
3	WT=2	2.1	0	0	1	0	4	5
4	WT=3	3.0	0	0	2	5	6	13
5	WT=4	4.0	0	0	2	3	1	6
6	WT=5	5.0	0	0	0	4	1	5
7	5<WT≤10	7.9	0	1	2	11	0	14
8	10<WT≤15	13.4	0	1	5	6	2	14
9	15<WT≤25	20.4	0	1	4	9	1	15
10	25<WT≤35	33.4	0	5	5	1	0	11
11	35<WT≤55	49.1	2	4	1	0	0	7
12	55<WT≤75	68.8	3	1	0	0	0	4

is described as follows: unacceptable ($k=1$), poor ($k=2$), fair ($k=3$), good ($k=4$) or excellent ($k=5$). The results of the survey are presented in Figure 1.3, where the percentage of passengers indicating the waiting time, processing time and space available to be unacceptable, poor, regular, good or excellent are shown. Particularly, for illustration of the methodology, the waiting time will be analyzed in detail. Table 1.8 illustrates the distribution of responses as a function of the waiting time experienced by passengers.

For the responses presented in Table 1.8, waiting times (WT) were measured for each passenger, prior to the interview. To facilitate the calculation, the 119 observed passengers have been separated into 12 groups of similar waiting times. It is possible to obtain the proportion of responses where the waiting time is assigned at or below category k. Let us denote these proportions of responses as p_{jk}, where j represents the group number, and k represents the category. Table 1.9 presents the proportions for the surveyed passengers at São Paulo/Guarulhos International Airport.

Table 1.9 Proportions (p_{jk}) of responses in group (row) j at or below category (column) k

Group	WT (min)	k – category				
		1 – Unac.	2 – Poor	3 – Fair	4 – Good	5 – Exc.
1	0.0	0.000	0.000	0.063	0.563	1.000
2	1.0	0.000	0.000	0.111	0.556	1.000
3	2.1	0.000	0.000	0.200	0.200	1.000
4	3.0	0.000	0.000	0.154	0.538	1.000
5	4.0	0.000	0.000	0.333	0.833	1.000
6	5.0	0.000	0.000	0.000	0.800	1.000
7	7.9	0.000	0.071	0.214	1.000	1.000
8	13.4	0.000	0.071	0.429	0.857	1.000
9	20.4	0.000	0.067	0.333	0.933	1.000
10	33.4	0.000	0.455	0.909	1.000	1.000
11	49.1	0.286	0.857	1.000	1.000	1.000
12	68.8	0.750	1.000	1.000	1.000	1.000

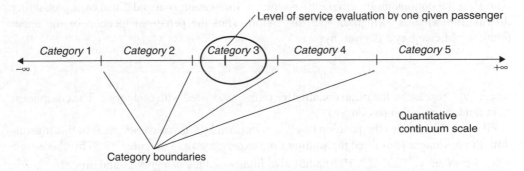

Figure 1.4 Illustration of the quantitative continuum scale

The proportion of responses represents a simplified LOS measure. It indicates the level of user satisfaction in each group. In group 7, only 21.4% of passengers rate the waiting time (7.9 min) as fair, poor or unacceptable; the great majority (78.6%) rate the waiting time as good or excellent. This LOS measure might be used by the management of an airport to assess the level of user satisfaction; however, it is not useful to precisely determine a quantitative LOS measure.

Let us define a level of service quantitative continuum ranging from $-\infty$ to $+\infty$. Values on the far negative side represent a "bad" level of service. Values on the far positive side represent "good" level of service. Zero represents a neutral position. Suppose this continuum can be divided into five regions, which represent each individual level of service category (Figure 1.4).

Each category has a lower and an upper boundary. In Figure 1.4 for instance, a given passenger has evaluated the level of service of a facility between the lower and upper boundaries of category 3. In the following paragraphs, only the upper boundary of each category will be considered as far as it concerns the methodology development.

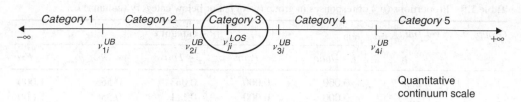

Quantitative
continuum scale

Figure 1.5 Location of category upper boundaries and LOS rating for passenger i

Suppose it is possible to obtain a quantitative LOS rating for the waiting time experienced. Consider that this rating v_{ji}^{LOS} can be defined as follows for a given passenger i:

$$v_{ji}^{LOS} = \mu_j^{LOS} + \varepsilon_{ji} \qquad (1.4)$$

where μ_j^{LOS} represents the mean LOS rating common to all passengers in group j, and ε_{ji} represents a deviation of a quantitative rating associated with a randomly selected passenger i in group j.

The position of a given category boundary is also assumed to be perceived at different points on the continuum by different passengers. Its location is also defined by a probability distribution with its own mean and dispersion. Thus the perceived location of the upper boundary of category k is given by:

$$v_{ki}^{UB} = \mu_k^{UB} + \varepsilon_{ki} \qquad (1.5)$$

where μ_k^{UB} represents the mean quantitative rating associated with category k. The component ε_{ki} is random based on passenger i.

Figure 1.5 illustrates the position of v_{ji}^{LOS}, as defined by a given passenger i. In this illustration, the passenger i has rated the waiting time experienced as fair (category 3) by choosing a value between v_{2i}^{UB} and v_{3i}^{UB}. He/she has also interpreted the category boundaries at v_{1i}^{UB}, v_{2i}^{UB}, v_{3i}^{UB} and v_{4i}^{UB}, as shown. It is worth noting that the upper boundary of category 5 is $+\infty$.

We assume the joint distribution of ε_{ji} and ε_{ki} to be bivariate normal, with means of zero, variances δ_j^2 and γ_k^2, and inter-correlation zero. In the absence of information to the contrary, it is usual to consider the variance γ_k^2 to be constant across all categories k; so we will assume that $\gamma_k^2 = \gamma^2$ for all k. Figure 1.6 illustrates the assumptions of normality for the distributions of v_{ji}^{LOS} and v_{ki}^{UB}.

The response of passenger i is assumed to be determined as follows. WTj (waiting time for passengers in group j) will be rated at or below point k for passenger i if:

$$v_{jki}^{\Delta} = v_{ji}^{LOS} - v_{ji}^{UB} = \mu_j^{LOS} - \mu_k^{UB} + \varepsilon_{ji} - \varepsilon_{ki} \leq 0 \qquad (1.6)$$

Clearly, v_{jki}^{Δ} is normally distributed, with mean

$$\mu\left(v_{jk}^{\Delta}\right) = \mu_j^{LOS} - \mu_k^{UB}, \qquad (1.7)$$

and variance

$$v\left(v_{jk}^{\Delta}\right) = \delta_j^2 + \gamma^2 = \sigma_j^2 \qquad (1.8)$$

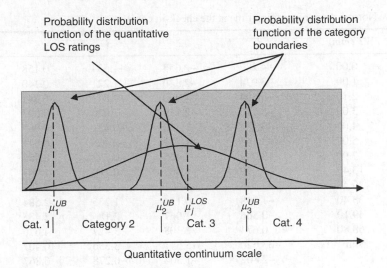

Figure 1.6   Illustration of the successive categories method for all passengers

Equation 1.6 can be illustrated using Figure 1.6. We note that v_{ji}^{LOS} is smaller than v_{3i}^{UR}. In this case $v_{jki}^{\Delta} = v_{ji}^{LOS} - v_{3i}^{UB} < 0$. So, WT$j$ is rated under category 3. Although it is very obvious, this equation will be very useful to the development of the model. The application of this equation to an integral of probability distribution, considering the mentioned assumptions, and after change of variables, can provide the following relation:

$$P_{jk} = \Phi\left[\left(\mu_k^{UB} - \mu_j^{LOS}\right)/\sigma_j\right] \tag{1.9}$$

Equation 1.9 represents the probability that a passenger will judge WT_j at or below category k. The inverse of this function is

$$\left(\mu_k^{UB} - \mu_j^{LOS}\right)/\sigma_j = \Phi^{-1}(P_{jk}) \tag{1.10}$$

Data from experimental design may be cast in the form of observed proportions p_{jk}, the proportions of judgments of WT_j at or below category k. Then according to the model,

$$\left(\mu_k^{UK} - \mu_j^{LOS}\right)/\sigma_j \cong [\Phi^{-1}(p_{jk})]$$

$$\cong (y_{jk}) \tag{1.11}$$

y_{jk} is the normal deviate corresponding to the proportion p_{jk} in the lower tail of the unit normal distribution.

Bock and Jones showed that the estimate $\underline{\mu}_k^{UB}$, $\underline{\mu}_k^{UB}$, can be determined as the average of the kth value of the standard normal deviates over all passenger groups j, that is:

$$\underline{\mu}_k^{UB} = \frac{1}{n}\sum_{j=1}^{n} y_{jk} \tag{1.12}$$

Table 1.10 Normal deviates: waiting time at the check-in counter

Group	WT (min)	1	2	3	4	Sum
1	0.00	−4.287	−2.653	−1.534	0.158	−8.316
2	1.00	−3.974	−2.340	−1.221	0.140	−7.394
3	2.10	−3.594	−1.960	−0.841	−0.841	−7.238
4	3.00	−3.773	−2.139	−1.020	0.097	−6.835
5	4.00	−3.184	−1.550	−0.431	0.967	−4.198
6	5.00	−3.161	−1.527	−0.408	0.841	−4.253
7	7.90	−3.099	−1.465	−0.792	0.457	−4.898
8	13.40	−3.099	−1.465	−0.180	1.067	−3.677
9	20.40	−3.135	−1.501	−0.431	1.501	−3.566
10	33.40	−1.748	−0.114	1.335	2.584	2.057
11	49.10	−0.566	1.067	2.186	3.435	6.123
12	68.80	0.674	2.308	3.427	4.676	11.087
	Sum	−33.620	−15.647	−3.336	10.407	−42.195
μ_k^{UB}		−2.802	−1.304	−0.278	0.867	
			$(-1.304 - 0.278)/2 = -0.791$			
μ_k^{UB} (normalized)		−2.011	−0.513	0.513	1.658	

According to the normal distribution, μ_k^{UB} will vary linearly with y, and so the estimate of μ_j^{LOS} and σ_j can be obtained by the regression line defined using these values of μ_k^{UB} as the dependent variables, and the y_{jk}, $k = 1, 2, ..., (m-1)$, for each j as the independent variables. The slope will be σ_j and the intercept on the μ_k^{UB} axis will be the value of μ_j^{LOS}. This last value is the mean LOS quantitative rating for group j.

Before proceeding with the calculations to obtain μ_j^{LOS}, let us summarize the necessary steps:

1. Separate the passengers into groups of similar waiting times. In the example, they were divided into 12 groups. Each of them has an average waiting time, denoted by WTj, where j is the group number ($j = 1, 2, ..., 12$).
2. Obtain the number of responses for each category in each group.
3. Calculate the proportions p_{jk}.
4. Calculate the normal deviates y_{jk}.
5. Calculate μ_k^{UB} as the average of y_{jk} over all groups for each category k.
6. Perform a regression analysis to obtain μ_j^{LOS}. For the regression, the independent variable should be $y_{jk;}$ The dependent variable should be μ_k^{UB}.

In the example proposed, steps 1–3 have been already undertaken. We now proceed to the calculation of the normal deviates y_{jk} (Table 1.10).

The μ_k^{UB}'s have been calculated as the average of the y_{jk}'s over all groups. The second last row of Table 1.10 calculates the mean of μ_k^{UB}'s for the lower and upper bounds of category 3. Its value (−0.791) represents the quantitative rating corresponding to the neutral position or indifference (mean of category 3). This value has been subtracted to the values of the originals μ_k^{UB}'s for obtaining the "normalized" μ_k^{UB}'s (last row Table 1.10).

Table 1.11   Necessary data for performing a regression analysis

Upper Bound of Category	Dependent Variable $\underline{\mu}_k^{UB}$	Independent Variable y_{7k}
1 – Unacceptable	−2.011	−3.099
2 – Poor	−0.513	−1.465
3 – Fair	0.513	−0.792
4 – Good	1.658	0.457

Table 1.12 μ_j^{LOS} s for each group (j)

Group (j)	μ_j^{LOS}	WTj (min)
1	1.64	0.00
2	1.57	1.00
3	1.97	2.10
4	1.52	3.00
5	0.84	4.00
6	0.89	5.00
7	1.20	7.90
8	0.71	13.40
9	0.62	20.40
10	−0.52	33.40
11	−1.49	49.10
12	−2.63	68.80

There is now enough data for obtaining the μ_j^{LOS}'s. We will illustrate the procedure for obtaining μ_7^{LOS}, which is the mean LOS rating for group 7 (waiting time = 7.9 minutes). In this case, a regression analysis must be performed between the two variables presented in Table 1.11.

The regression analysis provides the intercept of the curve, which is 1.20. This is the value of μ_7^{LOS}. The remaining μ_j^{LOS} (j=1 to 6, 8 to12) are presented in Table 1.12.

From Table 1.12 it can be seen that the mean quantitative LOS ratings become more negative as the waiting time increases; this represents the decreasing user satisfaction as the waiting time assumes greater values. It is possible to obtain a numeric function depicting the relationship between LOS and waiting times. The function obtained from a regression analysis performed using data from Table 1.12 is provided next:

$$\text{LOS} = 1.597 - 0.06(\text{WT})$$
$$R^2 = 0.96, \qquad F = 262.30$$

(1.13)

The curve corresponding to Equation 1.13 is represented by the line in Figure 1.7. The data points are represented by the dots. Equation 1.13 can be used to determine the level of service standards associated with the boundaries of categories. Table 1.13 shows the upper boundaries of categories 1–4.

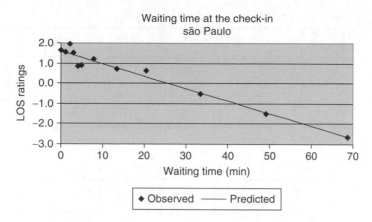

Figure 1.7   Data for waiting time at the check-in counter

Table 1.13   Category boundaries

Upper Bound of Category	$\underline{\mu}_k^{UB}$
1 – Unacceptable	−2.011
2 – Poor	−0.513
3 – Fair	0.513
4 – Good	1.658

Table 1.14   Proposed LOS standards

LOS	Waiting Time (min)
A	<1
B	1–17
C	17–34
D	34–58
E	>58

The substitution of the $\underline{\mu}_k^{UB}$ values of Table 1.13 into Equation 1.13 (as the LOS variable) provides the WT values corresponding to the upper boundaries of the categories. Table 1.14 shows the LOS standards calculated using this procedure.

For instance, LOS B was defined using waiting times corresponding to upper bounds of categories 3 and 4.

Finally, the conformity of the observed proportions of response in each category, designated $p_{jk} - p_{j,k-1}$ with those derived from the model designated $P_{jk} - P_{j,k-1}$, may be tested by computing a total χ^2 (chi-square) for the discrepancies between them:

$$\chi^2 = \sum_{j=1}^{n}\sum_{k=1}^{m-1} \frac{\left\{\left[\left(p_{jk} - p_{j,\,k-1}\right) - \left(P_{jk} - P_{j,\,k-1}\right)\right]N_j\right\}^2}{\left(P_{jk} - P_{j,\,k-1}\right)N_j} \tag{1.14}$$

To determine the degrees of freedom for the total χ^2, we note that there are $n(m-1)$ independent observed proportions (according to the assumptions stated before). From this total $2(n-1)$ degrees of freedom are consumed by the estimates of μ_j^{LOS} and σ_j not determined by the estimates of μ_k^{UB}, and $m-1$ are consumed by the estimates of μ_k^{UB}. Thus, the residual variation is on $(n-1)(m-3)$ degrees of freedom, and it is necessary to use not less than four categories and two objects if the model is to be tested.

(n is the number of groups: 12; and m is the number of categories: 5)

Equation 1.14 was applied and the chi-square value resulting was 13.476. The degrees of freedom are $(12-1)(5-3)=22$. In this case the chi-square value (13.476) is compared to 33.429 at 5% significance level (22 degrees of freedom). By this comparison we see that the model can be used for the LOS modeling.

1.5   Degree of Importance of Landside Components and Attributes

1.5.1   Introduction

Airport managers have to struggle with the decision of prioritizing resources. Although they are motivated to offer a reasonable level of service (LOS) to passengers, there is a growing worldwide tendency for cost reduction. In this scenario, an effort to determine the importance that passengers attribute to airport components will be useful, as it will be one indication of where airport managers should invest their limited resources such as funds, employees and their own attention. Evaluating the overall effectiveness of individual actions (e.g., improvement of check-in waiting time) requires evaluating the impact of individual components at the overall airport level of service (Correia et al., 2008).

This section presents a methodology to obtain the importance that passengers assign to the various components and attributes of an airport passenger terminal (APT). The Analytical Hierarchy Process (AHP) is employed in order to obtain quantitative weights representing the relative importance of components and their attributes (processing time, courtesy, etc.). Through the process of pair wise comparisons between the components and their attributes we can suppose that these weights could be further used to obtain a global LOS measure of an APT as a function of the LOS of individual components (check-in, departure lounge, etc.).

1.5.2   Selection of Components and Attributes

The Airports Council International level of service manual (ACI, 2000) was employed to pre-select the most important components and characteristics of components, according to the opinion of managers of 512 ACI airports. Final selection of variables to be included in the model was concluded with specialists of the Department of Air Transport and Airports of the Aeronautical Institute of Technology (Instituto Tecnológico de Aeronáutica: ITA) in meetings during May–July/2006. Table 1.15 presents the proposed variables indicated by the expert panel.

A questionnaire was developed so that passengers might compare the importance of one component (or attribute) over another component (or attribute). This process is called pair wise comparisons. For instance, passengers were asked to indicate the degree of importance of courtesy compared to comfort at the departure lounge. The scale provided to the user will be explained in the next section.

Table 1.15 Components and their attributes as proposed by an expert panel

Components	Characteristics
Parking	Courtesy, security, availability of parking spots
Departure Hall	Security, orientation, information, comfort, services
Check-in	Processing and waiting time, courtesy
Departure Lounge	Courtesy, comfort
Concessions	Courtesy, variety of stores

The number of factors to be analyzed should be kept into a minimum, in order to prevent passenger's confusion. Additionally, the questionnaire should be simple because passengers at the departure lounge usually do not have too much time to answer it. A pilot survey was undertaken to check the questionnaire. It was noticed that too many questions lead the passenger to answer the last ones without much consideration.

The management of São Paulo/Guarulhos International Airport did not allow the surveying of passengers about the importance of security screening and passport control. Thus, these components were not included in the analysis.

In addition to the passenger ratings of the importance components and their attributes, several socio-economic variables were collected during the survey:

• Gender; age; family income;
• airline; destination (national or international); purpose of trip (business, leisure, or combined);
• annual frequency of air trips; degree of familiarity to major Brazilian airports.

These socio-economic variables could be useful in a research to analyze groups of passengers that have common characteristics (same destination, similar family income, etc.). It might be the case that some groups of passengers have different perceptions of the importance of components and attributes.

1.5.3   The AHP – Analytical Hierarchy Process

The Analytical Hierarchy Process was first presented by Saaty (1980). It is one of the first multi-criteria decision methods. Its main objective is to represent the decision model in the most realistic way, including subjective and objective factors. Using this method, it is possible to structure a problem through hierarchical levels and make alternative comparisons through a quantitative importance scale. Table 1.16 presents the fundamental scale of the method.

According to Table 1.16, the maximum degree of importance of a given component or attribute is 0.90. That will happen when this component is extremely more important than another component. Additional methodological details can be found in Correia et al. (2007).

1.5.4   Descriptive Analysis of Passenger Responses

A brief descriptive analysis was accomplished using data obtained from passenger responses. In this case, the most common passenger groups are represented by: male passengers, age 21–30 years old, family income of US$10 000–20 000, frequency of travel – twice a year,

Table 1.16 Comparison among the scale used in this research with the scale of Saaty

Degree of Importance		Saaty Scale	Importance Degree
Component 1	Component 2		
0.90	0.10	9	Component 1 is extremely more important than Component 2
0.80	0.20	7	Component 1 is very important compared to Component 2
0.70	0.30	5	Component 1 is important compared Component 2
0.60	0.40	3	Component 1 is less important compared to Component 2
0.50	0.50	1	Both components have the same importance
0.40	0.60	1/3	Component 2 is less important compared to Component 1
0.30	0.70	1/5	Component 2 is important compared Component 1
0.20	0.80	1/7	Component 2 is very important compared to Component 1
0.10	0.90	1/9	Component 2 is extremely more important than Component 1

familiarity with São Paulo airports only (domestic airport and international airport), business passengers, flying GOL airlines, and flying to domestic destinations.

The next section presents the degrees of importance, which are valid for passengers at São Paulo/Guarulhos International Airport. The results might differ from the perceptions of passengers flying at different airports, especially if their socio-economic profile differs significantly from those surveyed.

1.5.5 Degrees of Importance of Components and Their Attributes

Figure 1.8 presents the weights indicating the degree of importance for the group of 103 interviewed passengers. They were obtained by application of the methodology previously presented in this paper. According to the Saaty scale (Table 1.16), the sum of values on a given level should be close to 0.90 (depending on the degree of consistency).

According to Figure 1.8, the check-in counter (0.33) is the most important component for departing passengers that have been interviewed at São Paulo/Guarulhos International Airport. The most important attributes for each component are presented in bold. For instance, security is the most important attribute for the parking component with a degree of importance of 0.48.

The values of Figure 1.8 could be further employed to obtain a global LOS measure for departing passengers at the APT. In this case, the AHP method must be used to rate the LOS for each component. Note that the degree of importance and rating are different things: for instance, a passenger might consider the check-in counter the most important component; however he/she might evaluate it to be the component with the relatively worst LOS at the APT, in comparison with the remaining components.

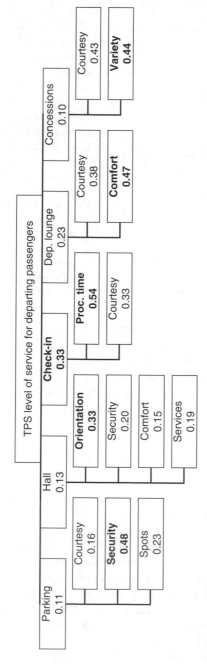

Figure 1.8 Weights of importance for components and its characteristics, São Paulo/Guarulhos International Airport

1.6 Conclusions

Airport level of service standards are an important input at the planning, design and operational stages. Nonetheless, there is no current handbook or standard methodology applicable to airports worldwide. Additionally, there is a lack of studies providing level of service standards considering user needs. This chapter provides several indicators for main airport terminal LOS components, as provided by academic studies and industry organizations. However, considering that level of service standards are very dependent on local passenger characteristics and behavior, we have proposed two methodologies that will be able to derive quantitative standards at any airport, by use of local passenger surveys and observations.

Although passenger surveys are always costly, their benefits surpass the costs to run them, especially considering that airport competition is increasing with the recent trends of airport decentralization, privatization, and globalization.

References

Airports Council International (2000) *Quality of Service at Airports: Standards & Measurements*. ACI World Headquarters, Geneva, Switzerland.

Ashford, N. and Wright, P. H. (1992) *Airport Engineering*, 3rd edn. John Wiley & Sons Inc., New York.

Bock, R. D. and Jones, L. V. (1968) *The Measurement and Prediction of Judgment and Choice*. San Francisco, Holden-Day.

de Barros, A. G. (2001) *Planning of Airports for the New Large Aircraft*. PhD. Dissertation, University of Calgary.

de Barros, A.G. and Wirasinghe, S.C. (1998) Sizing the Airport Passenger Departure Lounge for New Large Aircraft. *Transportation Research Record*, 1621, 13–21.

de Barros, A. G. and Wirasinghe, S. C. (2004) Sizing the Baggage Claim Area for the New Large Aircraft. *Journal of Transportation Engineering of ASCE*, 130 (3), 274–279.

Brunetta, L., Righi, L. and Andreatta, G. (1999) An Operations Research Model for the Evaluation of an Airport Terminal: SLAM (Simple Landside Aggregate Model). *Journal of Air Transport Management*, 5, 161–175.

Correia, A. R. (2009) *Evaluation of Level of Service at Airport Passenger Terminals*, 1st edn. LAP Lambert, Academic Publishing, Koln.

Correia, A. R., Galvão, M. and Wirasinghe, S. C. (2007) Degree of Importance of Airport Passenger Terminal Components and Their Attributes. *Airlines Magazine*, 37, 1–4.

Correia, A. R. and Wirasinghe, S. C. (2007) Development of Level of Service Standards for Airport Facilities: Application to São Paulo International Airport. *Journal of Air Transport Management*, 13 (2), 97–103.

Correia, A.R., Wirasinghe, S. C. and de Barros, A.G. (2008) Overall Level of Service Measures for Airport Passenger Terminals. *Transportation Research A*, 42 (2), 330–346.

Correia, A. R. and Wirasinghe, S. C. (2010) Level of Service Analysis for Airport Baggage Claim with a Case Study of the Calgary International Airport. *Journal of Advanced Transportation*, 44 (2), 103–112.

de Neufville, R. and Odoni, A. (2002) *Airport Systems: Planning Design and Management*, 1st edn. McGraw-Hill Book Company, New York.

Fernandes, E. and Pacheco, R. R. (2002) Efficient Use of Airport Capacity. *Transportation Research, Part A: General*, 36 (3), 225–238.

Forsyth, P. (1997) Price Regulation of Airports: Principles with Australian Applications. *Transportation Research E: Logistics and Transportation Review*, 33 (4), 297–309.

Gillen, D. and Lall, A. (1997) Developing Measures of Airport Productivity and Performance: an Application of Data Envelopment Analysis. *Transportation Research E: Logistics and Transportation Review*, 33 (4), 261–273.

Gosling, G. D. (1988) Airport Landside Planning Techniques: Introduction. *Transportation Research Record*, 1199, 1–3.

Hart, W. (1985) *The Airport Passenger Terminal*, 1st edn. John Wiley & Sons, Inc., New York.

Hooper, P. G. and Hensher, D. A. (1997) Measuring Total Factor Productivity of Airports – an Index Number Approach. *Transportation Research E: Logistics and Transportation Review*, 33 (4), 245–247.

Horonjeff, R. and McKelvey, F. X. (1994) *Planning and Design of Airports*, 4th edn. McGraw-Hill, New York.

Humphreys, I. and Francis, G. (2000) Traditional Airport Performance Indicators: A Critical Perspective. *Transportation Research Record*, 1703, 24–30.

IATA (1995) *Airport Development Reference Manual*. International Air Transport Association, Montreal.

Lemer, A. C. (1988) Measuring Airport Landside Capacity. *Transportation Research Record*, 1199, 12–18.

Mandle, P., Whitlock, E. and LaMagna, F. (1982) Airport Curbside Planning and Design. *Transportation Research Record*, 840, 1–6.

Martel, N. and Seneviratne, P. N. (1990) Analysis of Factors Influencing Quality of Service in Passenger Terminal Buildings. *Transportation Research Record*, 1273, 1–10.

Müller, C. (1987) *A Framework for Quality of Service Evaluation at Airport Terminals*. PhD Thesis, Institute of Transportation Studies, University of California, Berkeley.

Müller, C. and Gosling, G. D. (1991) A Framework for Evaluating Level of Service for Airport Terminals. *Transportation Planning and Technology*, 16, 45–61.

Mumayiz, S. A. (1985) *Methodology for Planning and Operations Management of Airport Passenger Terminals: A Capacity/Level of Service Approach*. PhD thesis, Department of Transport Technology, Loughborough University of Technology, Loughborough.

Ndoh, N. N. and Ashford, N. (1993) Evaluation of Airport Access Level of Service. *Transportation Research Record*, 1423, 34–39.

Saaty, T. L. (1980) *The Analytic Hierarchy Process*. McGraw Hill, New York. International, Translated to Russian, Portuguese, and Chinese, Revised editions, Paperback (1996, 2000), RWS Publications, Pittsburgh.

Seneviratne, P. N. and Martel, N. (1991) Variables Influencing Performance of Air Terminal Buildings. *Transportation Planning and Technology*, 16 (1), 1177–1179.

Seneviratne, P. N. and Martel, N. (1994) Criteria for Evaluating Quality of Service in Air Terminals. *Transportation Research Record*, 1461, 24–30.

Siddiqui, M. R. (1994) *A Statistical Analysis of the Factors Influencing the Level of Service of Airport Terminal Curbsides*. MSc thesis, Concordia University, Canada.

Thurstone, L. L. (1959) *The Measurement of Values*. University of Chicago Press, Chicago.

Transportation Research Board (1987) *Special Report 215: Measuring Airport Landside Capacity*. TRB, National Research Council, Washington D.C.

Wirasinghe, S. C. and Shehata, M. (1988) Departure Lounge Sizing and Optimal Seating Capacity for a Given Aircraft/Flight Mix – (i) Single Gate, (ii) Several Gates. *Transportation Planning and Technology*, 13, 57–71.

2

A Decision Support System for Integrated Airport Performance Assessment and Capacity Management

Konstantinos G. Zografos[A], Giovanni Andreatta[B], Michel J.A. van Eenige[C] and Michael A. Madas[D]

[A]Department of Management Science, Lancaster University Management School, Lancaster University, UK
[B]Department of Mathematics, University of Padova, Italy
[C]Environment and Policy Support Department, National Aerospace Laboratory (NLR), The Netherlands
[D]Transportation Systems and Logistics Laboratory (TRANSLOG), Department of Management Science and Technology, Athens University of Economics and Business, Greece

2.1 Introduction and Objectives

Notwithstanding the global economic crisis, long-term forecasts of air transport growth speak about a doubling in air transport demand in Europe by 2030 (Eurocontrol, 2008). Characteristically, although the planned capacity of 138 Eurocontrol Statistical Reference Area (ESRA) airports is planned to increase by 41% in total by 2030, the demand will still exceed capacity of the airport system by as many as 7.0 million flights in a high-growth scenario for 2030 (Eurocontrol, 2008). Under such a scenario, 14–39 European airports will need to operate at full capacity 8 h per day to accommodate only a part of the demand, similar to what most severely congested airports do now (Eurocontrol, 2008). A direct consequence of the mismatch between capacity and traffic growth is the increase of congestion and delays both in the air and on the

Modelling and Managing Airport Performance, First Edition. Edited by Konstantinos G. Zografos, Giovanni Andreatta and Amedeo R. Odoni.
© 2013 John Wiley & Sons, Ltd. Published 2013 by John Wiley & Sons, Ltd.

ground (at airports). Airports constitute the terminal nodes of a continuously expanding air transport network that should both efficiently and safely accommodate growing traffic.

The anticipated traffic growth in decades to come will again push capacity to the limit, thus triggering unprecedented levels of congestion with far reaching impact on the environment and the safety of operations. The latter concerns will pose serious challenges towards close airport performance monitoring and improvement. Airport decision makers should be able to cope with multiple – even conflicting – objectives and priorities assigned by various stakeholders regarding the multifaceted performance of the airport system such as the level of service offered to the travelling public, the efficiency of airport and air traffic management (ATM) operations, the quality of the surrounding environment, and the safety of the entire air transport system. The assessment of the airport performance requires a deep understanding of the manifold aspects of airport performance supported by advanced modelling capabilities and decision support systems, or tools for measuring it. Such decision aids should be diverse in that they should: (1) capture the behaviour of various entities (e.g. aircraft, passenger, baggage) processed through the system, (2) address different airport elements simultaneously (e.g. runway system, taxiway system, apron area, terminal), and (3) consider a large set of airport performance measures like capacity, delays, safety, security, noise and costs.

In order to deal with the multifaceted aspects of the airport decision making process, a wealth of decision support models and tools have appeared in both literature and practice (Odoni, 1991; Tosic, 1992; Odoni et al., 1997; Lucic et al., 2007; Correia et al., 2008; Long et al., 2009). Early modelling efforts developed rather focused applications (e.g. models, tools) both in terms of integration scope and degree of coverage. They basically constituted 'monolithic' modelling structures exhibiting either analytical or simulation modelling approaches with focused decision support capabilities mainly with view to a single airport performance measure, for example, runway capacity (not accounting for trade-offs). At the same time, early modelling efforts had a targeted/narrow coverage of specific elements of either airside (mainly runways) or landside. Since their early stages of development in the 1960s, airport performance models and tools have evolved substantially with a common orientation being the pursuit of more integration and expanded coverage capabilities. More recent research initiatives since 1990s attempted the integration of pre-selected and pre-existing tool configurations in order to model and evaluate simultaneously airport airside and landside and assess their interdependencies (Andreatta et al., 1999; Zografos and Madas, 2006). These efforts primarily suffered from the lack of a harmonized, fully integrated and automated computing environment needed to execute the various models, as well as limited trade-off analysis capabilities.

Despite the rich experience in both models and tools for airport performance assessment, modelling capabilities until 10 years ago or so addressed only partial aspects of the airport performance and exhibited several deficiencies: (1) they were concerned with specific flows or entities (e.g. aircraft, passengers, baggage), (2) they focused on specific airport elements (e.g. runway system, apron, terminal), (3) they considered one (or very few) airport performance indicator at a time (e.g. capacity, noise, safety, emissions), and (4) they were tailored for a specific level of decision making, either strategic, tactical or operational. At the outset, there was a clear lack of integrated modelling capabilities for assessing multiple performance measures simultaneously (and their trade-offs) for the airport system in its entirety (i.e. 'total airport'), that is, for both airport airside and landside simultaneously.

The latest developments in the airport modelling landscape involve the emergence of integrated platforms or systems[1]. Currently, there are a limited number of software products/decision support systems (mainly 'off-the-shelf') with integrated impact analysis capabilities for total airport operations. Most of them are purely simulation platforms basically integrating detailed simulation tools at a microscopic level. As a result, they do not exhibit macroscopic/aggregate analysis capabilities at the strategic decision making level with the use of analytical models. Furthermore, the existing, simulation based tools are quite complicated, rather data intensive, have a costly set up process for different airports, and require substantial tool familiarity and prior computational expertise. Another common feature for most of these tools is that they capture the airside-landside interaction, but still provide limited trade-off analysis capabilities, since they primarily focus on capacity and delay metrics. As a result, the basic modelling challenge remains, that is, to develop systems and tools that will not only capture the manifold aspects of airport performance in isolation, but will be also able to analyse, with reasonable effort, the various trade-offs and interdependencies among these performance measures, entities, or airport elements.

In response to the identified modelling needs, an integrated Decision Support System (DSS), the 'Supporting Platform for Airport Decision Making and Efficiency Analysis' (SPADE DSS), has been developed recently. The proposed system has the form of a computational platform that seamlessly integrates a variety of existing analytical models and simulation tools in order to capture the interdependencies among various measures of airport effectiveness (e.g. capacity, delays, level of service, noise, safety, costs and benefits) and enable performance trade-off analyses at various levels of detail (e.g. strategic, tactical/operational). Furthermore, the SPADE DSS allows decision makers and analysts to evaluate the efficiency of the entire airport complex simultaneously (including also interaction effects among airport elements). However, the most important and innovative element of the proposed modelling approach is the adoption of the 'use case' paradigm as the main building component of the system implementation structure. The use case driven implementation approach supports a problem or decision oriented approach that is capable of addressing airport planning decisions in a user-friendly manner and at a reasonable effort without requiring prior familiarity of the user with the selected tools (e.g. build baseline/'what-if' scenarios, prepare and exchange data sets, perform trade-off analysis).

The objective of this chapter is twofold:

1. To introduce the structure and constituting elements of the SPADE system; and
2. to demonstrate the decision support capabilities of the system under 'real-world' conditions by means of two manifestations of the system for strategic decision making (Athens International Airport) and operational/tactical decision making (Amsterdam Airport Schiphol).

The remainder of this chapter consists of three main sections. Section 2.2 provides an overall description of the high level structure of the SPADE DSS with special emphasis placed on the use case driven modelling concept. Section 2.3 provides a demonstration of two application instances of the system for strategic and operational/tactical decision making, respectively. Section 2.4 presents the concluding remarks and lessons learnt during the system implementation, whilst reporting some brief results from the evaluation of the system. Finally, the chapter is complemented by the acknowledgements and a list of reference sources.

2.2 SPADE DSS Description

2.2.1 Basic Modelling Concepts

The SPADE system introduces a modelling approach which integrates different, existing tools in a problem-oriented environment for addressing particular types of strategic or tactical questions in the decision maker's language. The basic innovative feature of the system is its capability to perform certain types of analyses in the form of pre-specified and pre-modelled questions that are frequently faced by airport decision makers and planners. These frequently asked questions (FAQs) examine the impact of certain airport management and planning decisions on various measures of airport effectiveness and have been elicited from a large panel of potential end users (SPADE-2, 2009; Zografos et al., 2013).

The basic modelling block of the SPADE DSS is the 'use case'. A use case is a 'structured modelling path' that is pre-built and pre-configured to deal with various decision support requirements expressed in the form of FAQs. Use cases are able to support the aforementioned, pre-specified set of decisions and perform integrated impact analyses with reference to multiple airport performance metrics in any airport. Practically, the user is guided by a wizard to select the level of detail (module), the desired type of analysis (use case), and specific question(s) (FAQs) to be addressed in the analysis. For example, a user might need to address airport capacity management/planning issues at the strategic/macroscopic level with particular emphasis on the assessment of capacity gains and the associated congestion relief due to the construction of a new runway.

Based on the aforementioned user selections, the system will then activate the appropriate tools in a user transparent manner; it will select which tools to run, it will feed the appropriate input (user defined and default) to each tool, it will execute the appropriate tool workflow, and finally it will store and display the requested output to the user. The use case driven approach offers some important benefits from the user point of view. First, it shields the user from the complicated tool world by adopting a problem oriented rather than tool driven approach. In a sense, it allows the user interface of the system to translate the 'mental map' of the decision maker (Gaines and Shaw, 1995; Chaib-Draa, 2002) into pre-specified modelling steps automatically performed by the system. Therefore, it de-emphasizes the importance of experience/familiarity with the use of specific tools in favour of expertise in the airport decision making domain. Second, it provides a 'holistic' view of airport operations by addressing both airside and landside simultaneously with view to multiple airport performance measures. Third, it supports trade-off and sensitivity ('what-if') analysis in a simple, quick, and user-friendly manner.

2.2.2   High Level Structure

The SPADE system was designed and built around the concept of use cases in the form of a 'suite of use cases'. Depending on the tools involved and the associated level of detail applied in the analysis, use cases were designated as 'strategic' or 'operational/tactical' and positioned under the strategic or the tactical/operational modules, respectively. The idea behind the separation of the two modules was based on the assumption that different profiles of users or circumstances may require different aggregation levels. For example, the strategic module may first provide a quick diagnosis tool for senior airport executives, who will then trigger a more detailed and focused investigation. The latter can be performed by specialized airport analysts/experts through the tactical/operational module.

From a logical viewpoint, the SPADE system architecture splits into two distinct design layers (i.e. horizontal/system wide, vertical/use case specific) and adheres to the widely-adopted Model-View-Controller (MVC) pattern (Buschmann et al., 1996). The high level design of the system's architecture was derived from the elicitation of requirements by

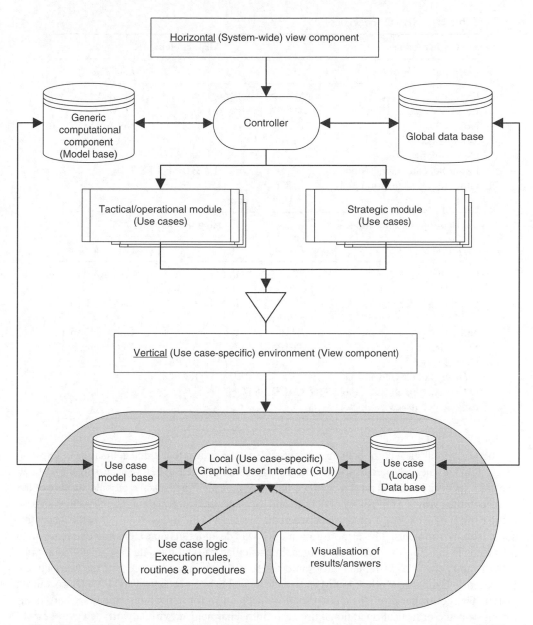

Figure 2.1 High level structure of the SPADE system [adapted from Zografos and Madas, 2007]

applying the Unified Process (Arlow and Neustadt, 2001), and resulted in a client-server approach. The horizontal/system-wide design layer includes the following main components: (1) the horizontal (system-wide) graphical user interface, (2) the generic computational component, (3) the generic input component (global data base), and (4) the system controller (Figure 2.1) (Zografos and Madas, 2007).

Table 2.1   SPADE model base

Strategic module	Airport elements	
Performance measure	*Airside*	*Landside*
Capacity	MACAD, ABS-ACD	SLAM
Level of Service / Delays	MACAD, ABS-ACD	SLAM
Noise	INM	Not Applicable
Safety/Risk	TRIPAC	Not Applicable
Cost-Benefit	CBM	
Flight Schedule Generation	FLASH, ABS-FSG	
Other (e.g. visualization tool)	LUCIAD ATC Playback	
Tactical/operational module	**Airport elements**	
Performance measure	*Airside*	*Landside*
Capacity	TAAM, RAMS Plus, SIMMOD	SAMANTA
Level of Service / Delays	TAAM, RAMS Plus, SIMMOD	SAMANTA
Noise	INM	Not Applicable
Safety/Risk	TRIPAC	Not Applicable
Cost-Benefit	CBM	
Flight Schedule Generation	TRAFGEN	
Other (e.g. visualization tool, runway occupancy estimator)	CAST, ESTOP	

The horizontal view component represents the user interface that enables the user to log-in to the system and make the necessary selections (e.g. module, use case) before accessing the vertical/ use case-specific environment. The generic input component represents the global data base of the system in the form of a repository of default and user-specified airport data. The SPADE data base is the central data exchange medium between the system and the integrated tools in the respective use case environments. The airport data model was designed as an extensible data base with interfaces to existing, external data sources. In particular, parts of the data base related to airport-specific and generic ATM-related information were designed in accordance with the Aeronautical Information Exchange Model (AIXM) 5.0 made available by Eurocontrol and FAA (Eurocontrol, 2010). The controller provides a coordination mechanism for activating, executing, and monitoring strategic or tactical/operational use cases into a harmonized virtual working environment.

The generic computational component (model base) constitutes the core of the system since it includes all models and tools involved in the various use cases of the vertical layer of the system (use case model base). The selection of the models/tools to be included in the SPADE model base (Table 2.1) was made with view to the elicited use case requirements, the compatibility and complementarity of tool characteristics and capabilities, as well as prior integration experience. Practically, the identified tools were classified and selected on the basis of: (1) the level of detail applied in the analysis (i.e. strategic, operational/tactical module), (2) the airport elements covered, and (3) the performance indicators addressed (Zografos and Madas, 2007; SPADE-2, 2009).

Based on the selection criteria mentioned previously, the model base includes both macroscopic – mostly analytical – tools such as MACAD (Stamatopoulos et al., 2004), SLAM (Brunetta et al., 1999), INM (FAA, 1997), and TRIPAC (NLR, 2013 based on the model described by Dutch Government, 2013), as well as microscopic – mostly detailed – simulation tools such as TAAM (Jeppesen, 2013), RAMS Plus (ISA, 2011), SIMMOD (FAA, 1989), SAMANTA (INCONTROL, 2011), and CAST (ARC, 2011). In addition to the various performance modelling capabilities, the model base features also traffic generation capabilities (i.e. TRAFGEN, FLASH, ABS-FSG) (SPADE-2, 2009) at both the aggregate and detailed levels. The latter are of high value to the end user since they can easily model the traffic-side conditions of alternative scenarios by generating flight schedules based on user's aggregate descriptions of the demand/traffic.

The vertical layer of the system shifts the focus to the specific type of analysis through the selection of a use case. Each use case has its own view component which inherits the properties and Graphical User Interface (GUI) principles of the system-wide view component. The use case-specific GUI implements the dialogues, interacts with the user, provides input wizards/data forms and visualizes the results/output. Within the vertical part, the business logic incorporates a set of execution rules, integration routines and procedures aiming to implement an automated back-office environment for the use case at hand. Finally, the vertical computational component 'borrows' tools from the generic model base, while the local (use case) data base retrieves input from the global data base.

An important feature of the SPADE system is the capability of its data model to handle both common, system wide data, stored in the global data base, and tool specific data, which are stored in files with specific formats within the use case (local) data base. Integrating the data requirements of the various tools has been a major challenge, which required the identification of common input requirements, the enforcement of data integrity and validation rules, and the reconciliation of different units of measurement and classification systems. Such common, system wide data can be directly retrieved from the global data base, are interchangeably used from the various use cases and hence need not be repeated or re-entered on the use case level. Typical categories of common data are airport topology/geometry characteristics, flight schedule data, as well as aircraft models/categories and their associated operational characteristics.

Another useful characteristic of the SPADE data model relates to the process streamlining for creating a new analysis, and, more specifically, alleviating the need to provide a complete data set for each new analysis. This is particularly important for detailed simulation tools, which require a high level of detail, involving substantial effort for data collection and entering. Moreover, many 'what-if' analyses rely on changing only a few pieces of information of an initial/baseline set of data. For these reasons, 'sensible defaults' are provided for every piece of information required by the system. As a result, the user is given the option to modify only a subset of data that are relevant to the type of analysis he/she wishes to perform.

2.2.3   Suite of Use Cases

The use cases that are currently supported by the SPADE system fall into the following three broad categories of analysis (Zografos and Madas, 2007):

- *Supply-Side Planning and Analysis*: the assessment of implications on airport operations (e.g. capacity, delays, noise, safety) as a result of supply-side interventions (e.g. construction of new or expansion of an existing terminal, runway, taxiway, apron stands, reconfiguration of existing or new infrastructural elements).

- *Impact Analysis from Changes in Operational Requirements, Procedures, Concepts, and Systems*: the assessment of implications on airport operations from the introduction of new types of aircraft (e.g. A3XX), procedures (e.g. reduced separation minima), concepts (e.g. Collaborative Decision Making), or technological capabilities (e.g. Advanced Surface Movement Guidance and Control System – A-SMGCS).
- *Demand-side (Traffic) Changes*: the assessment of implications on airport performance as a result of demand changes (e.g. fleet mix/characteristics, passengers' characteristics) or demand management measures (e.g. night bans, diversion of traffic to relieve airports) affecting the profile and/or volume of traffic.

The nine use cases currently embedded into the SPADE system are subdivided into the so-called 'strategic' and 'operational/tactical' use cases. The former provide decision support on a medium to long-term time horizon mainly through the use of macroscopic, low level-of-detail, mainly analytical models and tools. The latter provide decision support on a short- to medium-term horizon through the use of microscopic, detailed simulation tools. Use cases 1–3 are classified as strategic use cases with use cases 4–9 being classified under operational/ tactical use cases. The resulting list of use cases (UCs) accompanied by a brief description of their main decision support capabilities, performance measures, integrated tools, and typical users are presented in what follows (SPADE-2, 2009).

2.2.3.1   UC1: Airport Capacity Management

The objective of UC1 is to assess the impact of changes in airport infrastructure (e.g. expansion of runway system, apron area or check-in counters), operational procedures (e.g. security procedures, service times/disciplines or separation minima), and/or traffic volume or distribution (e.g. fleet characteristics, passengers' characteristics, traffic peaking or traffic growth) for a 'total' airport. It addresses performance indicators such as capacity, delay, level of service, noise, third-party risk, and cost-benefits, and aims to provide decision support to airport operators, airport analysts/consultants, policy making bodies (e.g. civil aviation authorities, slot coordinators), and airport associations or other bodies representing the collective interests of the industry (e.g. ACI Europe, IATA). UC1 integrates the following tools: (1) FLASH: flight schedule generator, (2) MACAD: airside capacity and delay tool, (3) SLAM: terminal capacity, delays and the level of service tool, (4) INM: noise model, (5) TRIPAC: third-party risk assessment model, and (6) CBM: cost-benefit model.

2.2.3.2   UC2: Match Capacity and Demand

The objective of UC2 is to provide a generic analysis for airport strategic planning problems: what is the airport's baseline performance? What is the effect on airport performance for a business as usual strategy? What is the effect on airport performance after expanding the runway system? Or, what is the effect on airport performance after the implementation of a demand management strategy? UC2 addresses performance indicators such as capacity, delay, noise, third-party risk and costs-benefits, and aims to provide decision support to airport operators, policy making bodies (e.g. civil aviation authorities, slot coordinators), and airport associations or other bodies representing the collective interests of the industry (e.g. ACI Europe, IATA). UC2 integrates the following tools: (1) ABS-FSG: flight schedule generator, (2) ABS-ACD: airside capacity and delay tool, (3) INM: noise model, (4) TRIPAC: third-party risk assessment model, and (5) CBM: cost-benefit model.

2.2.3.3   UC3: Airport Airside Analysis

The objective of UC3 is to assess the effects of changes in infrastructure/procedures from an airport airside perspective in order to highlight the airport elements that will be more congested or generate the highest delays. UC3 enables the analysis of a single airport layout based on the existing infrastructure and investigates how traffic increase may affect performance. In addition, it allows the comparison of how traffic increase may affect two alternative scenarios related to airside infrastructure and/or procedures. It addresses performance indicators such as capacity, flow, delays, noise and safety (in terms of bottlenecks), and aims to provide decision support to airport operators and air traffic service provides. UC3 integrates the following tools: (1) SIMMOD: fast-time airport airside simulator, (2) INM: noise model, and (3) Luciad ATC Playback: visualization tool for aircraft movements.

2.2.3.4   UC4: Fleet Characteristics Impact on Airport Operations

The objective of UC4 is to assess how potential changes in fleet characteristics (e.g. aircraft type, city pair and flight schedule) may affect airport operations in terms of airside capacity, efficiency, noise, as well as costs and benefits. It aims to provide decision support to airport operators, airlines, policy making bodies (e.g. civil aviation authorities and slot coordinators), and airport associations or other bodies representing the collective interests of the industry (e.g. ACI Europe, IATA). UC4 integrates the following tools: (1) TRAFGEN: traffic generator, (2) ESTOP: runway occupancy estimator, (3) TAAM: fast-time airport airside simulator (as alternative to RAMS Plus), (4) RAMS Plus: fast-time airport airside simulator (as alternative to TAAM), (5) INM: noise model, (6) CBM: cost-benefit model, and (7) CAST: visualization tool for aircraft movements.

2.2.3.5   UC5: Airport Capacity Utilization

The objective of UC5 is to analyse the impacts of changes in the external environment (e.g. airline alliances, new Schengen countries) and/or minor changes in terminal infrastructure (e.g. security or check-in procedures) or allocation rules (e.g. check-in and stands), with regards to capacity, passenger and baggage throughput, as well as costs and benefits. UC5 aims to provide decision support to airport operators and policy makers. It integrates the following tools: (1) SAMANTA: fast-time airport terminal simulator, (2) TAAM: fast-time airport airside simulator, and (3) CBM: cost-benefit model.

2.2.3.6   UC6: Airport Capacity Determination

The objective of UC6 is to assess the impacts of changes in traffic, operational usage of airside elements (e.g. runway configuration, taxiing speeds, separations), or weather (e.g. wind speed and direction, visibility, temperature). UC6 addresses performance indicators such as capacity, delays, and level of service, and aims to provide decision support to airport operators, airlines, policy making bodies (e.g. civil aviation authorities, slot coordinators), and airport associations or other bodies representing the collective interests of the industry (e.g. ACI Europe, IATA). It integrates the following tools: (1) TRAFGEN: traffic generator, (2) ESTOP: runway occupancy estimator, (3) TAAM: fast-time airport airside simulator, and (4) CAST: visualization tool for aircraft movements.

2.2.3.7   UC7: Airport Capacity versus Environmental Capacity

The objective of UC7 is to analyse the impacts of capacity and/or environmental constraints on airport performance. Typical questions are the following: what is the maximum environmental (noise-driven) capacity of the airport, what is the impact of a noise-optimized flight schedule, what is the maximum airside capacity of the airport, or what is the impact of an airside-optimized flight schedule? UC7 addresses performance indicators such as airside and terminal capacity, level of service, noise, third-party risk, as well as costs and benefits. It aims to provide decision support to airport operators, consultants, and policy makers. It integrates the following tools: (1) TRAFGEN: traffic generator, (2) TAAM: fast-time airport airside simulator (as alternative to SIMMOD), (3) SIMMOD: fast-time airport airside simulator (as alternative to TAAM), (4) SAMANTA: fast-time airport terminal simulator, (5) INM: noise model, (6) TRIPAC: third-party risk assessment model, and (7) CBM: cost-benefit model.

2.2.3.8   UC8: Taxiing Methodology

The objective of UC8 is to apply a taxiing methodology aiming to improve the airport performance with regards to capacity, delays, safety (e.g. number of conflicts), and efficiency based on a global quality factor. UC8 aims to provide support to airport operators, civil aviation authorities, air navigation service providers, and air traffic control authorities. It integrates the following tools: (1) TRAFGEN: traffic generator, (2) ESTOP: runway occupancy estimator, (3) TAAM: fast-time airport airside simulator, and (4) CAST: visualization tool for aircraft movements.

2.2.3.9   UC9: Impact of New Procedures and/or Equipment

The objective of UC9 is to assess the return on investment and operational benefits of introducing new procedures and/or airport equipment, with regards to airside and terminal capacity, delays, level of service, noise, third-party risk, as well as costs and benefits. For instance, what is the impact of changes in night regimes, separation minima or aircraft turnaround times. It aims to provide support to airport operators, airlines, civil aviation authorities, and policy makers. It integrates the following tools: (1) TAAM: fast-time airport airside simulator, (2) SAMANTA: fast-time airport terminal simulator, (3) INM: noise model, (4) TRIPAC: third-party risk assessment model, and (5) CBM: cost-benefit model.

It is interesting to observe here that there is some overlapping in the types of analysis supported by certain use cases. This functional 'redundancy' has been purposely maintained mainly due to the fact that different use cases apply different tool combinations for performing each type of analysis. In effect, the type of analysis supported by one use case can be further elaborated/granularized or even aggregated through the use of another, more detailed or more aggregate, use case depending on the level of detail required in the analysis. The complementary utility of use cases becomes even more important when considering the possibility to compare or validate the results obtained for the same type of analysis from different use cases and different/alternative tools involved in the same type of analysis (modelling 'redundancy'). Finally, the modelling 'redundancy' increases the flexibility from the potential user's point of view since it allows the selection of those use cases involving (commercial or proprietary) tools to which the user has already direct license or access.

2.3   SPADE DSS Applications

This section provides a demonstration of two exemplary application instances of the SPADE system. The demonstration aims to present some indicative decision support capabilities of the system through an application for strategic decision making (Athens International Airport) and a second one for the operational/tactical level of decision making (Amsterdam Airport Schiphol).

2.3.1   SPADE DSS Application for Strategic Decision Making

An application of the SPADE system pertaining to the strategic level of decision making will be discussed in this section. This application is based on a test study performed with the Use Case 1, titled 'Airport Capacity Management (ACM)' with Athens International Airport (AIA) acting as the airport demonstration test-bed[2]. The study aims to demonstrate the main decision support capabilities of Use Case 1 dealing with the management of airport airside and landside capacity vis-à-vis existing or forecasted traffic (demand-side) fluctuations.

2.3.1.1   Description of Use Case 1 (Airport Capacity Management)

Use Case 1 (UC1), titled 'Airport Capacity Management (ACM)', applies to the strategic level of decision making and considers issues related to airport airside and landside analysis, including capacity, delays, level of service, safety, noise, and cost-benefit performance indicators. The specific types of 'Frequently Asked Questions (FAQs)' supported by UC1 involve the assessment of the impact of: (1) changes/interventions in the physical airside and/or landside infrastructure (FAQ1) (e.g. construction of new or expansion of existing runway, taxiway, apron stands, check-in counters, security control points), (2) changes in operational procedures or standards (FAQ2) (e.g. reduced separation minima, runway operating policy), and (3) fluctuations/changes in the distribution, properties and/or volume of traffic (FAQ3) (e.g. changes in fleet mix/characteristics, application of demand management measures).

The use case is intended to fulfil the decision making needs of three broad types of potential users. First, high level airport decision makers aiming to assess the future performance of an airport under forecasted levels of traffic with major emphasis on measures required to cope with sharp traffic increases. Second, civil aviation authorities concerned with the identification of the elements constraining airport capacity, the identification of bottlenecks and major sources of delays, and the implications of regulatory decisions or new operational standards (such as reduced separation minima). Third, national slot coordinators interested in monitoring the efficiency of slot allocation strategies, in determining declared airport capacity, and in assessing the impact of different declared capacity levels in conjunction with different levels of traffic.

The core tools integrated in UC1 are MACAD (Stamatopoulos et al., 2004), SLAM (Brunetta et al., 1999), and FLASH (SPADE-2, 2009). MACAD is a macroscopic, analytical model providing an overall assessment of the capacity and delays of the airside operations (i.e. runway system, taxiway system, apron area). It uses a stochastic analytical model for the estimation of the capacity envelope of the runway system based on a modified version of the LMI model (Wingrove et al., 1995) in conjunction with a slightly modified version of the DELAYS model (Malone, 1995) for computing runway delays numerically. A full-scale validation of MACAD tool has been performed in several European airports with capacity and delay figures being estimated very close to observed (actual) values (Stamatopoulos et al., 2004). SLAM is a macroscopic, analytical model for estimating capacity and level of service (according to IATA standards) for passenger and baggage in airport terminals. It is designed to assess alternative configurations and/or service policies of the various processing, holding, and flow facilities in a terminal. Furthermore, SLAM has been also successfully validated in several European airports (Brunetta et al., 1999). It is worth noting that MACAD and SLAM have been successfully integrated (Andreatta et al., 1999; Zografos and Madas, 2006; Zografos et al., 2013) within the framework of several European R&D Projects (e.g. TAPE, OPAL, SPADE, SPADE-2).

The integration of MACAD and SLAM (Andreatta et al., 1999; Zografos et al., 2013) allows the capacity and delay analysis of the airside and landside operations simultaneously at a macroscopic level, and is achieved through a computational loop between MACAD and SLAM. This loop supports the exchange of delays computed by each model and analyses the propagation of delays from the airside to the landside and vice versa for a specific daily flight schedule. In particular, both MACAD and SLAM run an initial daily flight schedule generated by FLASH based on user-specified characteristics of traffic (e.g. volume, hourly distribution, aircraft mix). For each departure flight, if the landside delay computed by SLAM is larger than the airside delay computed by MACAD, it is added to the original scheduled time of the flight. The computational loop closes with a second run of MACAD based on the new departure times (updated by the delay increments computed by SLAM and the first run of MACAD), to generate 'total' airside and landside delays. The rationale behind this approach is to identify whether landside and/or airside operations affect the scheduled departure times of a specific flight, and subsequently to estimate the propagation effects on the timely processing of the flight. This is particularly important especially under abrupt traffic peak or disruption conditions (e.g. severe weather) that may shift delays from the air on the ground (airside) and subsequently in the terminal with the opposite being also possible.

If no other performance trade-offs are involved in the analysis (i.e. 'basic run'), the use case execution stops and proceeds directly with the display of scenario output. Otherwise, part of the output (practically the flight schedule and the associated delays) of the integrated MACAD-SLAM-MACAD run is directly channelled to post-processing tools such as INM, TRIPAC, and CBM in order to perform noise, safety, and/or cost-benefit analyses (SPADE-2, 2009). These tools post-process the flight schedule generated by FLASH along with the updates (i.e. delays) computed by MACAD and SLAM in order to calculate noise and safety figures, while also estimating the cost of delays for various airport stakeholders (e.g. airports, airlines).

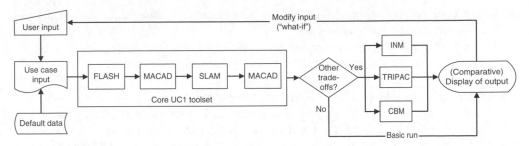

Figure 2.2   Use Case 1 workflow

In a final step, the user is able to perform with minimal effort some simple sensitivity tests in the form of 'what-if' scenarios aiming to examine the relationship between changes in airport capacity determinants or traffic parameters and selected measures of airport performance. Finally, the user is also provided with the option to analyse simultaneously and trace easily differences or the impact of changes introduced in the form of 'what-if' scenarios. This is achieved through output wizards being able to display output on a comparative basis for multiple scenarios simultaneously. Figure 2.2 presents the tool workflow and the sequence of major events taking place in the context of UC1.

The definition, set-up, and execution of a specific scenario with the tool workflow presented above requires a long list of input parameters that can be either provided directly by the user or alternatively retrieved from default data sets that are pre-stored into the system's global data base. The user has the flexibility to select only those input parameters for which he/she desires to specify input, while all remaining input will be then populated with default data pertaining to the specific airport and scenario under consideration. This functionality provides much flexibility to the user and reduces substantially the cumbersome input specification process that is particularly applicable to data-intensive tools. Overall, the UC1 input requirements can be classified under the following broad categories of data:

Infrastructural elements of the airport (supply-side)
• Number and type(s) of runway set(s),
• properties of the runway set(s) such as separation minima, taxi times, runway occupancy times, aircraft approach speeds and so on,
• runway configuration(s),
• number, type, and properties of apron stands (e.g. stand preparation time for the next aircraft by type of stand, stand vacation time in case of departure delays),
• number, types, service times and properties of check-in counters,
• number, types, service times and properties of security screening points, and
• number, types, and properties of gate lounge areas.

Operational rules and procedures
• Turnaround times by type of aircraft, type of flight, and type of apron stand,
• separation standards by type of aircraft and runway set,
• hourly operational policy of runway configuration(s),

- type of operation (i.e. optimal mix, arrivals only, departures only, even mix, arrivals and 'free' departures) for each runway set,
- check-in service time and properties by group of check-in counters, and
- security service time and properties by group of security screening points.

Characteristics and volume of airport traffic (demand-side)

This category of data involves basically input related to the total volume, hourly distribution, or profile characteristics of traffic for both arrivals and departures within the day. Practically, these data are reflected on the daily flight schedule prepared by FLASH based on input provided by the user such as the split between arrivals and departures, the hourly distribution of arrivals and departures within the day, the type of flight (e.g. Intra/Extra-Schengen), the airline operating the flight, the aircraft model etc.

Based on the list of UC1 input parameters and the tool workflow discussed previously, the following types of output are produced at various levels of aggregation:

- Daily capacity versus demand figures for both arrivals and departures,
- capacity envelopes (i.e. maximum throughput of the runway system in terms of arrivals versus departures per unit of time) for both runway set(s) and runway configuration(s),
- airside and landside (i.e. overall) delay figures for both arrivals and departures,
- overall airside and apron delays by type of flight (i.e. intra-Schengen, extra-Schengen), as well as airside delay figures by cause (e.g. due to runway congestion, apron congestion),
- level of service tables (according to IATA standards) for various terminal processing and holding facilities (check-in, security, gate lounges),
- performance (e.g. queue length) of various terminal facilities,
- noise levels and contours, population/houses within noise contours,
- third-party risk (i.e. probability of accident within a specific area), risk contours, population/ houses within risk contours, and
- cost- benefit data by aircraft type, origin-destination, airline, as well as the airport as a whole.

2.3.1.2 Scenario Description

The demonstration of the SPADE system for the strategic level of decision making is based on a test study with Athens International Airport (AIA) acting as the demonstration test-bed airport. AIA is the largest and busiest airport in Greece (approximately 40.7% of total passenger movements and 43.4% of total aircraft movements in Greece for 2009) (HCAA, 2010) and aims to establish a strategic international presence as the largest airport hub in south-eastern Europe. Passenger traffic in 2009 was approximately 16.2 million passengers (63% of which were international passengers), ranking AIA in ACI Europe Airport Group 2[3] and among the 30 busiest airports in Europe in terms of passenger traffic. In 2009, AIA offered direct scheduled passenger services to 113 destinations (80 international) in 52 countries, serviced by 70 airlines and handled approximately 210 000 aircraft movements. Even amid unfavourable industry conditions, Athens International Airport achieved satisfactory traffic development results outperforming most of its European counterparts, with the number of flights recording a significant annual growth in 2009 (+5.4%), and passenger traffic witnessing a marginal loss (−1.5%) (AIA, 2010).

The Athens International Airport currently has two passenger terminals (and one cargo/ freight terminal) connected by an underground pedestrian tunnel, the Main Terminal and the Satellite Terminal. Both terminal buildings cover a total space of 180000 m² and provide 24 passenger-boarding bridges, 144 check-in counters, 8 security screening points, and 11 luggage claim conveyor belts (7 for Intra-Schengen flights and 4 for Extra-Schengen flights). The airside infrastructure includes two parallel independent runways with segregated operations, that is, one runway is dedicated to arrivals and the other to departures. Furthermore, in addition to contact stands/bridges, the airport has also 65 remote apron stands. An aerial layout of AIA is presented in Figure 2.3.

Two alternative scenarios were developed for demonstration purposes. The main objective of these scenarios was to investigate the 'behaviour' of the airport system, identify bottlenecks, and assess various measures of its performance in response to increased traffic conditions in the long run. Scenario 1 basically reflects the current situation since it assumes the existing airport infrastructure for both airside and landside, as well as current operational policies, rules, and procedures (supply side). It involves FAQ3 dealing with the assessment of impacts on various airside and landside performance aspects (e.g. delays, level of service, noise) as a result of certain traffic conditions on the demand-side. The latter pertains to a daily flight schedule generated by FLASH and reflects the thirtieth busiest day for 2009 (710 aircraft movements). Scenario 2 assumes the same infrastructure and operational policies, but it introduces (again through FAQ3) an optimistic traffic growth forecast in the long run (after the end of the concession period in 2026). In particular, it assumes a notable traffic increase (approximately 20% increase in the number of aircraft movements on top of the typical day schedule of 2009) that could be the outcome of a 'hubbing' scenario for Athens International Airport (south-eastern Mediterranean hub).

2.3.1.3 Discussion of Results

This section presents some indicative results[4] for Scenarios 1 and 2 discussed in the previous section. The demonstrated scenarios aim to provide decision support for a senior airport planner/decision maker on a medium to long-term horizon by: (1) assessing the multifaceted aggregate impacts of traffic growth on the airport airside and landside operations, and (2) providing a quick diagnosis tool that identifies specific bottlenecks, inefficiencies, or areas/elements that may require a more detailed and focused investigation.

Initially, Figure 2.4 presents the runway capacity envelope for AIA's two parallel independent runways according to the existing infrastructure and operational rules/procedures (applied in both Scenarios 1 and 2). It is important to note here that Scenarios 1 and 2 assume the same airport infrastructure and operational rules/procedures, hence capacity results are identical (and presented once). Departure and arrival operations are segregated, while the departures only and arrivals only capacity are estimated at 30 and 26 movements, respectively. As already mentioned, AIA has two parallel, independent runways each dedicated to a specific type of movement (i.e. segregated operations), such that there is no trade-off (i.e. 'free departures') between arrivals and departures.

Another important aspect for an airport decision maker is to associate capacity levels with delay outcomes. Based on the estimated capacity levels, the system computes moderate average delays of 4–6 min for Scenario 1 with delays being deteriorated (8–10 min) in the traffic growth Scenario 2 (Figure 2.5). Delays exhibit similar patterns in

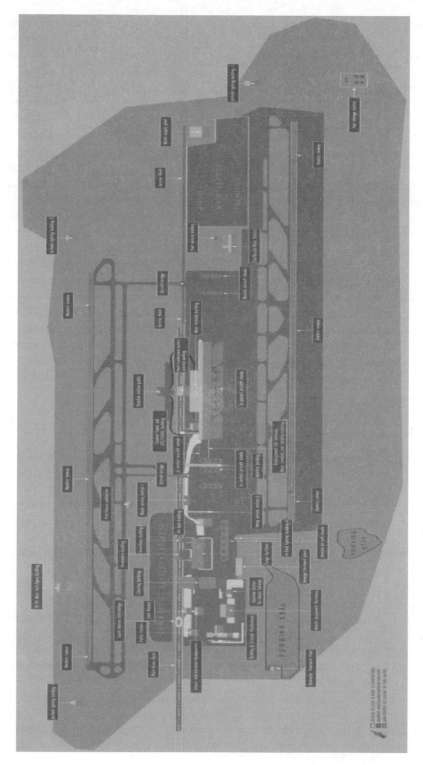

Figure 2.3 Athens International Airport layout (www.aia.gr)

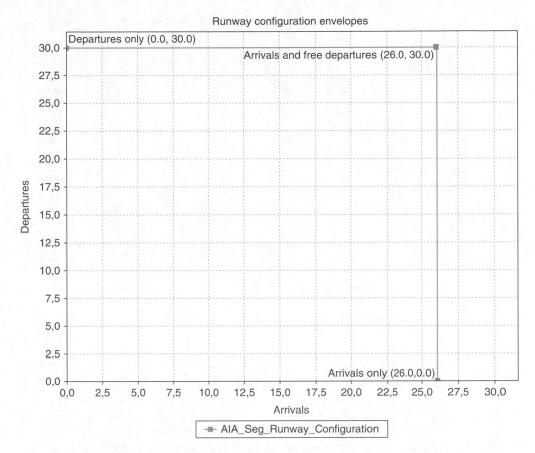

Figure 2.4   AIA runway capacity envelope (Scenarios 1 and 2)

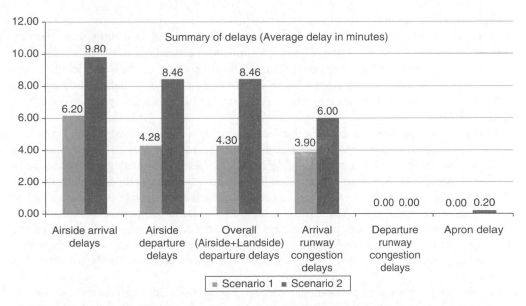

Figure 2.5   Aggregate delay statistics (Scenarios 1 and 2)

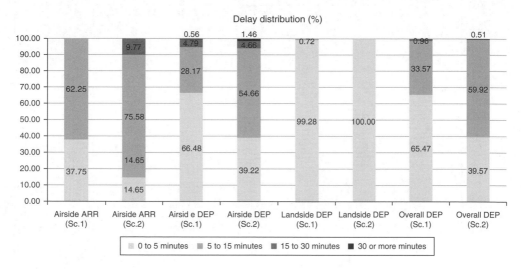

Figure 2.6   Delay distribution statistics (Scenarios 1 and 2)

both scenarios with delays being mainly accumulated on the airside and particularly the arrival process. The main reason behind airside arrival delays seems to be runway congestion in the arrival process with apron congestion having a negligible effect on delay accumulation. Overall, landside does not seem to be a constraining factor since it hardly contributes in the departure delays[5] that are due to airside bottlenecks. The latter findings signify an important added value of the SPADE system in that it particularly detects bottleneck areas/elements and their impact on the performance of the airport system in its entirety.

Having traced the main problematic areas/elements, the airport decision maker might need to drill further down into the delay patterns and their evolution throughout the day (Figures 2.6 and 2.7). In particular, Figure 2.6 confirms the existence of bottlenecks mainly in the airside arrival process for both scenarios with only few exceptional delays in the airside departure process. Moreover, there is an apparent trend of delays moving from 0–5 min (Scenario 1) to 5–15 min (Scenario 2). Again, landside does not appear to contribute at all in delays.

At a more detailed level of analysis, the decision maker is also provided with the option to further investigate the delay patterns and explore possible roots of congestion by contrasting traffic with capacity. For example, a tangible evidence of the airside bottlenecks appearing in the arrival process for Scenario 2 is provided in Figure 2.7. It presents the hourly distribution of arrival delays on the airside vis-à-vis the scheduled number of arrival movements per hour for the optimistic traffic growth scenario (Scenario 2). It is important to observe that although congestion and delays are currently (Scenario 1) kept at reasonably low levels (based on the existing infrastructure and traffic levels), the airport would suffer from severe congestion for a major part of the day (07:00–17:00) under a traffic growth scenario like the one modelled in Scenario 2.

As far as arrivals are concerned, demand (i.e. scheduled movements) would be constantly close (or equal) to nominal arrival capacity (26 arrivals) after the first morning

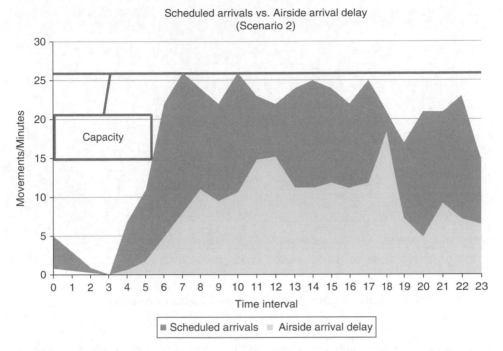

Figure 2.7   Hourly distribution of airside arrival delays versus scheduled arrivals (Scenario 2)

peak, hence resulting in airside arrival delays of more than 10 min on average for each hour between 08:00 and 19:00. Arrival delays seem to closely follow the arrivals' profile, but with a time lag of 1–2 hours between the time when the demand rate peaks and the time that delays reach their peak, respectively. Notable delays build up throughout the day, while delays increase rapidly during the evening peak when the system operates at capacity for a significant time interval. The latter provides a clear indication of the saturation of runway capacity and highlights the temporal propagation effect of delays, which is basically due to the lack of a time buffer to allow a full airport system recovery between or during arrival waves.

The airport decision maker has also the option to analyse the performance of the landside/terminal facilities such as check-in counters, security control points, and gate lounges. Figure 2.8 presents some exemplary check-in queue (i.e. queue length) statistics for Scenario 2. One can easily spot three spikes in the queue length, in early morning, early and late afternoon. By crossing these data with the opening times of the relevant check-ins, it can be easily found that the queue build up is due to the early presentation of passengers before the actual opening of the check-ins. Certainly, it is the responsibility of the terminal management to decide if this queue profile is acceptable or further action should be taken. In the latter case, UC1 could be used to assess the impact of an earlier opening of the check-ins on the queue spikes.

On the other hand, Figure 2.9 provides a summary of the level of service (LOS) statistics according to IATA standards (from A-excellent to F-unacceptable)[6] (IATA, 2004) for the

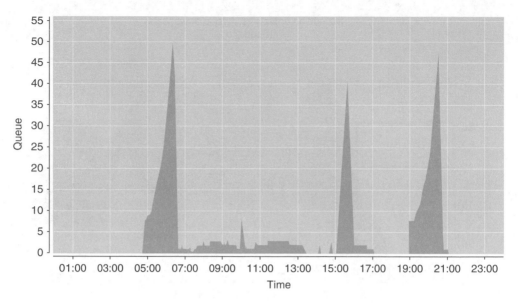

Figure 2.8   Hourly check-in queue statistics (Scenario 2)

terminal processing and holding facilities for both Scenarios 1 and 2. As normally expected, Figure 2.9(a) confirms that Scenario 2 will slightly deteriorate the landside level of service due to increasing traffic conditions. However, the deterioration seems to be quite modest: the percentage of facilities experiencing a lower level of service than in Scenario 1 is less than 2%. In Figure 2.9(b), one can see that the time spent in a facility is much more sensible to the increase in traffic. In the worst case, the time spent in the busiest check-in increases by 12.5% from 96 min in Scenario 1 to 108 in Scenario 2, while for the busiest security facility it increases by 8.6% from 93 to 101 min. This is somehow to be expected from elementary Queuing Theory since queuing time is growing more than linearly when demand rate increases.

Figure 2.9(a and b) provides a kind of summary situation of the landside as a whole. The decision maker might be interested to obtain similar statistics for every single terminal facility of interest. Figure 2.10 is a portion of a bigger table, where each row corresponds to a single facility such as the cluster of check-ins serving a specific airline, while each column defines the time interval. Each cell reports the worst (or the average, if so desired) LOS during the time period specified by its column in the facility identified by its row. The LOS table (partially) reported in Figure 2.10 allows easily detection of which specific facilities and time periods are the most critical and do deserve better and closer scrutiny.

Finally, congestion, delays, and level of service aspects represent one primary outcome of airport operations especially in case of temporal mismatches between demand and capacity. On the other hand, during the last two decades there has been a continuously increasing concern for externalities of airport operations other than delays; typical examples are environmental (e.g. noise, emissions) or safety/risk aspects. In that end, the SPADE system incorporates noise and safety assessment tools that enable the decision maker to obtain easily (without requiring integration efforts or time-consuming and data-intensive tool

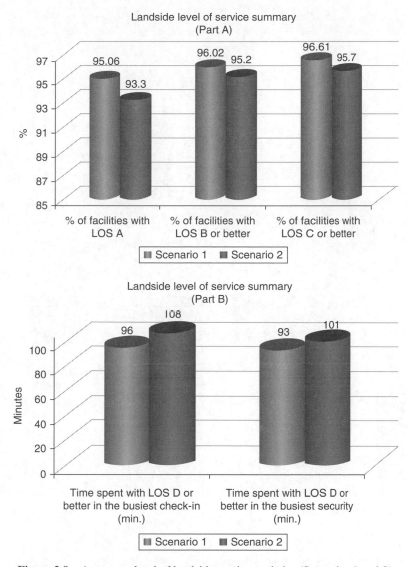

Figure 2.9   Aggregate level of landside service statistics (Scenarios 1 and 2)

set-up times) some first insights into the impact of airport operations on noise, external risk/ safety, as well as costs and benefits to airlines and the airport in its entirety. For example, Figure 2.11 provides an overview of indicative noise and external risk output in view of the conditions that occur in Scenarios 1 and 2. It particularly depicts noise levels (in dB), noise contours and affected areas (in km²) in the airport vicinity, as well as third-party risk levels (probability of accident within a specific area), risk contours and affected areas (in km²) within risk contours. Most importantly, one can directly trace the deteriorative impact of increased traffic (Scenario 2) in the form of increase in the affected surface/area (in km²) at various levels of noise (in dB) and risk (probability of accident).

Full screen	10:00	11:00	12:00	13:00	14:00	15:00	16:00	17:00	18:00	19:00	20:00	21:00	22:00
CheckinAF	A	A	A	A	A	C	A	A	A	A	A	A	A
CheckinBA	A	A	A	F	F	A	A	A	A	A	A	A	A
CheckinCO	A	A	A	A	A	A	A	A	F	A	A	A	A
CheckinCY	A	A	A	A	A	A	A	A	A	A	A	A	A
CheckinDL	A	A	A	A	A	A	A	A	A	A	A	A	A
CheckinGeneral	A	A	A	A	A	A	A	A	A	A	A	A	A
CheckinIB	A	A	B	A	A	A	F	A	A	A	A	A	A
CheckinKL	A	A	A	A	A	A	A	A	A	A	A	A	A
CheckinLH	E	A	A	A	A	F	F	A	A	A	A	B	A
CheckinLX	A	A	A	A	A	A	A	A	A	F	F	A	A
CheckinOA	A	A	A	A	A	A	A	A	A	A	A	A	A
CheckinOS	A	A	A	A	A	A	A	A	A	A	A	A	A
CheckinSK	A	A	A	A	A	A	F	F	A	A	A	A	A
CheckinSU	F	A	A	A	A	F	F	A	A	A	A	A	A
MarrPassportControl	A	A	A	A	A	A	A	A	A	A	A	A	A
MDepPassportControl	A	A	A	A	A	A	A	A	A	A	A	A	A

Figure 2.10 Detailed level of landside service statistics (Scenario 1)

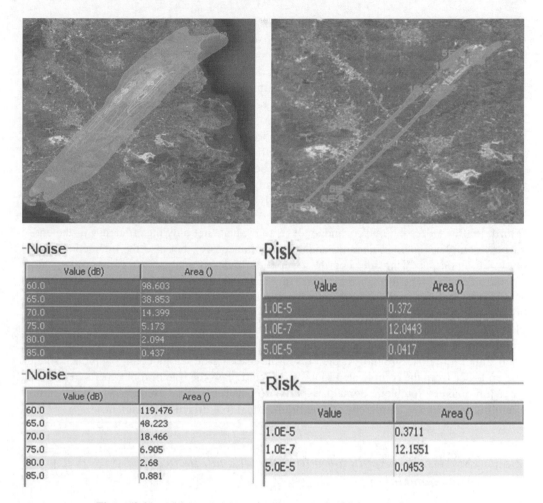

Figure 2.11   Noise and risk output (Scenario 1 top, Scenario 2 bottom)

2.3.1.4   Conclusions

The previous sections presented an application of the SPADE system pertaining to the strategic level of decision making. This application was based on a test study performed with UC1, titled "Airport Capacity Management (ACM)" with Athens International Airport (AIA) acting as the airport demonstration test bed. UC1 deals with the assessment of the impact of changes in airside or landside infrastructure, operational procedures or standards, and/or the distribution/volume of traffic. Moreover, it addresses various measures of airport performance and enables trade-off analysis between capacity, delays, level of service, safety, noise and cost-benefit indicators. UC1 produces output related to the aforementioned performance indicators at various levels of detail (e.g. aggregate, detailed). Aggregate output includes daily capacity versus demand figures, overall (airside and landside) delay figures for both arrivals and departures, airside delay figures by cause (e.g. due to runway congestion, apron congestion), runway capacity envelopes, as well as landside summary and level of service figures. On the other hand, detailed output contains results such as delay distributions, hourly statistics, facilities' statistics and performance, list of flights delayed, detailed external risk/safety and noise output (including maps), as well as detailed cost-benefit output for airlines and the airport as a whole.

Several advantages of the instantiation of the SPADE system for strategic decision making (UC1) can be identified. First, it builds on existing modelling capabilities by integrating tools (e.g. SLAM, MACAD, INM) that have been already extensively validated in several airports worldwide (Brunetta et al., 1999; Stamatopoulos et al., 2004; Long et al., 2009). Second, it serves as a quick diagnosis tool for senior airport executives/decision makers since it detects easily (and macroscopically) bottlenecks, inefficiencies, or problematic areas for which a more detailed and focused investigation might be necessary. As a matter of fact, the strategic module of the SPADE system integrates mostly macroscopic models that are simple and quick, less data-intensive and have a less costly set up process for different airports (at the expense of level of detail though) as compared to most of the existing, simulation-based tools. Third, it allows decision makers and analysts to evaluate not only the efficiency of the entire airport complex (both airside and landside simultaneously), but also the interaction effects between airside and landside. Fourth, it supports trade-off analyses by considering a large spectrum of airport performance measures (e.g. capacity, delays, level of service, noise, safety, costs and benefits). The latter signifies a basic contribution of the SPADE system as compared to the traditional, isolated use of tools or even previous integration attempts outside the SPADE environment since it alleviates the end user (e.g. airport executive/analyst) from the cumbersome and rather complicated technical process of integrating tools and exchanging their data on the background. At the outset, the use case-driven modelling approach shields the end user from the complicated tools' world and plays the role of tool experts in assisting decision makers (problem domain experts) to answer several frequently arising airport planning questions in a problem-oriented rather than tool-driven environment.

2.3.2   SPADE DSS Application for Operational/Tactical Decision Making

In this section, an application of the operational/tactical decision making functionality of the SPADE system will be discussed. This application is based on a test study performed with Use Case 7, titled "Airport Capacity versus Environmental Capacity", and Use Case 9, titled "Impact of New Procedures and/or Equipment", with Amsterdam Airport Schiphol as the airport demonstration test-bed[7]. The study's objective is to demonstrate the main decision support capabilities of Use Case 7, assessing the effects of a change in the air traffic volume on airport performance, and the main decision support capabilities of Use Case 9, dealing with the introduction of new or improved operational procedures on airport performance. The two use cases will be applied sequentially: the application of Use Case 7 will be followed by the application of Use Case 9. On the basis of the application of Use Case 7, the impact of a (forecasted or anticipated) short- to medium-term increase in air traffic is assessed with regards to delay, noise and third-party risk. Such an increase is expected to have some negative impacts on airport performance (e.g. in terms of delays and noise). The application of Use Case 9 will enable the investigation to what extent the (expected) negative impacts of this increase (assessed through Use Case 7) can be alleviated or reduced to a minimum by assessing the impact of a reduction in aircraft turnaround times. A reduction in turnaround times could be, for instance, the result of more efficient airport operations at the apron due to the introduction of improved Collaborative Decision Making (CDM) by the main airport stakeholders involved (e.g. airlines, airport, ground handlers).

2.3.2.1 Description of Use Case 7 (Airport Capacity versus Environmental Capacity) and Use Case 9 (Impact of New Procedures and/or Equipment)

As mentioned earlier, the application of the SPADE system pertaining to the operational/ tactical level of decision making is based on a test study performed with Use Case 7 and Use Case 9, which are further described next.

Use Case 7

Use Case 7 (UC7), titled "Airport Capacity versus Environmental Capacity", considers issues related to airport airside and landside analysis, including capacity, delays, level of service, third-party risk, noise, and cost-benefit performance indicators. The specific types of FAQs supported by UC7 are: (1) what is the maximum environmental (noise) capacity of the airport? (FAQ1); (2) what is the impact of a noise optimized flight schedule on airport airside performance, terminal performance, third-party risk and economics? (FAQ2); (3) what is the impact of additional flights to airport airside performance, terminal performance, noise, third-party risk and economics? (FAQ3); (4) what is the maximum airside capacity of the airport? (FAQ4); and (5) what is the impact of an airside capacity optimized flight schedule on terminal performance, noise, third-party risk and economics? (FAQ5).

The use case is intended to support the decision making needs of three broad types of potential users. First, airport management/flight management involved in constructing flight schedules to assess the airport performance under short- to medium-term forecasts of air traffic with major emphasis on identifying bottlenecks and satisfying environmental (mainly noise) regulations. Second, national government concerned with airport and environmental regulations. Third, research institutes and consultancy agencies interested in investigating the future performance of the airport under anticipated levels of traffic in order to seek innovative ways to deal with sharp traffic increases and more severe environmental regulations.

The core tools integrated into UC7 are TRAFGEN (SPADE-2, 2009), TAAM (Jeppesen, 2013), and SAMANTA (INCONTROL, 2011). TRAFGEN is a tool used to generate flight schedules. A flight schedule generated by TRAFGEN is then processed by TAAM to simulate the airport airside processes (including the runway system, taxiway system, apron area, and the terminal manoeuvring area) with regards to capacity and delays of airside operations. It supports the examination of alternative airport configurations, processes, and operating procedures and rules. TAAM generates a modified flight schedule that incorporates the airside delays. This modified flight schedule is processed by the simulation tool SAMANTA to assess the airport terminal performance with regards to passenger and baggage throughput, delays and level of service, enabling the investigation of alternative configurations and/or operations of the various terminal facilities (e.g. check-in, security, lounges, and handling systems). TAAM and SAMANTA have been validated for various airports and have been successfully integrated within the framework of several European R&D projects (e.g. OPAL, SPADE, SPADE-2).

The integration of TRAFGEN, TAAM and SAMANTA allows the simultaneous analysis of capacity, delay and level of service of the airport airside and terminal operations at

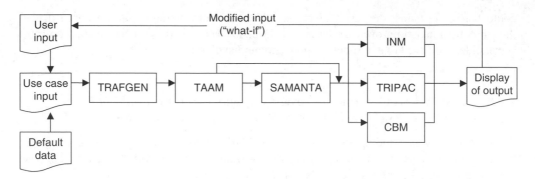

Figure 2.12   Use Case 7 workflow

microscopic level. If no other performance trade-offs are involved in the analysis, the use case execution stops and proceeds with the display of the output. Otherwise, part of the output (mainly the flight schedule, delays, and trajectory data) of the integrated TRAFGEN-TAAM-SAMANTA run is directed to tools such as INM (for noise assessments based on the flight schedule and trajectories generated by TAAM), TRIPAC (for addressing third-party risk based on the trajectories generated by TAAM), and CBM (for the calculation of cost-benefits for various airport stakeholders, using the airport airside and terminal delays generated by TAAM and SAMANTA).

Once the appropriate set of integrated tools has performed its computations, the results of the study are displayed to the user, enabling him/her to examine the relationships between changes in air traffic and the selected airport performance measures. Figure 2.12 displays the tool workflow in the case that all airport performance measures are considered.

Similar to UC1 (see Section 2.3.1.1) the definition, set-up, and execution of a specific scenario for UC7 with the tool workflow presented above require a long list of input parameters. These parameters can be either provided directly by the user or alternatively retrieved from default data sets that are pre-stored into the system's global data base. The user has the flexibility to select only those input parameters for which he/she desires to specify input, while all remaining input will be populated with default data associated with the specific airport and scenario under consideration. This functionality provides much flexibility to the user and reduces substantially the cumbersome input specification process that is particularly applicable to some data-intensive tools used in UC7. Overall, the UC7 input requirements can be classified under the following categories of data:

- Airport layout,
- airspace route and sector layout,
- airport usage, operational and service characteristics,
- air traffic control rules,
- aircraft performance characteristics,
- aircraft trajectories and routes, and SIDs/STARs/route selection,
- air traffic and fleet data,
- passenger and baggage data,
- aircraft noise parameters,
- population density data (i.e. data of population living in the vicinity of the airport),

- risk parameters, and
- cost-benefit parameters.

The main categories of output of UC7 are:

- Airport airside capacity, delay and efficiency,
- airport terminal capacity, delay and efficiency (passengers and baggage),
- noise (e.g. noise levels at specific locations, (surface of) noise contours and number of people living within noise contours),
- third-party risk (e.g. probability of an accident within a specific area, (surface of) risk contours and number of people/houses within risk contours), and
- economy (e.g. cost-benefit for the airport and cost for airlines).

The output can be displayed at various levels of detail. These levels range from key (or aggregated) values for capacity, cost-benefit, efficiency/delay, noise and third-party risk, to detailed performance indicators for a single airport facility, to noise or third-party risk contours for a specific noise level or third-party risk level, respectively, and even to the actual output files generated by individual tools.

Use Case 9

Use Case 9 (UC9), titled "Impact of New Procedures and/or Equipment", considers the impact of implementing new procedures or equipment at an airport on return on investment and operational benefits (e.g. in terms of efficiency and level of service), but also on environmental aspects such as noise and third-party risk. The analyses in UC9 are built around three FAQs that deal with the impact of: (1) a change in the night regime time slot (FAQ1), (2) the application of reduced separation minima (FAQ2), and (3) a change in aircraft turnaround times (FAQ3).

   The use case is intended to support the decision making needs of four broad types of potential users. First, airport management concerned with assessing the impact of the installation and use of new airport equipment, and the effects of the implementation of new airport procedures. Second, airline fleet procedure/equipment managers involved in dealing with the impact of the implementation of new flight procedures and airport equipment on airline operations. Third, civil aviation authorities involved in the regulation aspects associated with the implementation of new airport procedures and equipment. Fourth, departments of environmental policy (at national and regional levels) dealing with the impact of new procedures and equipment on environmental regulation.

   UC9 deals with new procedures and equipment, but not with changes in the air traffic volume and composition, and therefore, it does not incorporate any traffic generator tool like TRAFGEN. However, with the exception of TRAFGEN, the same (core) tools are integrated into UC9 as into UC7. Moreover, the tool workflow of UC9 is identical to the one of UC7, once the tool TRAFGEN is omitted in Figure 2.12.

   Similar to UC1 and UC7, the definition, set-up, and execution of a specific scenario in UC9 with the tool workflow require input parameters that can be either provided directly by the user or alternatively retrieved from default data sets that are pre-stored into the system's global data base. Further, UC9 has the same categories of input and output as use case UC7, and its output can be presented at the same levels of detail as in use case UC7.

2.3.2.2 Scenario Description

The demonstration of the SPADE system for operational / tactical decision making is based on a test study with Amsterdam Airport Schiphol (AAS) acting as the demonstration test-bed airport. AAS is the main airport in the Netherlands, and it is located at about 15 km to the south-west of the centre of Amsterdam. In 2010, the total number of air transport movements was about 386 000 corresponding to 45.1 million passengers and making use of a network consisting of 301 destinations (Schiphol Group, 2011), ranking AAS as the fifth-largest airport in Europe in terms of air transport movements and number of passengers. In terms of cargo, AAS is ranked as the third largest in Europe, with more than 1 512 000 tons of air cargo handled in 2010 (Schiphol Group, 2011).

At the airside, AAS is one of Europe's most complex airports. It has five main runways (and a secondary runway for general aviation). Figure 2.13 shows the runway layout. The runway

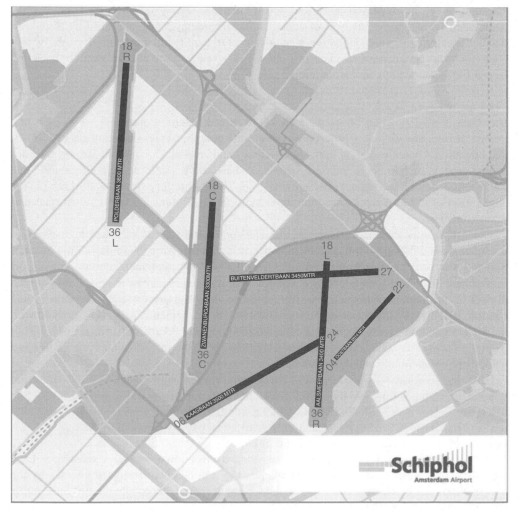

Figure 2.13 Amsterdam Airport Schiphol runway layout

configuration enables AAS to cope with variations in wind direction without a significant impact on capacity. During outbound peaks, two main runways for departure and one main runway for arrival operations are normally used. During inbound peaks, two main runways for arrival and one main runway for departure operations are used. At the landside, AAS has a single terminal building for passenger arrivals and departures of both domestic and international flights. The terminal can be accessed by road (car, taxi or bus) and rail (through an underground railway station that is connected to the Dutch railway network).

Three scenarios were developed for demonstration purposes. The main objectives of these scenarios were, firstly, to investigate the impact of an increase in air traffic on airport performance (including the identification of bottlenecks and the assessment of several types of performance measures resulting from this increase), and, secondly, to assess how a reduction in aircraft turnaround times might limit the (expected) negative effects of this traffic increase on airport efficiency. Scenario 1 is the baseline scenario with a single day of air traffic, which is based on a peak day in 2007 with about 1340 aircraft movements as displayed in Figure 2.14. It reflects the airport infrastructure and operational rules, procedures and policies in 2007.

Scenario 2 is a modified version of Scenario 1 and pertains to the application of FAQ3 of UC7 (what is the effect of additional flights on the overall airport performance?). It represents a potential short- to medium-term increase in air traffic by in total 2.7% compared to Scenario 1, to enable airport management to identify any bottleneck in the airport processes and check for a violation of environmental regulations due to this change in traffic. The results of Scenario 2 could lead airport management to examine ways accommodating this increase and limiting any resulting negative impact on, for instance, the efficiency of airport operations, the level of service to the travelling public, and the quality of the surrounding environment.

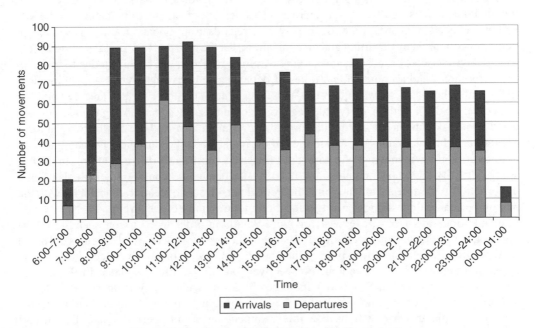

Figure 2.14 Hourly distribution of traffic (Scenario 1)

Scenario 3 is a modified version of Scenario 2 and is an application of FAQ3 of UC9 (i.e. a change in aircraft turnaround times). This application can be considered as a potential, short-to medium-term means to alleviate the (expected) negative impacts of the increased traffic in Scenario 2 on airport performance in terms of, for instance, delay and level of service to passengers. A reduction in aircraft turnaround times could be the result of the application of Collaborative Decision Making (CDM). CDM could yield more efficient airport operations at the apron, for instance, through information sharing by airport users. Such a reduction could be of specific interest to an airport like AAS, since AAS is a major, global hub airport with many transfer passengers, but also of interest to airlines operating at AAS. In Scenario 3, the impact of a reduction in aircraft turnaround times by 10% (as compared to Scenario 2), reflecting the resulting improved efficiency of apron processes due to the implementation of CDM, is assessed with regards to delay and level of service.

2.3.2.3   Discussion of Results

This section presents some indicative results[8] of the AAS test study. The demonstrated scenarios aim to provide decision support to, among other airport stakeholders, airport management on a short- to medium-term horizon by: (1) assessing the integrated impact of an increase in traffic (i.e. comparing Scenario 2 with 1) and (2) assessing the potential efficiency improvement of reduced aircraft turnaround times to limit the negative impacts of traffic increase (i.e. comparing Scenario 3 with 2).

The discussion starts with the presentation of the airport performance in the baseline scenario (i.e. Scenario 1), especially concerning airport airside performance. As mentioned in Section 2.3.2.1, the SPADE system displays the results of a scenario at various levels of detail. First, it presents the key or aggregated values of performance indicators. Some key values are listed in Figure 2.15 for Scenario 1 with regards to airport airside efficiency (average delays encountered by departing and arriving aircraft at the gate due to processes at the airside (e.g. delays caused by runway congestion or resulting from delayed pushback due to traffic in the apron area) and average ground times of departing and arriving aircraft – including taxi times and delays, and delays in line-up), noise (legal noise contour: surface area and number of people inside), and third-party risk (legal risk contour: surface area). A legal noise contour and a legal risk contour are or could be used for legislation, that is, these could act as national regulatory reference values applied to the airport under study. For instance, Dutch regulation defines contours within which the number of houses is constrained (by regulation). In the test study, a 55 dB legal noise contour and a 10^{-6} risk contour are considered, with risk being the probability of decease of a notional/theoretical person due to an aircraft accident, with the person permanently residing at a single location in the vicinity of the airport.

In addition, the user can request for more details in order to focus on specific airport processes or elements, or specific performance indicators. For Scenario 1, Figures 2.18 and 2.19 provide more details by exploring the passengers flow in one of the check-in areas and by examining specific noise contours, respectively, as will be discussed later. In Scenario 2, the air traffic analysed in Scenario 1 increases by 2.7%, without any further change to the airport and/or its environment and processes, resulting in about 1375 aircraft movements. Obviously, this increase will have a direct impact on airport performance. The airport management can first have a glance on the impacts of this increase and then focus on specific

Efficiency

Average departure Gate delay:	34.43	Seconds per movement
Average arrival Gate delay:	0.0	Seconds per movement
Average departure Taxi time:	477.21	Seconds per movement
Average arrival Taxi time:	537.62	Seconds per movement

Noise

Legal noise contour: ❶	55.0	*dB*
Noise contour surface:	147.623	km²
Number of people within noise contour:	50484	People

Risk

Legal noise contour: ❶	1.0E−6	*decease probability per year*
Risk contour surface:	0,4	km²

Figure 2.15   Key values (Scenario 1)

Efficiency

Average departure Gate delay:	44.45	Seconds per movement
Average arrival Gate delay:	0.0	Seconds per movement
Average departure Taxi time:	784.45	Seconds per movement
Average arrival Taxi time:	572.27	Seconds per movement

Noise

Legal noise contour: ❶	55.0	*dB*
Noise contour surface:	156.364	km²
Number of people within noise contour:	72677	People

Risk

Legal noise contour: ❶	1.0E-6	*decease probability per year*
Risk contour surface:	0,4	km²

Figure 2.16   Key values (Scenario 2)

performance indicators into more detail. The key values as displayed in Figure 2.16 (which are the same as for Scenario 1 in Figure 2.15) provide a global view for the airport management on the impact of the traffic increase.

As expected, a direct comparison of Figures 2.15 and 2.16 confirms that the overall airport efficiency deteriorates due to this traffic increase. In general, the computed delays for

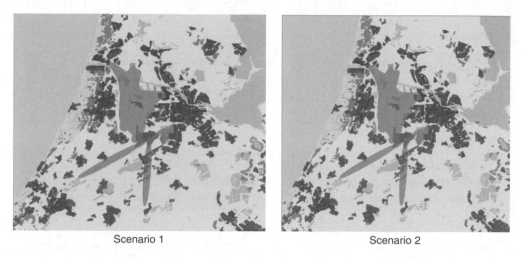

Scenario 1   Scenario 2

Figure 2.17   Noise: legal noise contour (set at 55 dB(A) in the test study)

Scenario 1 deteriorate in Scenario 2 at the airport airside. This deterioration mainly manifests itself in the departure processes rather than the arrival processes. A reason for this could be that the airport airside operates more often at maximum capacity in departure peak periods in Scenario 1, so that the effect of a few additional flights in these periods is already significant. These findings illustrate that the exploration of the key values for efficiency as provided by the SPADE system enables some first diagnosis of the main bottleneck areas/elements and their impact on the overall efficiency of the airport system.

For the airport management, noise is another important aspect of airport performance. Environmental impacts of aviation are receiving increasing attention, and many airports are already subject to noise regulations. The key values for noise provide direct information to the airport management as to whether noise regulations have been met or not in terms of the number of people exposed to the 55 dB(A) noise level, which is a specific, regulatory noise level applied in this test study. Moreover, by comparing these key values with those of Scenario 1, airport management can assess the rate at which the surface of the legal noise contours increases and at which the number of people living inside this noise contour increases. In particular, the increase in the latter suggests that city districts with relatively many inhabitants are located quite close to the airport, since the fairly limited increase in surface area of the legal noise contour by 6.0% results in an increase of 44% in the number of people, while the location and shape of the contour do not change significantly as depicted in Figure 2.17. These values imply that an increase in air traffic should be accompanied by noise abatement procedures to reduce the impact on the population living in the vicinity of the airport. Finally, the key values displayed in Figures 2.15 and 2.16 indicate that the small traffic increase has no significant impact on third-party risk. Hence, third-party risk regulations do not seem to pose restrictions on a small growth in traffic at AAS.

Having identified the main problematic areas, airport management might need to focus on the details. By focusing on the various airport processes and elements, airport management can identify the airport processes/elements constituting bottlenecks in the total airport process and assess the level of service offered to passengers. Figure 2.18 displays the passengers flow

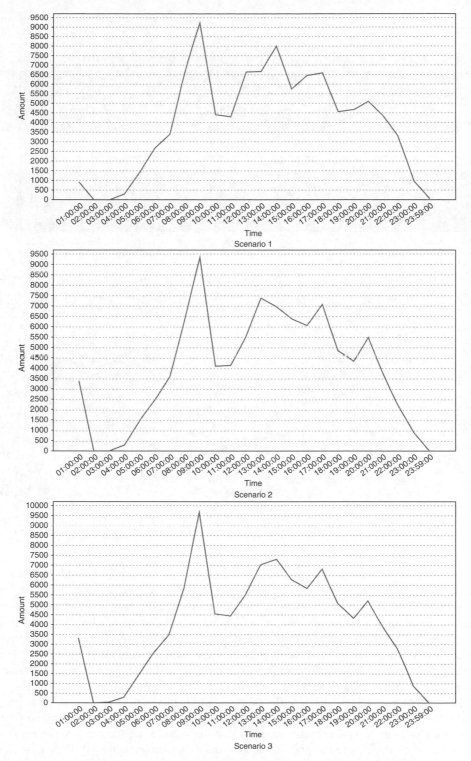

Figure 2.18 Passengers flow (in number of passengers) at Check-in Facility 1 for each of the scenarios

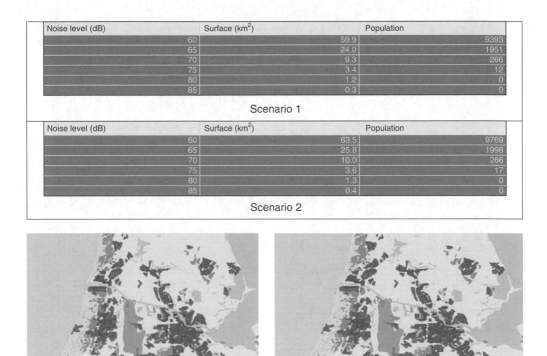

Noise level (dB)	Surface (km^2)	Population
60	59.9	9393
65	24.0	1951
70	9.3	266
75	3.4	12
80	1.2	0
85	0.3	0

Scenario 1

Noise level (dB)	Surface (km^2)	Population
60	63.5	9769
65	25.8	1998
70	10.0	266
75	3.6	17
80	1.3	0
85	0.4	0

Scenario 2

Scenario 1   Scenario 2

Figure 2.19   Noise: surface area and population in specific noise contours, and 60 dB(A) noise contour

in one of the check-in areas for Scenarios 1 and 2. It appears that the fairly moderate increase in traffic does not have a significant impact on the passengers flow in this check-in area. Moreover, if the current level of service is considered to be appropriate, no additional measures concerning resources or procedures have to be taken in order to – at least – maintain the current service levels in this check-in area. In other words, the level of service offered to passengers in this area does not deteriorate with the passenger increase in Scenario 2. These findings enable airport management to further analyse actual bottlenecks in order to ensure an efficient passenger handling and, hence, a seamless passenger travel experience.

It is evident that the increase in traffic is expected to enlarge the noise exposure to people in the vicinity of the airport. Figure 2.19 directly confirms this expectation by displaying noise results for Scenario 2. Again, it also suggests that the suburbs are located fairly close to the airport, since the increase in the population exposed to, for instance, the 60 dB(A) noise level is 4.0% and that of the surface area of the associated noise contour 6.0%, exceeding the increase in air traffic, while the location and shape of the contour hardly change. In other words,

Efficiency		
Average departure Gate delay:	44.45	Seconds per movement
Average arrival Gate delay:	0.0	Seconds per movement
Average departure Taxi time:	784.45	Seconds per movement
Average arrival Taxi time:	572.27	Seconds per movement
Scenario 2		
Average departure Gate delay:	38.82	Seconds per movement
Average arrival Gate delay:	0.0	Seconds per movement
Average departure Taxi time:	804.84	Seconds per movement
Average arrival Taxi time:	572.4	Seconds per movement
Scenario 3		

Figure 2.20   Key values (Scenario 2: top and Scenario 3: bottom)

these figures suggest that airport management may add flights with caution, because noise regulations might be violated. In that case, an increase in air traffic should be accompanied by additional measures to alleviate external impacts on the environment, safety/risk and so on.

The application of UC7 demonstrates certain decision support capabilities of the SPADE system as a means of assessing the impact of a traffic increase. The results revealed some negative effects due to this increase, such as the deterioration of delays and the noise exposure to the surrounding community. On the one hand, these effects might render AAS less attractive to passengers and airlines. On the other hand, these effects might impose governmental constraints, since the legal noise and/or risk contours may go beyond legal limits. Hence, under the present conditions of the test study, airport management should raise the question: how to limit some of these negative impacts; for instance, the impact on delay and level of service?

Many measures could be investigated to reduce negative impacts of an increase in traffic, but only a few might be implemented by the airport operator and its users on a short- to medium-term timeframe. One such measure could be the further enforcement of Airport Collaborative Decision Making (CDM), through an improved cooperation between all airport actors by sharing information, as supported by (among others) Eurocontrol, ACI Europe and IATA (Eurocontrol, 2012). Airport CDM aims to improve the overall efficiency of airport operations, with a particular focus on the aircraft turnaround and pre-departure sequencing processes, resulting in a reduction in aircraft turnaround times. The latter could make AAS again more attractive to both passengers and airlines, and subsequently generate more airport revenues.

UC9 provides the capability to assess the impact of reduced aircraft turnaround times and present them in the same format as for UC7. Aircraft turnaround times are expected to have neither a major impact on the overall noise impact and third-party risks, nor on the handling of departing passengers inside the airport terminal (see for instance Figure 2.18 on passengers flow in one of the check-in areas: the number of passengers passing through the check-in area per hour does not change significantly). The presentation of the indicative results of this part of the test study is therefore limited to the key values of airport airside efficiency as displayed in Figure 2.20.

The key values indicate that a 10% reduction in aircraft turnaround times would lead to a significant reduction (about 28%) in the average departure gate delay. However, there seems to be some minor negative side-effects with regards to the average taxiing time for departing aircraft. Although the latter results have not been investigated into depth, a reason for these side-effects might be that shorter turnaround times (without any change to the airport and/or any further change to its environment and processes) effectively result in an increased arrival rate at the departure runways. All these findings demonstrate that the SPADE system enables airport management to explore whether a specific reduction in aircraft turnaround times would result in the demanded improvement in the airport performance or whether such a reduction should be part of a broader range of enhancements in various airport processes.

2.3.2.4 Conclusions

The previous subsections presented an application of the SPADE system associated with the operational/tactical level of decision making. This application was based on a test study conducted with two use cases: (1) UC7, titled 'Airport Capacity versus Environmental Capacity', and (2) UC9, titled 'Impact of New Procedures and/or Equipment', with Amsterdam Airport Schiphol acting as the airport demonstration test-bed. In this application, UC7 dealt with the assessment of the impact of a short- to medium-term increase in air traffic on the overall airport performance, including airport airside and landside efficiency, noise, and third-party risk. UC7 provided the results of these performance indicators at various levels of detail, ranging from key (or aggregated) values to detailed values for specific airport processes/facilities, and contours for a specific noise or third-party risk level. UC9 analysed the impact of a decrease in aircraft turnaround times (for instance, as a result of improved apron processes due to the introduction of CDM), and produced the same types of output as UC7.

The application of the SPADE system through UC7 and UC9 demonstrated the system's capability to enable airport stakeholders and policy makers to perform integrated impact analyses and trade-off studies. Similar to UC1, various advantages can be identified. First, it builds on existing modelling capabilities by integrating extensively validated and widely used tools, which have been applied to several airports worldwide. Second, it provides detailed decision support to airport management to investigate (potential) changes in the air traffic volume and composition as well in airport processes, enabling the detection of possible problematic areas (e.g. bottlenecks or violation of regulations) and allowing for a timely quest for ways to efficiently and effectively alleviate these problematic areas. Third, it enables airport management to evaluate the interaction between processes at the airport airside and the airport terminal, and perform trade-off studies. This illustrates an essential contribution of the SPADE system as compared to the traditional, isolated use of tools. Finally, the use case approach allows users to focus on the airport question at hand rather than the complicated technicalities of integrating tools, collecting and providing data for each of the tools, and exchanging data between tools.

2.4 Conclusions

Stakeholders and policy makers involved in the airport decision making process are asked to make decisions, draw policy directions, and operate in a quite complicated environmental, institutional, and operational setting. They often face challenging decision making questions with strong interdependencies and often conflicting objectives (trade-offs).

Existing tools can successfully address all decision making levels and provide decision support to various airport studies, but they practically exhibit limited applicability and usefulness to high level decision making in airport planning and design, on the grounds that their use requires in-depth technical knowledge and expertise. In practice, the decision maker is interested in obtaining an insight and examining alternative (even approximate) solutions and 'what-if' scenarios into particular strategic or tactical problems. However, existing tools require substantial familiarity and prior knowledge to tackle with their use, build scenarios addressing the study objectives, and prepare the appropriate data sets pertaining to the given study and scenario.

To this aim, the SPADE DSS allows the user to obtain the required analysis results without the burden of having to master each tool technicalities. It is like allowing to 'driving a car' without the necessity of being a 'mechanical engineer'. The latter can be achieved by the use case-driven architecture applied in SPADE. In the SPADE context, a use case is a 'wizard-type', integrated use of a set of tools assisting airport-domain experts in addressing frequently arising airport development, planning or design problems without requiring familiarity with the tools themselves. This concept enables the user to perform the analysis under consideration through 'pre-structured' and built-in, 'wizard-type' navigation aids in a single run by shielding the user from the complicated model and tool world and thus enabling him/her to focus on the real question to be addressed. Besides, another important advantage brought by SPADE is its capability of dealing with the study of essential trade-offs faced by the airport decision makers, capability that is not offered by any individual tool.

At the outset, the key benefit of the SPADE system is its strong and positive impact on two key strategic aspects: (1) the improvement of the airport decision making process quality through an integrated and systematic impact analysis, for example, of capacity, delay, noise, safety, and (2) the homogenization and rationalization of this process at a European level by addressing a standard set of questions or use cases related to airports. Furthermore, it is important to stress the fact that by integrating different tools that are each concerned with a different part of the airport complex, like, for instance, a tool devoted to airside (aircraft) and another devoted to landside (passengers and baggage), one obtains hints, insights and conclusions that would not be achievable by running the individual tools in isolation. For example, the influence and the implications of airside delays on landside operations could not be detected nor properly interpreted without the seamless integration of all concerned tools.

A proof of the SPADE operational concept and system development has been already provided through the successful demonstration and evaluation of the SPADE suite of use cases on the basis of selected test scenarios at major European airports acting as demonstration test-beds (e.g. Athens, Malaga, Amsterdam-Schiphol, Naples). Based on the system demonstration and evaluation process, it was concluded that SPADE fulfils the users' expectations and elicited user requirements, it produces results at a good approximation of reality, and can be efficiently used to model and support real-world airport planning problems. Most importantly, potential end users of the system and airport stakeholders confirmed the importance of the key subject addressed by the system (the integration of tools for obtaining a 'total' airport view and providing support in trade-off analyses), and yielded constructive and valuable feedback/suggestions for future improvements. Directions for enhancements and further research efforts pertain mainly to the increase of complementarity among use cases, the examination of advanced technological capabilities enabling the synchronization of the SPADE database with external databases or other sources available to the end users, and the development of more sophisticated modelling capabilities for traffic generation purposes.

Acknowledgements

Part of the research work presented in this chapter has been supported by the European Commission (Directorate General Energy and Transport), under contracts TREN/04/FP6AE/S07.29856/503207 and TREN/06/FP6AE/S07.58054/518362, 'Supporting Platform for Airport Decision Making and Efficiency Analysis (SPADE I+II)' Research Projects (2004–2009) and the Hellenic General Secretariat of Research and Technology (GSRT), under contract SPADE-2/NC (NC-1375-01).

Notes

1   A more detailed discussion is provided in (Zografos et al., 2013).
2   The results presented in this study are only indicative and are discussed for demonstration purposes without implying any explicit endorsement or other commitment from the side of Athens International Airport.
3   Between 10–25 million passengers per year.
4   The presented results are based on exemplary data and assumptions for Athens International Airport and should be interpreted as indicative types of output used only for demonstration purposes rather than an actual planning or assessment exercise.
5   Landside delays are computed only for departures.
6   These standards reflect basically measures such as space (square meters) available per passenger, ease of flow/speed and so on.
7   The results presented in this study are only indicative and are discussed for demonstration purposes without implying any explicit endorsement or other commitment from the side of Amsterdam Airport Schiphol.
8   The presented results are based on exemplary data and assumptions for Amsterdam Airport Schiphol and should be interpreted as indicative types of output used only for demonstration purposes rather than an actual planning or assessment exercise.

References

Andreatta, G., Brunetta, L., Odoni, A. R., et al. (1999) A Set of Approximate and Compatible Models for Airport Strategic Planning on Airside and Landside. *Air Traffic Control Quarterly*, 7(4), 291–319.

Airport Research Center (ARC) (2011) *CAST Overview*. [Online] Website: http://www.airport-consultants.com [Accessed 19 February, 2013].

Arlow, J. and Neustadt, I. (2001) *UML and the Unified Process: Practical Object-Oriented Analysis and Design*. 1st Edition, Addison-Wesley Professional.

Athens International Airport (AIA) S.A. (2010) *AEROSTAT Handbook 2009*. Athens International Airport, Athens, Greece.

Brunetta, L., Righi, L. and Andreatta, G. (1999) An Operations Research Model for the Evaluation of an Airport Terminal: SLAM (Simple Landside Aggregate Model). *Journal of Air Transport Management*, 5(3), 161–175.

Buschmann, F., Meunier, R., Rohnert, H., et al. (1996) *Pattern-Oriented Software Architecture: A System of Patterns*. Volume 1, John Wiley & Sons, Ltd., Chichester.

Chaib-Draa, B. (2002) Cognitive Maps: Theory, Implementation and Practical Applications in Multiagent Environments. *IEEE Transactions on Knowledge and Data Engineering*, 14(6), 1–17.

Correia, A. R., Wirasinghe, S. C., and de Barros, A. G. (2008) A Global Index for Level of Service Evaluation at Airport Passenger Terminals. *Transportation Research Part E*, 44(4), 607–620.

Dutch Government (2013) *Regulations for Civil Airports.* http://wetten.overheid.nl/BWBR0026564/volledig/geldigheidsdatum_15-02-2013#Bijlage2 (In Dutch) [Accessed 15 February, 2013].

Eurocontrol (2008) *Long-Term Forecast: IFR Flight Movements 2008–2030.* v1.0, Forecast prepared as part of the Challenges of Growth 2008 project, Brussels.

Eurocontrol. (2010) *Aeronautical Information Exchange Model (AIXM).* [Online] Available from: http://www.aixm.aero/public/subsite_homepage/homepage.html [Accessed 19 February, 2013].

Eurocontrol (2012) *European Airport CDM.* [Online] Available on: http://www.euro-cdm.org [Accessed 19 February, 2013].

Federal Aviation Administration (FAA) (1989) *SIMMOD Reference Manual.* AOR-200, Office of Operational Research, Washington, D.C.

Federal Aviation Administration (FAA) (1997) *INM Technical Manual.* Version 5.1, Report No. FAA-AEE-97-04, Washington, D.C.

Gaines, B. R. and Shaw, M. L. G. (1995) Concept Maps as Hypermedia Components. *International Journal of Human-Computer Studies*, 43(3), 323–361.

Hellenic Civil Aviation Authority (HCAA) (2010) *Bulletin of Air Transport Statistical Data.* [Online] Available from: http://www.hcaa.gr/content/index.asp?tid=15 (In Greek) [Accessed 19 February, 2013].

INCONTROL Simulation Solutions (2011) *Enterprise Dynamics (ED) Brochures: The ED SAMANTA and ED Airport Suite.* [Online] Software Brochure available on: http://www.incontrolsim.com [Accessed 19 February, 2013].

International Air Transport Association (IATA). (2004) *Airport Development Reference Manual.* 9th Edition, Montreal – Geneva.

ISA Software (2011) *RAMS Plus Simulation Solutions: What is RAMS Plus?.* [Online] Software Brochure available on: www.ramsplus.com [Accessed 19 February, 2013].

Jeppesen (2013) *Total Airspace and Airport Modeler (TAAM).* [Online] Brochure available on: http://ww1.jeppesen.com/industry-solutions/aviation/government/total-airspace-airport-modeler.jsp [Accessed February 19, 2013].

Long, D., Hasan, S., Trani, A. A. and McDonald, A. (2009) *Catalog of Models for Assessing the Next-Generation Air Transportation System.* LMI Report NS802T2 for NextGen Airportal Project, Logistics Management Institute (LMI), VA.

Lucic, P., Ohsfeldt, M., Rodgers, M. and Klein, A. (2007) *Airport Runway Capacity Model Review.* Research Report by CSSI and Air Traffic Analysis, Inc. for FAA ATO-P Performance Analysis and Strategy.

Malone, K. M. (1995) *Dynamic Queuing Systems: Behavior and Approximations for Individual Queues and Networks.* PhD Dissertation, Operations Research Center, Massachusetts Institute of Technology, Cambridge, MA.

NLR (2013) *Third Party Risk Analysis.* [Online] http://www.nlr.nl/capabilities-iii/third-party-risk/ [Accessed February 19, 2013].

Odoni, A. R. (1991) *Transportation Modelling Needs: Airports and Airspace.* Volpe National Transportation Systems Center, Cambridge, MA.

Odoni, A. R., Bowman, J., Delahaye, D., et al. (1997) *Existing and Required Modelling Capabilities for Evaluating ATM Systems and Concepts.* Modelling Research under NASA/AATT, International Center for Air Transportation, Massachusetts Institute of Technology (MIT), Cambridge, MA.

Schiphol Group. (2011) *Annual Report 2010.* Schiphol Group, Schiphol, The Netherlands.

SPADE-2 Consortium. (2009) *Supporting Platform for Airport Decision-Making and Efficiency Analysis: Use Case Descriptions Reports.* Series of Technical Reports prepared within the framework of the SPADE-2 project funded by the European Commission, Directorate General for Energy and Transport, Brussels, Belgium.

Stamatopoulos, M. A., Zografos, K. G. and Odoni, A. R. (2004) A Decision Support System for Airport Strategic Planning. *Transportation Research Part C*, 12(2), 91–117.

Tosic, V. (1992) A Review of Airport Passenger Terminal Operations Analysis and Modelling. *Transportation Research Part A*, 26A(1), 3–26.

Wingrove, W. E., David, A. L., Kostiuk, P. F. and Hemm, R. V. (1995) *Estimating the Effects of the Terminal Area Productivity Program.* LMI Report NS301R3, Logistics Management Institute (LMI), VA.

Zografos, K. G.and Madas, M. A. (2006) Development and Demonstration of an Integrated Decision Support System for Airport Performance Analysis. *Transportation Research Part C*, 14(1), 1–17.

Zografos, K. G.and Madas, M. A. (2007) Advanced Modeling Capabilities for Airport Performance Assessment and Capacity Management. *Transportation Research Record*, Aviation 2007, 60–69.

Zografos, K. G., Madas, M. A. and Salouras, Y. (2013) A Decision Support System for Total Airport Operations Management and Planning. *Journal of Advanced Transportation*, 47(2), 170–189.

3

Measuring Air Traffic Management (ATM) Delays Related to Airports: A Comparison between the US and Europe

John Gulding[A], David A. Knorr[B], Marc Rose[C], Philippe Enaud[D] and Holger P. Hegendoerfer[D]

[A]*Federal Aviation Administration, Office of Performance Analysis, USA*
[B]*Office of International Affairs, Federal Aviation Administration, USA*
[C]*MCR LLC, USA*
[D]*Performance Review Unit, EUROCONTROL, Belgium*

3.1 Introduction

Managing delay and the predictability of flight times is critical to an airline's success. In the US alone, the Air Transport Association (ATA) estimates that in 2010 delays cost airlines more than $6.4 billion (Airlines for America, 2012). Costs to passengers are estimated to be another $17 billion (Brody Guy, 2010). Airport capacity constraints at airports are a primary source of delay in both the US and Europe (EUROCONTROL and FAA, 2012). Airports play a substantial role in absorbing delays from airspace constraints and other airports. The management of delays absorbed, both at airport gates and on taxi-ways, can impact airline costs and environmental emissions. Airports can also work with Air Traffic Control (ATC) services to influence delays taken in the air at congested airports. Understanding the magnitude and causes of delay associated with airports is critical to making the right investment decisions about technologies, procedures, and airport slot controls.

This chapter compares airport associated delays in the US and Europe. The primary focus is on ATM-related delays that can be influenced by Air Traffic Management (ATM) actions.

Modelling and Managing Airport Performance, First Edition. Edited by Konstantinos G. Zografos, Giovanni Andreatta and Amedeo R. Odoni.

ATM works to minimize delays through efficient use of available airport and airspace capacities under varying weather conditions. ATM plays a primary role in managing where delays are absorbed. The location where delay is absorbed can impact fuel burn, efficiency, and the safety of the air transportation system.

Establishing common definitions of delay is a key requisite to meaningful comparisons. In this chapter, US and Europe delays are compared for both departures and arrivals. Best practices in ATM at airports are also discussed. While demand management at airports is an additional tool for managing delays, analysis of strategies for constraining demand are outside the scope of this work.

For the purposes of this chapter, an ATM-related airport delay is defined as delay where ATM has an impact in managing the operations. Delays that occur inside the airport terminal (e.g., due to security checks) or in preparing aircraft for departure (fueling, baggage handling, etc.) are not considered as part of the ATM manageable pool.

In order to improve comparability of data sets, the analyses in this chapter were limited to controlled (IFR) flights from or to the 34 busiest airports in both the US and Europe. Data sources for this chapter come from airlines, ATM systems and airports. In the US, airlines are required to provide performance data to the US Department of Transportation if they have at least 1% of the total domestic scheduled-service passenger revenue flights (plus others who provide data voluntarily). In Europe, the Central Office for Delay Analysis (CODA) collects data from airlines on a voluntary basis. Currently some 60% of scheduled airline flights are providing data including more than 71% at the main 34 airports[1]. Both FAA and EUROCONTROL have databases on a flight by flight basis from their respective Traffic Flow Management Systems.

3.2   Operations at the Main 34 US and European Airports

This section provides general background on the main 34 airports in the US and Europe. From a standpoint of comparability while US airports are generally bigger, movements per runway are similar (see Table 3.1).

Table 3.1 also highlights that European airports serve 21% more passengers per runway. Figure 3.1 confirms that the average aircraft size (seats) is also larger in Europe.

Table 3.1   Passenger/operations indicators for the 34 main airports

Main 34 airports	**Europe**		**US**		*Difference US vs Europe*
	2010	*vs 2008*	*2010*	*vs 2008*	
Average number of annual IFR movements per airport ('000)	237	−9%	389	−6%	+64%
Average number of annual passengers per airport (million)	24	−3%	31	−3.1%	+29%
Passengers per IFR movement	102	+6%	80	3.1%	−22%
Average number of runways per airport	2.5	0%	4.1	0.7%	+64%
Annual IFR movements per runway ('000)	95	−9%	96	−6.7%	+1%
Annual passengers per runway (million)	9.7	−3%	7.7	−3.8%	−21%

Source: EUROCONTROL PRC/FAA ATO

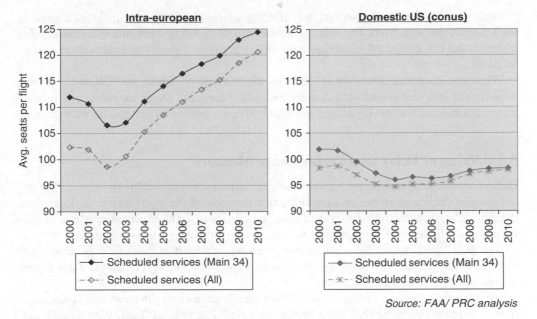

Figure 3.1 Average seats per scheduled flight (2000–2010)

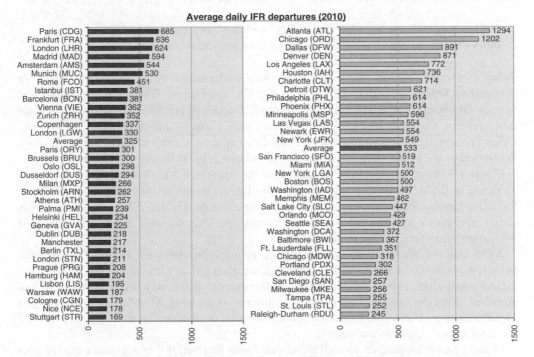

Figure 3.2 Average daily IFR departures at the 34 main airports (2010)

Figure 3.2 highlights the average daily IFR departures for each of the main 34 airports in the US and Europe in 2010. The average number of IFR departures per airport is considerably higher in the US, compared to average daily IFR departures at the 34 main airports in Europe in 2010[2].

While the economic downturn in 2009 has eased capacity constraints, US and European airports are still challenged with demand exceeding capacity. Delay measures can provide important insights into this imbalance and highlight opportunities for better delay management.

3.3 Value of Delay as a Performance Measure

Delays provide key information for valuing and prioritizing operational performance. "Delay" has several definitions in the aviation field, as it depends on the referenced time. In general, delay is defined as "later or slower than expected or desired". The expected or reference time is commonly linked to airline scheduled times, and performance is often expressed in terms of an "on-time" percentage. Scheduled times often include built-in inefficiencies, commonly referred to as "buffers" or "padding", that can be improved upon with runways, technologies, or procedures that increase capacity.

For economic analysis, airlines and ANSPs normally define delay in terms of an unimpeded time to capture system inefficiencies already included in scheduled times. With nominal reference times, delays are based on flight times when the system is not congested (discussed later in the chapter). This chapter attempts to create meaningful comparisons of ATM-related performance at airports in the US and Europe in two focus areas:

1. Delays versus airline schedules including "on-time" performance; and,
2. delays versus unimpeded reference times with a particular focus on ATM.

Delay is a common link between the five Key Performance Areas (KPAs) used by Air Navigation Service Providers (ANSPs) in the US and Europe. The Civil Air Navigation Service Organization (CANSO) promotes these KPAs as best practices in ANSP performance measurement. The KPAs are Safety, Cost-effectiveness, Capacity, Efficiency, and Environment. Delay is normally the result of demand exceeding capacity or the inefficient use of capacity. When delays are not properly managed, safety risks increase (Murphy and Shapiro, 2007).

Figure 3.3 shows a conceptual framework for the analysis of ATM-related delays at airports by phase of flight. Measuring performance with respect to airline schedules is not sufficient for understanding the underlying capacity and other constraints in the ATM system. Comparing actual operations to unimpeded reference times provides valuable information to ATM decision-makers on areas for improvement.

ATM may not always be the root cause for an imbalance between capacity and demand (which may also be caused by airlines, airports, weather, military training, noise and environmental constraints, airport scheduling, etc.). Depending on the way traffic is managed and absorbed along the various phases of flight (airborne vs ground), ATM has a different impact on airspace users (time, fuel burn, costs), the utilization of capacity (en-route and airport), and the environment (gaseous emissions).

In most cases, keeping an aircraft at the gate saves fuel but, if it is held and capacity goes unused, the cost to the airline of the extra delay may exceed the savings in fuel cost by far.

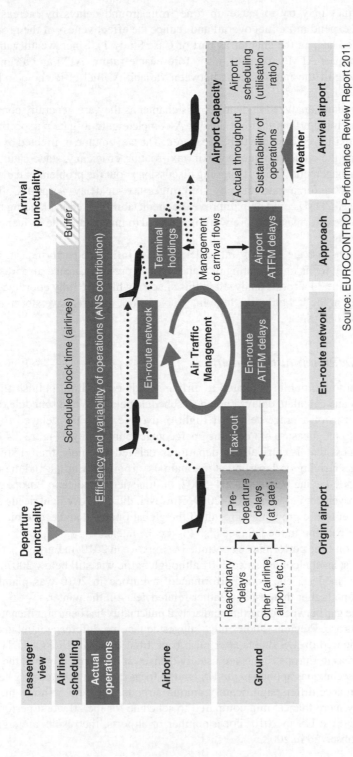

Figure 3.3 Conceptual framework for measuring ATM-related service quality

Source: EUROCONTROL Performance Review Report 2011

Additionally, airlines may try to make up time from ground delays by increasing flight speeds which can expend more fuel overall and reduce the effectiveness of the ground delay program (addressed later in the chapter as part of Case Study 1). Since weather uncertainty will continue to impact ATM capacities in the foreseeable future, ATM and airlines need a better understanding of the interrelations between variability, fuel efficiency, and capacity utilization.

Taxi-in delays are not analyzed in detail in this chapter as they are generally considered to be much less of a problem than taxi-out delays. A complete gate-to-gate perspective would, however, require taxi-in performance to be measured. The taxi-in time is affected by a number of factors including the choice for runway exit, taxi-routing efficiency, stand saturation, and stand allocation management. There is ongoing discussion about the problem of early-arriving aircraft having to wait for a gate, thereby suffering a taxi-in delay. Airport Collaborative Decision Making (A-CDM) could help improve the predictability of actual landing times, and thus improve taxi-in performance. More work is required to understand the influence of ANS on this flight segment.

Weather conditions and the level of traffic demand also influence performance. In Europe there has been a great focus on regulating demand through pre-coordinated airport slots. This is the case for 30 of the 34 airports analyzed in this chapter, which are fully coordinated (IATA Level 3). Currently in the US only the three major New York airports have slot management programs related to delays.

3.3.1   On-Time/Punctuality Measures

Most delay known to the public is related to airline schedules. Airlines publish flight times between city pairs and calculate delays known as schedule delays. While schedule delays can be calculated in terms of average delay per flight or average delay per delayed flight, most passengers relate to the measure of "on-time" percentage (punctuality). Figure 3.4 compares on time percentages, that is, arrivals or departures delayed by more than 15 min versus schedule, for flights arriving or departing at the main 34 airports in Europe and the US.

After a continuous decline between 2004–2007, on-time performance in Europe and in the US shows an improvement between 2008–2009. However, this improvement needs to be seen in a context of lower traffic growth as a result of the global financial and economic crisis and increased schedule padding in the US (see also Figure 3.7 later).

While in the US on-time performance continued to improve in 2010, in Europe, performance dropped to the worst level recorded since 2001 although traffic was still below 2007 levels and traffic growth was modest. The poor performance in Europe in 2010 was mainly due to industrial actions and higher than usual weather-related delay in the winter.

Prior to 2010, the gap between departure and arrival punctuality has been significant in the US and quasi-nil in Europe. Previous studies have suggested that the difference is related to additional delay experienced on US flights after push back from the gate (EUROCONTROL and FAA, 2012). Whereas in Europe flights are usually delayed at the departure gate, the air traffic flow management techniques applied in the US tend to focus more on the gate-to-gate phase.

On-time performance differs significantly among airports. Figure 3.5 shows the share of arrivals delayed by more than 15 min compared to schedule for the 20 worst performing airports in Europe and the US in 2010. For a number of airports, the results are significantly different to those observed in 2008.

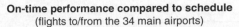

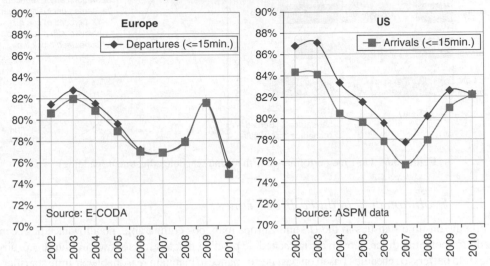

Figure 3.4 On-time performance compared to airline schedule (2002–2010)

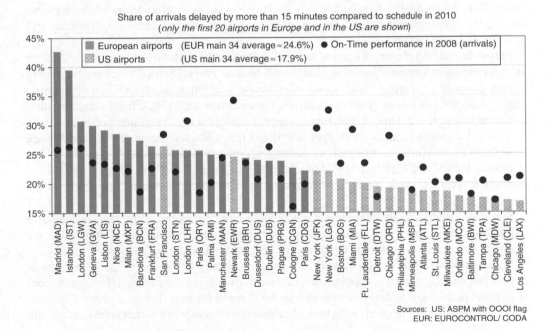

Figure 3.5 Arrival punctuality (airport level)

Late flights at individual arrival airports are influenced by previous flight-legs (reactionary delay) and performance at departure airports. Punctuality may be more a function of network performance as opposed to performance at individual airports.

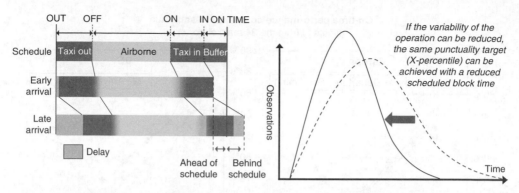

Figure 3.6   On-time performance with airline schedule adjustment (padding)

3.3.2   Evolution of Scheduled Block Times

While the on-time statistics in the US and Europe may look comparable, no allowance has been made for potential differences in airline schedule padding practices. As shown in Figure 3.6, airlines add buffers to schedules based on past performance to maintain reasonable on-time statistics.

Measures of delay relative to airline schedules do not provide insight into where delays occur. Airline schedules may include added time (padding) to account for anticipated delays. In addition, delays relative to schedule may occur due to airline-caused delays while preparing for departure (baggage handling, fueling, etc.), as well as due to carried-over (reactionary) delay from previous legs.

Schedule padding can mask the real amount of delay in the system. For example, consistent arrival holding at London Heathrow (LHR) will be built into scheduled block times. Adding one extra minute to airline block times would have a resulting additional cost of approximately US$50+ per minute (Ball et al., 2010). For an airline with 1100 flights a day, the additional minute in its schedule would have an approximate cost of $20 million per year.

Figure 3.7 shows changes in average scheduled flight block times in the US and Europe from 2000 to 2010 using a relative normalized performance measure (DLTA[3]).

Between 2000 and 2010, scheduled block times remained relatively stable in Europe while in the US, average block times have increased by some 3 min between 2005–2009 before declining in 2010.

Seasonal effects are also visible with scheduled block times being on average longer in winter than in summer. US studies by the former Free Flight Office have shown that the majority of increase is explained by stronger winds on average during the winter period.

Some of the differences between the US and Europe may be related to the differences in airport slot management. It is important to point out that while pre-coordinated slot management at airports helps manage delay, over-constraining demand for new flights can have economic costs. More analysis is needed in the area of delays versus supporting higher levels of demand.

3.3.3   Delays by Phase of Flight

For most commercial airline flights, airlines collect data on actual times for gate delay, taxi-out, airborne, and taxi-in. ANSPs are able to further subdivide airborne times into

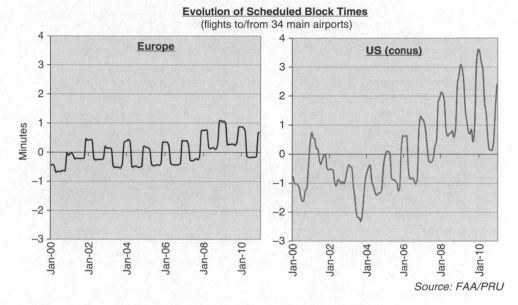

Figure 3.7 Evolution of scheduled block times (2000–2010)

departure, cruise, and terminal area phases. Subdividing flight times into phases enables a better understanding of the excess fuel burn associated with delays.

Different ATM strategies concerning the flight phase where delay is absorbed have been used in the US and Europe. The result of those strategies and external conditions are reflected in the data by flight phase. Figure 3.8 is a summary of performance by phase of flight using a relative normalized performance measure (DLTA) which compares city pair performance to its own average over time. This particular measure is useful for analyzing trends but does not provide information on delay relative to unimpeded times.

The calculations highlight where the changes in flight times have occurred by flight phase. In both the US and Europe, departure and total delay improved significantly in 2008 and 2009 due to the economic downturn resulting in a drop in demand.. While in the US performance continued to improve in 2010, total delays in Europe increased again significantly, mainly due to exceptional events (industrial actions, extreme weather). To a lesser degree, taxi-out times are also decreasing post-2008. In the US, airborne times have continued to be relatively higher than in pre-2008 periods.

Note that in Europe, the change in total delay correlates well with changes in gate delays (see Table 3.4 later in the chapter). In the US, gate delays are the major driver of change, but the other phases (taxi-out, en-route cruise and taxi-in) are also changing over time. The increase in total gate-to-gate duration in the US in Figure 3.8 correlates well with the increase in airline scheduled block times in the US in Figure 3.7 which is due to the fact that airlines use previously observed block-to-block times to build their schedules for the next season. The data suggest that Europe's approach for managing congestion consists primarily of absorbing most of the delay at the departure gate, whereas in the US, delay management is more dynamically carried out in all phases of a flight. While gate delay can be more fuel efficient, it can lead to

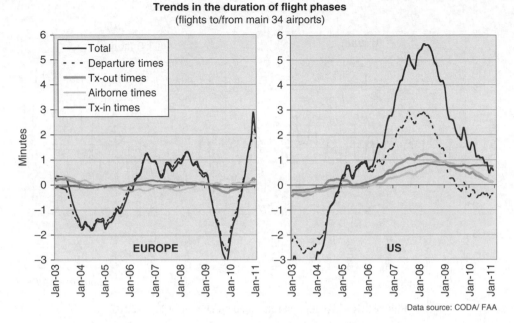

Figure 3.8   Trends in the duration of flight phases (2003–2010)

under-utilized capacity, if the strategy is overused. Furthermore, if an aircraft has to speed up in the air to make up for time lost on the ground, the overall fuel burn can be higher.

The large increases in US delays prior to 2008 are driven by over-scheduling at congested airports in the New York-Philadelphia area. These increases in relative flight times eventually resulted in the imposition by the FAA of scheduling (slot) constraints in April 2008 at Newark (EWR) and Kennedy (JFK) airports, while maintaining existing restrictions at La Guardia (LGA).

3.4   ATM-Related Operational Performance at US and EuropeanAirports

In the previous section it was shown how punctuality measures may have hidden ineffi-ciencies due to schedule padding on the part of airlines. To gain more insight into performance, this section considers ATM-related delays at airports compared to unimpeded times. Examining delay pools by phase of flight can be useful for identifying best practices and facilitating consistent comparisons between airports. The underlying framework for meaningful US and European comparisons is based on breaking out flight phases and relating delay to unimpeded flight times. Figure 3.9 gives a graphic representation of these flight phases.

Delay, when compared to *unimpeded times*, identifies the cost of congestion or route inef-ficiency. Unimpeded flight times can be viewed as the constraint-free flight times (without traffic or weather constraints). Unimpeded times can be calculated for individual flight phases

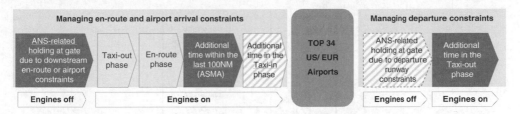

Figure 3.9 Conceptual framework for the evaluation of ATM-related delays at airports

to support further insights into where inefficiencies/constraints are occurring and where delays are absorbed. In this section "delay" is used as an indicator of a potential inefficiency in the system. However, it must be remembered that the reference time in this calculation is based on a hypothetical single flight scenario.

Common definitions of unimpeded times and delay by phase of flight are needed for meaningful comparisons and benchmarking best practices. The following comparisons between Europe and the US are based on joint work by the FAA and EUROCONTROL's Performance Review Unit (PRU) (EUROCONTROL and FAA, 2012).

Table 3.2 summarizes benefit pools by phase of flight. Predictability refers to the consistency with which delays occur. ATM-related holdings at the departure gate occur for a small percentage of flights, whereas delays in the gate-to-gate phase (taxi, en-route, airborne terminal holdings) occur across all flights. Thus, the former are less predictable than the latter.

Airports (together with ATM and airlines) can influence the efficient absorption of necessary delay. Overall some 80% of all ATM-related delays are either absorbed at or around airports (gate, taxi-out, and terminal airborne). In the US, taxi-out delays account for half the total ATM-related delay. Airports clearly play a key role in the performance of the ATM system as simply moving delays from the taxiways to the gate can save costs and reduce CO_2 emissions. Comparing delays between two sets of airports operating under differing conditions and procedures can serve as a virtual "testbed" for observing best practices in ATM, as well as for the scheduling of demand.

The remainder of this chapter focuses on comparisons of ATM-related delay associated with managing departure, arrival, and en-route constraints on flights from/to the main 34 airports in Europe and the US. ATM cannot reduce the majority of delay in the system on a daily basis, but can manage how delay is absorbed. Managing where delays are absorbed throughout a flight can save costs and emissions. Table 3.3 highlights the focus areas to be discussed in the remainder of the chapter, organized by constraint area and location of absorbed delay.

Whenever demand exceeds airport arrival capacity, the resulting delays need to be distributed along the trajectory. For both the US and Europe delay absorption starts near the arrival airport in the form of holding or vectoring patterns. Delay is managed in the airspace close to the airport in order to keep pressure on the runways. In most cases, the primary goal of ATM and airlines is to utilize all available runway capacity. Failure to keep pressure on the runways may result in missed slots or excess spacing, events which will drive up total delay. The risk associated with "runway pressure" is that, when the number of aircraft close to the arrival airport exceeds the "holding" capacity of the terminal airspace, having more aircraft close-in may have safety implications, as well as result in excessive fuel burn. The task of absorbing the right amount of delay at the right place constitutes a major challenge for ATM.

Table 3.2 Summary of delays by phase of flight (2010)

Estimated benefit pool actionable by ATM for a typical flight (2010) (flights to/from the main 34 airports)		Estimated additional time (avg. per flight in min)		Predictability (% of flights affected)		Fuel burn	Est. excess fuel burn (kg)[13]	
		EUR	US	EUR	US	engines	EUR	US
Holding at gate per departure (only delays >15 min included)	en route-related	1.8	0.05	5.7%	0.1%	OFF	≈0	≈0
	airport-related	1.2	1.0	3.3%	1.6%	OFF	≈0	≈0
Taxi-out phase (min per departure)[14]		4.9	5.0	all		ON	73 kg	75 kg
Horizontal en-route flight efficiency[15]		2.1–3.8	1.3–2.5	all		ON	176 kg	114 kg
Terminal areas (min. per arrival)[16]		2.5	2.45	all		ON	103 kg	100 kg
Estimated benefit pool actionable by ATM		≈12.5–14.2	≈9.8–11.0				352 kg	289 kg

Table 3.3 Constraint locations versus location where delay is absorbed at airports

Location of absorbed delay	Managing airport arrival constraints	Managing en-route constraints	Managing runway departure constraints
GROUND Departure gate	delay is absorbed at the gate at the departure airports	delay due to en-route constraints is moved back to gate at the departure airport	delay due to runway constraints at the departure airport is moved back to the gate
Taxi-out	delay is absorbed in the taxi out phase at the departure airports	En-route constraints managed during taxi-out	Delay mainly due to departure demand exceeding runway capacity is absorbed in the taxi-out phase
Taxi-in	*delay is absorbed in the taxi-in phase (mainly related to gate availability and variability of arrivals). Currently not managed by ATM*		
AIRSPACE	*delay is absorbed in the airspace around the arrival airport*		
Case Studies	*Case Study 1: Improving arrival delays with Controlled Times of Arrival (CTAs)*		*Case Study 2: Improving Taxi delays with airport CDM (Virtual taxi queues)*

3.4.1   Managing En-Route and Arrival Constraints at the Departure Gate

In Europe when traffic demand is anticipated to exceed the available capacity in en-route control centers or at an airport, ATC units may call for "Air Traffic Flow Management (ATFM) regulations". Aircraft subject to ATFM regulations are held at the departure airport according to "ATFM slots" allocated by the Central Flow Management Unit (CFMU) in Brussels, Belgium.

In the US, ground delay programs (GDP), airspace flow programs (AFP) and ground stops (GS) are mostly used in case of severe capacity restrictions at airports when other ATFM measures, such as Time Based Metering (TBM) or Miles in Trail (MIT) en-route are not sufficient. The Air Traffic Control System Command Centre (ATCSCC) applies Estimated Departure Clearance Times (EDCT) to delay flights prior to departure. Most of these delays are taken at the gate.

Table 3.4 compares ANS-related departure delays attributable to en-route and airport constraints on flights to/from the main 34 airports in each region. For comparability reasons, only EDCT and ATFM delays larger than 15 min were included in the calculation.

The share of flights affected by ATFM/EDCT delays due to en-route constraints differs considerably between the US and Europe. In Europe, flights are as much as 50 times more likely to be held at the gate for en-route constraints. US constraints occur primarily during the season of convective weather that may block parts of the airspace. In Europe, by contrast, en-route constraints are driven for the most part by demand exceeding the capacity of individual en-route sectors and result in holding flights on the ground. As stated previously, some delays related to en-route constraints in the US are managed during the taxi-out phase, but no specific data are available on the impacts of this practice.

For delays related to airport constraints, the percentage of delayed flights at the gate in the US is closer to the European figure.

The application of ground delays only when other ATFM measures (TBM, MIT) are not sufficient in the US leads to a lower share of flights affected by EDCT delays but higher delays per delayed flight than in Europe. More analysis is needed to see how higher delays per delayed flight are related to moderating demand with "airport slots" in Europe.

The change between 2008 and 2010 is consistent with overall trends between the US and Europe. In the US, delay declined with traffic whereas in Europe, delay increased largely due to off nominal events, mainly due to industrial actions and extreme weather in winter.

3.4.2   Managing Arrival Constraints within the Last 100 NM

This section estimates the level of inefficiencies due to airborne holding, metering, and sequencing of arrivals. Figure 3.10 shows the arrival radar tracks on a day in February 2008 at London Heathrow (LHR), Frankfurt (FRA) and Paris (CDG). It illustrates how local ATM strategies affect arrival flows within a radius of 100 NM at those three major European airports. Whereas at London Heathrow (LHR) the majority of the approach operations take place in close proximity to the airport (circular holdings), at Frankfurt (FRA) and Paris (CDG), the sequencing of arrival traffic starts much further out.

The actual transit times within the 100 NM ring are affected by a number of ANS and non-ANS related parameters including flow management measures (holdings, etc.), airspace design, airport configuration, aircraft type, pilot performance, environmental restrictions, and in Europe, to some extent the objectives agreed by the airport scheduling committee when declaring the airport capacity.

Table 3.4 ANS-related delays at the gate due to constraints en-route or at the arrival airport

Only delays >15 min. are included			En route related delays >15 min. (EDCT/ATFM)			Airport related delays >15 min. (EDCT/ATFM)		
(flights to/from main 34 airports)		IFR flights (M)	% of flights delayed >15 min.	delay per flight (min.)	delay per delayed flight (min.)	% of flights delayed >15 min.	delay per flight (min.)	delay per delayed flight (min.)
US	2008	9.2	0.1%	0.1	57	2.6%	1.8	70
	2010	8.6	0.1%	0.05	44	1.6%	1.0	66
Europe	2008	5.6	5.0%	1.4	28	3.0%	0.9	32
	2010	5.0	5.7%	1.8	32	3.3%	1.2	36

Source: EUROCONTROL PRC/FAA ATO

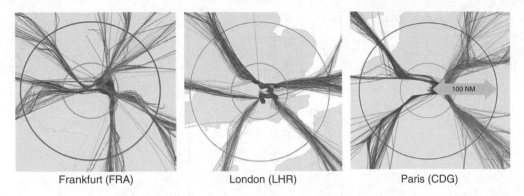

Frankfurt (FRA)   London (LHR)   Paris (CDG)

Figure 3.10   Impact of local ATM strategies on arrival flows

In order to evaluate tactical arrival control measures (sequencing, flow integration, speed control, spacing, stretching, etc.), irrespective of local ATM strategies and shape of the terminal maneuvering area (TMA), a standard "Arrival Sequencing and Metering Area" (ASMA) with a radius of 100 NM around each airport was defined (see radius applied in Figure 3.10). The "additional" time is used as a proxy for the level of inefficiency within the last 100NM. It is defined as the average additional time beyond the unimpeded transit time for each airport.

Figure 3.11 shows the estimated additional time within the last 100 NM for the US and Europe in 2008 and 2010. For clarity only the 20 most penalizing airports of the 34 main airports are shown[4].

In Europe, London Heathrow (LHR) is a clear outlier[5], having by far the highest level of additional time within the last 100 NM, followed by Frankfurt (FRA).

The US shows a more uniform picture. Despite a substantial improvement compared to 2008, there is still a notable difference when it comes to the airports in the greater New York area, which show the highest level of inefficiencies within the last 100 NM in 2010.

Due to the large number of variables involved, the direct ANS contribution towards the additional time within the last 100 NM is difficult to determine. One of the main differences of the US air traffic management system is the ability to maintain pressure on the airport while moving some delay to en route airspace with time based metering and miles-in-trail spacing.

In Europe, the support of the en-route function is limited and rarely extends beyond the national boundaries. Hence, most of the sequencing is done at lower altitudes around the airport. Additional delays beyond what can be absorbed around the airport are taken on the ground at the departure airports (see Section 3.4.1).

In both the US and Europe, there are alternative approaches for absorbing airborne delay for arrival constraints. Pilots, using aircraft Flight Management Systems (FMS), have the potential to absorb necessary delay more efficiently.

3.4.2.1   Case Study 1: Improving Arrival Management with Controlled Times for Arrivals (CTAs)

Aircraft arriving at an airport are subject to delays, just like departing ones. During peak periods, arrivals will queue up in the terminal area driving up fuel costs and impacting the

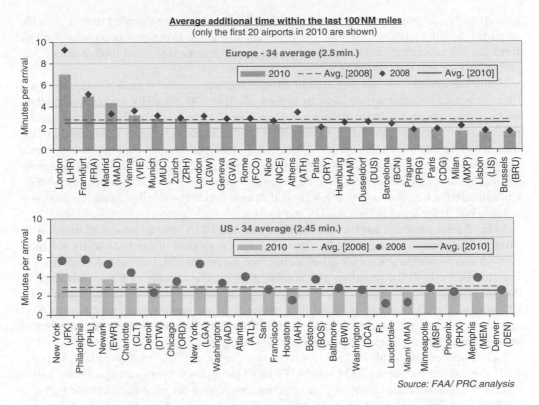

Figure 3.11   Estimated average additional time within the last 100 NM

environment. For flights arriving or departing at main airports, arrival delays in the terminal area account, on average, for up to 25% of the "excess" fuel burned per flight (see Table 3.2). However, there is good reason why this delay is absorbed near the airport during peak periods: the cost of "missing" an arrival slot would increase delays for all following aircraft. But ATM and aircraft technologies now make it possible to absorb much of the "necessary" delay outside the terminal area, while still keeping pressure on the runways.

In the US, there is widespread use of the Traffic Management Advisor system, which helps manage flow rates into the terminal areas, because absorbing delay at high altitudes reduces terminal area congestion and can reduce fuel burn. This system gives controllers information on how much delay should be absorbed by each aircraft before crossing the boundary into the terminal area. In addition, controllers have the ability to move necessary delay (through miles-in-trail restrictions) more than 100 miles from the terminal area.

In Europe, there is a similar management system called the Arrival Manager (AMAN), which also helps control flows into the terminal areas. However, moving delay back to en-route altitudes may either cause en-route congestion or require coordination with other ANSPs. For the most part, there is limited sharing of delay absorption between ANSPs although cooperation is increasing[6]. London Heathrow (LHR) is an example of a congested airport where minimal delay can be absorbed in en-route airspace. As a result, holding patterns (stacks) have been established around the airport. Frankfurt (FRA) airport uses lateral maneuvering of aircraft within terminal airspace to absorb delay. While sometimes absorbed further from the airport, US delay taken within 100 NM is similar to Europe overall (as previously shown). Both the US and Europe have opportunities for improvements that could further reduce fuel and environmental impacts (CO_2).

Reducing unnecessary queuing around airports requires increased predictability of arrival times to ensure available airport capacity is used efficiently. Additionally, procedures to establish a virtual queuing priority system would also need to be in place. With a prioritization procedure for accepting aircraft at an arrival fix outside an airport, flights would not have the incentive to rush to the terminal area to obtain a "place in the queue", only to experience significant holding or other forms of delay absorption. NEXTGEN[7] and SESAR[8] planning address the issue of four dimensional trajectories "contracted" between airlines and ANSPs. Coordination of trajectories between ATM and airlines will help solve the excess terminal area holding occurring today in both the US and Europe.

Several pilot programs have tested the ability of aircraft to meet target arrival times using Flight Management Systems (FMS)[9]. By using the capabilities of the FMS to fly the most cost-efficient speed to absorb necessary delay, fuel savings would increase by comparison to the current methods for vectoring and holding aircraft. The concept of Controlled Time of Arrival (CTA) sets up a process where pilots are given a time to reach the terminal area meter fix, where aircraft enter the terminal area airspace. If all arrivals participated in such an approach, flights could efficiently absorb delay using FMS capabilities through cost effective speeds and eliminate the excess time spent at lower altitudes.

Delta Airlines has established a CTA process for flights into Atlanta. This process is essentially carried out outside the ATM system. Delta's Operations Center is using a system called "Attila™", which estimates arrival times to the terminal area for most flights flown by Delta Airlines and their regional carrier Atlantic Southeast Airways. Arrival times are calculated using business parameters, defined by Delta, to prioritize high value flights. Adjustments are made by the Operations Center that communicates electronically assigned arrival times to pilots. Adjustments are limited to 2 min faster or slower to accomplish reordering, allowing efficient use of capacity and reductions in terminal area delay. In Atlanta, Delta Airlines is the primary carrier so they are able to apply these adjustments without major disruptions from/to other carriers. These small adjustments to speed have not been a concern to controllers, as anecdotal evidence indicates controller workload is reduced by the spreading out of the arrival demand.

Table 3.5 highlights the estimated savings associated with adjusting desired flight times to a fix outside Atlanta airport. Note that most of the savings is associated with moving aircraft forward to less congested periods. The intent is to start arrival peaks sooner to reduce overall delay.

It should be pointed out that in Atlanta's case, Delta is the primary carrier and can arbitrate the priority of aircraft to help achieve priorities and reduce congestion. Clearly there are times when flights from other airlines and actions by controllers impede Delta's objectives but overall improvement has been achieved. In cases where multiple airlines are competing for service, a more robust solution would be needed with the ANSPs assuming the role of ensuring equitable treatment of all airlines. In all cases where CTAs are used, air traffic controllers have the ultimate responsibility for safe separation in today's system.

Around the world, several ANSPs have worked with Airlines and Airports to create arrival times that are provided to pilots in an effort to manage terminal congestion and save fuel. The

Table 3.5   Summary of Attila delay and fuel savings (2009)

Month	Days Operational	Flight time saved (minutes)	Fuel saved (gallons)	A0* Improvement		Average dwell savings (min.)
				%	No.	
Jan-09	28	8169	165 939	3.3	293	–
Feb-09	27	15 519	306 377	5.5	473	–
Mar-09	24	16 477	335 304	6.2	504	1.0
Apr-09	28	20 624	353 981	6.1	690	0.6
May-09	25	9023	158 299	4.3	328	0.1
Jun-09	29	25 664	447 530	6.6	878	0.6
Jul-09	28	23 659	414 610	7.6	863	0.8
Aug-09	24	19 709	347 212	9.7	894	1.1
Sep-09	28	22 150	404 662	7.6	887	0.7
Oct-09	27	20 153	373 822	5.0	603	0.7
Nov-09	29	29 603	539 558	6.8	927	0.7
Dec-09	30	18 735	345 094	5.3	669	0.6
Totals	*327*	*229 485*	*4 192 386*	–	*8010*	–

Notes: Atlantic Southeast Airways (ASA) became operational on March 24, 2009
*"A0 Improvement" refers to change in flights arriving before the scheduled time

challenge is to keep pressure on runways to maximize throughput while minimizing system fuel burn. Success has been mainly achieved when smaller numbers of airlines are involved and where the last hour of a flight is managed within a single country's (or ANSP's) airspace. Complexity is increased when a high number of airlines are involved and assigned delay is managed across ANSP borders. More work must be done in this area to assure an equitable process with full participation by airlines for arrivals into large airports with complex surrounding airspace.

In addition to reducing delay due to arrival runway constraints, improving the predictability of arrival times with CTAs may also reduce gate constraints and associated taxi-in times.

3.4.3   Managing Departure Runway Constraints – A Look at Taxi-Out Delay

The FAA and EUROCONTROL have developed comparable methodologies for approximating the measurement of taxi-out efficiency. Many airports around the world are now collecting more refined surface movement data where taxi delays can be explicitly identified.

The analysis of taxi-out efficiency in the next sections refers to the period between the time when the aircraft leaves the stand (actual off-block time/push back) and the take-off time (wheels up). The additional time is measured as the average additional time beyond an unimpeded reference time. In principle, the time an aircraft leaves the stand is approximated with the Gate-Out time recorded by the aircraft.

3.4.3.1   Managing Departure Constraints on the Taxiways

The taxi-out phase and hence the performance measure is influenced by a number of factors such as take-off queue size (waiting time at the runway), distance to runway (runway

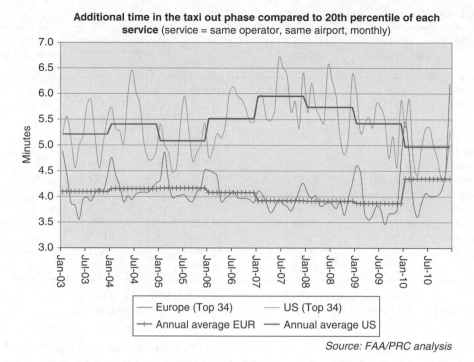

Additional time in the taxi out phase compared to 20th percentile of each service (service = same operator, same airport, monthly)

Source: FAA/PRC analysis

Figure 3.12 Additional times in the taxi-out phase (system level) (2003–2010)

configuration, stand location), downstream restrictions, aircraft type, and remote de-icing to name a few. Of the aforementioned causal factors, the take-off queue size[10] is considered to be the most important one (Idris et al., 2001).

In the US, the additional time observed in the taxi-out phase also includes delays due to local en-route departure and miles-in-trail restrictions. In Europe, the additional time might also include a small share of ATFM delay, which is not taken at the departure gate or some delays imposed by local restriction such as Minimum Departure Interval (MDI). Detailed data related to en-route delay absorbed during the taxi-out phase is currently not available.

To gain a better understanding of the US and European taxi delay differences, two distinct methodologies were applied. The first approach is simpler and allows for application of a consistent methodology in diverse environments. The method uses the twentieth percentile of each service (same operator, airport, etc.) as reference for the "unimpeded" time and compares it to the actual times. This can be easily computed with US and European data. Figure 3.12 shows the additional time in the taxi-out phase compared to the twentieth percentile of each service.

Two interesting points can be drawn from Figure 3.12:

• On average, additional times in the taxi-out phase appear to be higher in the US with a maximum difference of approximately 2 min more per departure in 2007. Between 2008 and 2010, US performance improved continuously which narrowed the gap. In Europe, performance remained relatively stable but showed a notable deterioration in 2010, which was mainly due to off nominal events (industrial actions and extreme weather in winter).

- Seasonal patterns emerge, but with different cycles in the US and in Europe. Whereas in Europe the peaks during the winter months are most likely due to weather conditions, in the US the peaks in the summer are most likely linked to congestion.

The comparison of additional times by airport in Figure 3.13 is based on the respective official methodologies of each group. Although some care should be taken when comparing the two indicators due to differing methodologies, Figure 3.13 tends to confirm the trends seen in Figure 3.12. Historically, there have been higher average additional times in the taxi-out phase in the US than in Europe (6.2 min per departure in US compared to 4.3 min per departure in Europe in 2008). However, the difference is much smaller in 2010 which is due to the increased focus on taxi-out efficiency in the US and a deterioration of performance in Europe. For reasons of clarity, only the 20 most penalizing airports of the 34 main airports are shown.

The observed differences in inefficiencies between the US and Europe reflect the different flow control policies and the absence of scheduling caps (pre-coordinated airport slot control) at most US airports. An additional factor is that the US Department of Transportation collects and publishes data for on-time departures. This practice motivates the airlines to focus on getting off-gate on time as often as possible.

The impact ANSPs have on reducing overall delay is marginal when runway capacities are constrained. Data on the magnitude of taxi delays are useful, however, for developing policies and procedures geared towards keeping aircraft at the gate longer thereby reducing fuel and emissions. Overall Europe appears to have made the management of taxi delays a higher priority than the US. At many European airports there are built-in procedures to minimize taxi queues. In the US there have been initiatives to reduce taxi queues through queue management at select airports including JFK and Boston (Simaiakis et al., 2011). Departing aircraft are sequenced by managing the pushback times. Shorter queues improve departure predictability (Bhadra et al., 2011).

3.4.3.2 Managing Departure Runway Constraints at the Gate

Moving potential taxi delays back to the gate is a common practice in Europe. However, a common data source that enables comparisons between Europe and the US is not yet available. A number of individual ANSPs within Europe are collecting this information, but more work is needed to build a consistent data source. In the US, keeping an aircraft at the gate in response to runway constraints is not a common practice. The reasons include the scarcity of gates at many airports, as well as the aforementioned emphasis of the US Department of Transportation on departure on-time statistics. Case Study 2 next highlights some of the efforts underway in Europe and the US, which are focused on managing runway departure constraints at the gate.

3.4.3.3 Case Study 2: Managing Taxi-Out Delay with Airport CDM

As shown in the previous section, the US has experienced higher taxi delay than Europe on average. Taxi delays represent an opportunity to reduce airline fuel costs and improve the environment. In Europe, Airport Collaborative Decision Making (A-CDM) is focused on improving

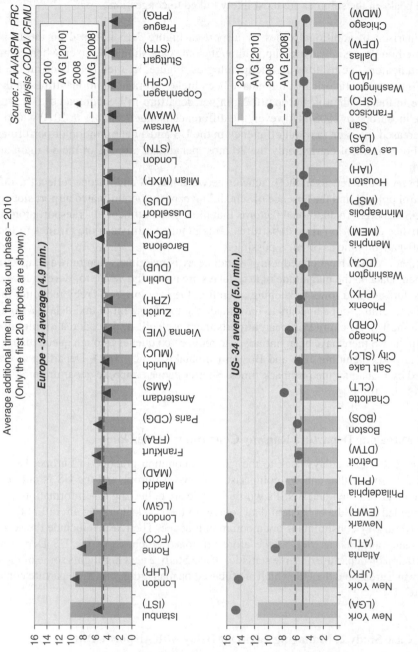

Figure 3.13 Average additional time in the taxi-out phase

information sharing as aircraft arrive and depart at airports. A-CDM improves the predictability of turnaround time and projections of demand to key outbound departure points. Improving predictability helps efficient use of available capacity in the system. A-CDM supports better predictions of runway taxi queues. Predictions are based on taxi time estimates from each gate area to each runway. Through these improved predictions expected taxi queues can be estimated and managed. When queues are projected to exceed an agreed number of flights, aircraft can be held at the gate (or in a holding area) with engines off until the queue is reduced. A-CDM supports a "virtual queue" as aircraft keep their taxi position without having to wait at the runway. This process is jointly executed by airlines, airports, and ANSPs. Currently, several European airports are using the A-CDM process and more are in the implementation phase.

Zurich Airport (ZRH) implemented an early form of A-CDM focused on exchanging real-time data between the airport operator, the airlines, and other service providers including baggage handling, catering, and aircraft cleaning services. At Zurich, the airport manages movement areas and virtual queuing for runways.

Munich Airport (MUC) was the first airport to implement the full European version of A-CDM in 2007. The robust sharing of airport, airline, and ATM data is reported to be very successful. Improvements include better management of airport and airline resources reducing turnaround time and overall delays. It has been estimated that taxi-out times have been reduced by 1 min per flight (DFS and Munich Airport, 2010). In addition, the A-CDM system provides data to the EUROCONTROL Central Flow Management Unit on aircraft departure times for improved estimates of en route sector loading.

Beyond the four fully A-CDM compatible airports (Brussels: BRU, Frankfurt: FRA, Munich: MUC, and Paris: CDG) operational in 2011, another six airports were planned to be fully implemented by end of 2012.

In the US, there has been a more recent focus on the taxi-out delay problem. This has included the imposition of regulations like the "3-hour Rule" (U.S. DoT, N JO 7110.524, 2010), a Passenger Rights law that requires airlines to compensate passengers if they are delayed for more than 3 on an airport's taxiway. More recently, the FAA Air Traffic Organization (ATO) has been exploring processes similar to the European A-CDM approach. One of the challenges in the US is high gate utilization and the limited availability of common use gates, which can impact a flight's ability to stay at the gate during busy periods. Several studies are underway in the US addressing virtual taxi queues as taxi delays represent between 10–25% of the total delays of flights arriving or departing at the main 34 US airports. The solution is based less on technology and more on devising processes that airlines can follow and be assured of equitable access to runways without having to experience long queues. Working with the Massachusetts Institute of Technology (Simaiakis, 2009), Boston Logan Airport has instituted a program to keep aircraft at the gate to reduce taxi congestion and fuel emissions. The program provides air traffic controllers in the tower with an optimal rate for gate departures to assure maximum throughput while managing time in the taxi queue.

Where gate availability could be an issue with virtual queues, other options for holding in ramp or taxi areas with engines off can be explored. In some cases there may be a need for demand management measures like airport slot controls to spread out demand. Part of the reason why taxi-out delays in Europe have been historically less than in the US is related to the pre-coordinated airport slot process in the strategic phase. Almost all of the main 34 airports in Europe have pre-coordinated airport slots versus only three among the main 34 in the US. Interestingly, the three major airports in the New York area that are subject to slot constraints still have the highest taxi delays in the US. It has been difficult to measure what impact the slot constraints at the New York airports have had, because demand has decreased since 2008 when the constraints were imposed.

JFK is the one US airport currently managing the departure process in order to reduce taxi delay through "virtual queues". The need for action became obvious during winter operations in recent years when long taxi-out queues would form causing some aircraft to return for de-icing prior to take-off. Returning for de-icing meant considerable delay (the aircraft would lose its spot in the departure queue), as well as added expense. This was too costly for airlines and presented a clear need for change. The need for departure management was reinforced with the closure in early 2010 of the airport's longest runway due to the resurfacing project at JFK. The Port Authority of New York and New Jersey (PANYNJ) and airline stakeholders then decided to establish a temporary departure control system based on the concept of "virtual taxi queues". The overall objective was to reduce the total number of aircraft queuing on taxiways, as well as reduce taxi-out delay (though gate delay may increase). It is important to note that, at this stage, there is no significant ATC or FAA coordination with the new departure management system at JFK. This contrasts sharply with what happens in Europe, where ANSPs are strongly involved with the European version of A-CDM.

Beginning on March 1, 2010, a single contractor, designated by PANYNJ, operates the PASSUR Aerospace[11] system that manages the taxi slot allocation process across all ramp areas and terminals at JFK. Airlines are responsible for updating PASSUR with the most current flight schedule information on the day of departure. With OAG and updated information from airlines, PASSUR assigns departure slot times (within 15-min intervals) to each flight up to 2 h in advance. The 15-min interval is the expected time window within which a given aircraft is supposed to enter the active movement area, that is, "cross the spot". PASSUR operators use historic data on runway throughput in different configurations and weather conditions to set the number of available slots in each time period. Available capacity is allocated according to the flight schedule. Roughly speaking, the three largest operators at JFK, American, Delta, and Jetblue, are allocated approximately 70–75% of available slots over the course of the operating day. Other airlines (including many international carriers) receive the remaining 25–30%, with allocations in any time period depending on schedule and aircraft readiness. Airlines are free to re-sequence the slots of their own flights and to exchange slots with other airlines, but the swaps must be confirmed by the departure management contractors via a chat facility in the PASSUR program. If a flight is delayed, because of mechanical or other problems, an airline can substitute a later flight of its own into the vacated slot. If slot time is much later than the scheduled departure time for a flight, the airline has the option to delay boarding so passengers can spend more time in the terminal. Most exchanges are proposed within individual ramp towers, and are usually among code-sharing partners. The departure management contractors also try to sequence aircraft by giving particular flights using certain fixes priority slot times in a way that maximizes runway throughput. However, since there is no coordination between PASSUR and ATC, such sequences are not binding on ATC or airlines. ATC ground controllers still process departing aircraft on a first-come-first-served basis when aircraft arrive at the "spot". However, the order in which aircraft show up at the spot and the timing of their requests to enter the taxiway system have already been controlled and influenced by the PASSUR contractors. This is the key difference compared to the previous system.

PASSUR aims to have between 6–8 aircraft in the active movement area waiting for take-off at any given time. (PASSUR does not control the size of the queue of arriving aircraft.) Due to the reduced number of aircraft on taxiways, it is then much easier to change runway configuration when needed, therefore reducing delay further. The system has also become more flexible over time. Overall, airlines seem satisfied with this virtual queue approach and PANYNJ is looking to expand use of this system to a year-round operation. However, it has been difficult to evaluate accurately the performance of PASSUR, due to the declining number of flights at JFK. Traffic in March and April of 2010 was down by between 5–10% compared to 2009. While delays have declined significantly, changes in underlying conditions, like demand and weather, complicate the assessment of the impact of surface traffic management improvements, such as PASSUR.

Looking forward, there are clearly opportunities for improvements by using data recorded by systems like the ASDE-X system[12] at New York JFK airport (Bhadra et al., 2011).

3.5 Summary and Conclusion

Measuring performance with respect to airline schedules (e.g., by recording lateness vis-à-vis scheduled arrival times) is not sufficient for understanding the underlying capacity and other constraints in the ATM system. Measuring delay versus unimpeded times provides valuable information to ATM decision-makers on the "total inefficiency" of the current system, also known as "improvement pools". Delay segregated by phase of flight is especially useful for highlighting where inefficiencies are occurring. Comparing the US and European ATM systems has highlighted best practices and opportunities for focused improvements. Specifically, reductions in taxi delays in the US may be achievable using European best practices for managing queues. For improved absorption of arrival delay, Europe may learn from US best practices. With virtual queues applied to departure runways, as well as arrival meter fixes, improvements can be made in the management of airport delays to reduce airline costs and reduce fuel burn.

Given the constraints in effect on any given day of operations, it is possible to influence where delay is absorbed through proper management of air traffic. Achieving large reductions in overall delay requires increasing effective capacity or constraining demand. Constraining demand at an airport has other economic impacts that must be considered.

In all cases, the comparison of the same measures of performance in different environments such as the US and Europe can be a key element in developing meaningful best practices for efficient airport management.

Notes

1 As of January 2011, airlines which operate within the European airspace more than 35 000 flights per year are required to submit the data on a monthly basis according to Regulation (EU) No 691/2010.

2 Figure 3.2 only shows IFR flights. Some airports – especially in the US – have a significant share of additional VFR traffic.

3 The Difference from Long-Term Average (DLTA) metric is designed to measure changes in time-based (e.g., flight time) performance normalized according to selected criteria (origin, destination, aircraft type, etc.) for which sufficient data are available. It provides a relative change in performance without underlying performance driver.

4   Due to data issues, the calculation of the reference times in the US and Europe is based on slightly differing methodologies which are expected to produce rather similar results. The methodologies are documented in the *2010 U.S./Europe Comparison of ATM-related Operational Performance* (EUROCONTROL and FAA, 2012).

5   Performance at London Heathrow airport (LHR) is influenced by decisions taken during the airport scheduling process regarding average holding in-stack.

6   For example, there is cooperation between DFS and Austro Control for Munich (MUC) arrivals.

7   The Next Generation Air Traffic Control System (NextGen) is a comprehensive overhaul of the US National Airspace System (NAS) to provide new capabilities that make air transportation safer and more reliable while improving the capacity of the NAS and reducing aviation's impact on the environment.

8   The mission of the SESAR (Single European Sky ATM Research) Joint Undertaking is to develop a modernized air traffic management system for Europe. This future system will ensure the safety and fluidity of air transport over the next 30 years, will make flying more environmentally friendly and reduce the costs of air traffic management.

9   Examples of CTA uses include: (1) Airservices Australia provides CTAs to pilots on a regular basis for flights coming from the South Pacific to Sydney Airport; (2) UK's National Air Traffic System (NATS) has held trials where flights have absorbed specified delays en route in exchange for bypassing holding queues; (3) for the first arrivals after the 6 a.m. curfew at Zurich airport, Swiss International Air Lines is giving aircraft pilots CTAs at the fix outside the airport to prioritize flights and reduce ATM holding/vectoring.

10   The queue size that an aircraft experienced was measured as the number of take-offs that took place between its pushback and take-off time.

11   PASSUR Aerospace builds decision support tools using PASsive SURveillance data for airlines and airports.

12   Airport Surface Detection Equipment, Model X.

13   Fuel burn calculations are based on averages representing a "standard" aircraft in the system. (Taxi ≈ 15 kg/min, Cruise. ≈ 46 kg/min, TMA holding 41 kg/min).

14   The estimated inefficiencies in the taxi-out phase refer only to departures from the main 34 airports. If all flights to/from the main 34 airports were considered, the "inefficiency" per flight would be lower because departures from less congested airports to the main 34 airports were included.

15   The horizontal flight efficiency figures relate to the distance between the 40 NM radius at the departure and the 100 NM radius at the arrival airport. The range in horizontal en route flight efficiency relates to direct route extension (A–D)/G which assumes the need to maintain a route structure in the TMA area and the en route extension (A–G)/G which assumes that all the route structure including TMA can be improved. Europe/US differences in the average distance would lead to different results, as the "inefficiency" is measured as a percentage of the great circle distance. For comparability reasons, the estimated additional time calculation was based on an average great circle distance of 450 NM for the US and Europe.

16   The estimated inefficiencies in the last 100 NM refer only to arrivals at the main 34 airports. If all flights to/from the main 34 airports were considered, the "inefficiency" per flight would be lower because arrivals at less congested airports from the main 34 airports were included.

References

Airlines for America (2012). *Annual and Per-Minute Cost of Delays to U.S. Airlines*. [online]. Available at: <http://www.airlines.org/Pages/Annual-and-Per-Minute-Cost-of-Delays-to-U.S.-Airlines.aspx > [Accessed February 19, 2013].

Bhadra, D., Knorr, D., and Levy, B. (2011). *Benefits of Virtual Queuing at Congested Airports Using ASDE-X: A Case Study of JFK Airport*. 9th USA/Europe Air Traffic Management Research and Development Seminar (ATM2011), Berlin, Germany, June 2011. [pdf]. Available at: <http://www.atmseminar.org/seminarContent/seminar9/papers/36-Bhadra-Final-Paper-4-13-11.pdf> [Accessed February 19, 2013].

Ball, M., Barnhart, C., Dresner, M., et al. (2010). *Revised Release: Total Delay Impact Study: A Comprehensive Assessment of the Costs and Impacts of Flight Delay in the United States* [pdf]. Available at: <http://www.isr.umd.edu/NEXTOR/rep2010.html> [Accessed February 19, 2013].

Brody Guy, A. (2010). *Flight delays cost $32.9 billion, passengers foot half the bill*. [online]. UC Berkley News Center, 18 October. Available at: <Newscenter.berkeley.edu/2010/10/18/flight_delays >[Accessed February 19, 2013].

DFS Deutsche Flugsicherung GmbH/ Flughafen Muenchen GmbH (2010). *Airport CDM Munich, Results 2009*.[pdf]. Available at: <http://www.euro-cdm.org/library/airports/munich/airport_cdm_munich_results_2009.pdf> [Accessed February 19, 2013].

EUROCONTROL Performance Review Commission and the FAA Air Traffic Organization System Operations Services (2012). *US/Europe Comparison of ATM-Related Operational Performance 2010*. [pdf] Brussels: EUROCONTROL/FAA. Available at: <http://www.eurocontrol.int/documents/useurope-comparison-atm-related-operational-performance-2010> [Accessed February 19, 2013].

Idris, H., Clarke, J.P., Bhuva, R., and Kang, L. (2001). *Queuing Model for Taxi-Out Time Estimation. Submitted to ATC Quarterly, September 2001*. [pdf]. Available through DSpace@MIT at <http:// http://dspace.mit.edu. handle/1721.1/37322> [Accessed February 19, 2013].

Murphy, D. and Shapiro, G. (2007). *A Quantitative Model for En Route Error Rate Analysis. Proceedings of the USA-Europe ATM 7th Seminar, July 2007, Barcelona, Spain*. [pdf]. Available at: <http://www.atmseminar.org/seminarContent/seminar7/papers/p_020_S.pdf> [Accessed February 19, 2013].

Simaiakis, I., Balakrishnan, H., Khadilkar, H., et al. (2011). *Demonstration of Reduced Airport Congestion Through Pushback Rate Control*. [pdf]. Available through DSpace@MIT at <http:// http://dspace.mit.edu/handle/1721.1/60882> [Accessed February 19, 2013].

Simaiakis, I. (2009). *Modeling and control of airport departure processes for emissions reduction*. [pdf]. Available through DSpace@MIT at <http://dspace.mit.edu/handle/1721.1/58289> [Accessed February 19, 2013].

U.S. Department Of Transportation, FAA Notice N JO 7110.524 (2010). *Enhancing Airline Passenger Protections (Three-hour Tarmac Rule)* [pdf]. Available at: <www.faa.gov/documentLibrary/media/Notice/N7110.524.pdf> [Accessed February 19, 2013].

4

Forecasting Airport Delays

David K. Chin, Alius J. Meilus, Daniel Murphy, and Prabhakar Thyagarajan
Federal Aviation Administration, Office of Performance Analysis, USA

4.1 Introduction

The central task of any Air Navigation Service Provider (ANSP) is to provide safe and efficient air travel. Implicit in this task is the maximization of the use of the available capacity of the air transportation system. Meanwhile, policy makers concerned with the experiences of consumers often focus on delays as a measure of how well the aviation system is functioning.

A key difficulty for an ANSP is that they are responsible for only the capacity of the system, and, in the case of the Federal Aviation Administration's (FAA) Air Traffic Organization (ATO), do not control demand. Since policy makers are focused on consumer outcomes (delays) as a measure of system performance, it is necessary to understand how demand and capacity intersect to cause delays. Furthermore, it is sometimes difficult to directly measure those factors over which the ANSP does exert some control, such as airport throughput, and any results reported by the ANSP after the fact may be viewed with skepticism and risk the appearance of being self-serving.

Therefore, the FAA began an effort to project delays six months in advance. This effort serves two purposes. First, if delays can be projected sufficiently far in advance, the FAA can take steps to mitigate these delays, either through operational adjustments or through administrative actions. Second, the FAA can warn policy makers and consumers ahead of time, so that consumers can adjust their behavior or expectations accordingly.

4.2 Historical Example – JFK Summer 2007

The genesis of this FAA effort stems from the experiences at JFK in the summer of 2007. The summer of 2007 was perhaps the worst period for delays in the United States. Much of the delay can be attributed to a significant increase in scheduled operations at JFK.

Modelling and Managing Airport Performance, First Edition. Edited by Konstantinos G. Zografos, Giovanni Andreatta and Amedeo R. Odoni.

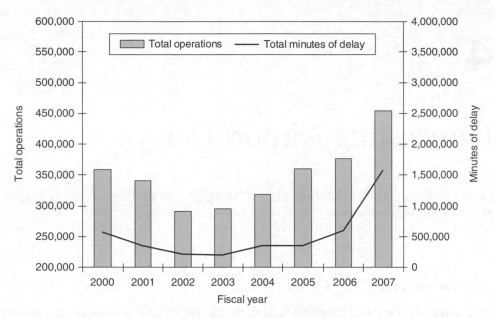

Figure 4.1 Operations and OPSNET delays at John F. Kennedy International Airport (JFK) for fiscal years 2000–2007, showing a correlation between increased operations and delay

Between Fiscal Year (FY) 2006 and FY2007, OPSNET-reported operations at JFK increased by over 75 000 operations, or more than 20%. (The Operations Network (OPSNET) is the FAA's official data source for air traffic operations and delay counts.) Correspondingly, the number of minutes of delay associated with OPSNET delays attributed to JFK increased by over 150%, as shown in Figure 4.1. In addition, the operations of the adjoining New York airports, EWR and LGA, are strongly coupled to JFK because the airports share airspace. As a result, EWR and LGA experienced increased delay. The delays at the New York area airports in turn cascaded throughout the entire National Airspace System (NAS).

Early in 2007, many in the aviation industry knew delays would increase that summer. In the spring of 2007, the FAA tried to warn the public that delays were likely to be particularly bad in the summer. In remarks that were given extensive coverage in the popular press, then FAA Administrator, Marion Blakey, said "We know we're going to have a tough summer," (Hughes, 2007) and "Our best advice is to allow extra time." (Levin, 2007).

However, without detailed modeling, the Agency was unable to anticipate how bad the summer delays would be. In July, over a quarter of flights departed JFK more than 1 h late. Once catastrophic delays materialized, airlines attempted to shift blame to the FAA and other parties. James May, then CEO of the Air Transport Association, testified before the House Transportation and Infrastructure Subcommittee on Airline Delays and Consumer Issues as follows:

> The easy response to delays is to say that scheduled airline operations are the cause, but that response overlooks the many other factors discussed above, especially the capacity supply

factors. ATA and its members are particularly concerned about the inability of the FAA to deliver services at published levels. (May, 2007)

Forced to defend the FAA's performance, in a speech at the Aero Club of Washington D.C. on September 11, 2007, former Administrator Blakey said "...airlines need to take a step back on the scheduling practices that are at times out of line with reality." (Blakey, 2007).

These delays, and the recriminations that followed, might have been avoided had the Agency been aware of the severity of the delays that would result from that schedule. The FAA decided that a more pro-active approach to monitoring airline scheduling was needed, and in its 2009 strategic planning document, the FAA Flight Plan (Federal Aviation Administration, 2009), the agency created two initiatives: "Identify airports forecasted to have chronic delay in the next six months"; and "Mitigate forecasted delays with congestion action teams composed of FAA, airports and operators." This chapter describes the identification initiative.

4.3 Delay Forecasting Methodology

Two main factors drive the accuracy of delay forecasting: accurate demand estimation and an accurate delay model. In the following section we will describe how airline schedules are used to project demand.

In the subsequent section, we will discuss two different models that are used to convert demand estimates into delay estimates. The first of these models uses parameterized relationships between annual operations and average delays for individual airports. The second model incorporates capacity limits into a detailed, system-wide model of delays. The two models require different forms of projected demand, and unique requirements for these inputs will be described in the appropriate modeling section.

4.3.1 Projected Demand

Accurate demand projection is essential to this exercise. Airlines provide schedules up to a year in advance of the day on which a flight is operated. While this information is extraordinarily useful, there are several limitations which we will address next.

4.3.1.1 Airline Schedule Evolution

First, we are attempting to project delays 6–12 months in advance. While airline schedules are first published up to a year in advance, schedules continue to be adjusted until 1–2 months in advance. There are several reasons why airline schedules change. One major driver is that airlines can only speculate what the economic conditions will be six months from now. In Figure 4.2, we show how rapidly airline schedules changed in response to spiking fuel costs and the onset of the financial crisis in the fall of 2008. A distinguishing feature of schedule adjustments driven by economics is that they tend to affect all days of the week in similar ways.

Even when not faced with dramatic economic changes, airlines adjust their schedules to reflect the number of tickets sold on flights. Airlines typically offer a complete schedule for

sale several months in advance, even if they do not expect to operate all flights when the final
schedule is made. A good example of this effect occurs every year on the American holiday of
Thanksgiving. Thanksgiving Day falls on the fourth Thursday in November, and while most
Thursdays are high traffic days, Thanksgiving is one of the least-traveled days of the year in
the United States. In the Spring and Summer, the advertised schedule for Thanksgiving does
not look appreciably different from any other Fall Thursday. However, over the course of the
Fall, the number of flights scheduled for Thanksgiving progressively falls, as shown in
Figure 4.2. The final schedule offered is 33% below a typical Fall Thursday. Other reasons for
changes in schedules are changes in the availability of aircraft or crews or changes in the
economics of the work force and equipment, for example, new labor contracts.

At some point in advance of the month during which a flight is operated, an airline's schedule
will cease to be adjusted. The length of this stable period varies by airline, because it is driven
by factors such as the agreements that airlines have made with their employee groups. For
example, airlines have to schedule pilots to operate their flights, and their pilot agreements may
require the airline to allow the pilots six weeks during which the pilots can "bid" on the flights.
In this case, the schedules must be finalized six weeks ahead of time. Other planning and
operational considerations lead airlines to finalize schedules one to two months in advance.

Taking into consideration the dynamic nature of airline schedules, we have concluded that
while airline schedules provide a good indicator of where demand is going, they are not quan-
titatively reliable more than three months in advance. Therefore, we have collaborated with
the FAA's Office of Aviation Policy and Plans (APO) to supplement the airline schedules with
a short-term, economics-based forecast of demand for each airport. APO is responsible for
generating the FAA's Terminal Area Forecast (TAF). The TAF contains projections of traffic

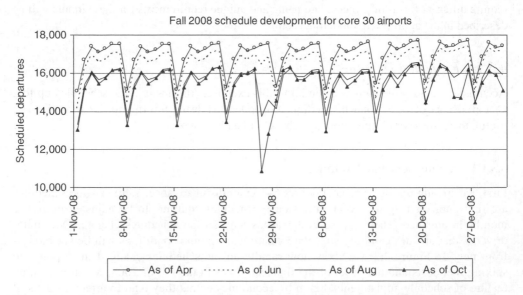

Figure 4.2   Scheduled departures at the US's Core 30 airports for November 2008–December 2008.
Four different published schedules are shown as of the beginning of April, June, August, and October
2008. The data show that the airlines reduced their schedules overall as the economic crisis of 2008
unfolded, and also show how schedules for the American holiday of Thanksgiving evolve

from one year to 15 years into the future. APO has devised a methodology to provide our office with a similar product that is focused on the near-term of 6–12 months, on a monthly basis. We believe that the combination of the extant airline schedules and APO's near-term forecast provides us with a reasonable guide for near-term demand.

4.3.1.2 Unscheduled Traffic

While scheduled airlines operate the majority of the flights into the busiest airports in the United States, there are a significant number of flights which are "unscheduled". These flights include General Aviation (GA) aircraft, but also include cargo and other types of commercial operators. At the moment, we do not take unscheduled traffic into account when using the parameterized, annualized traffic model, but we do supplement the demand input data for the detailed model with historical unscheduled traffic.

4.3.2 Annual Service Volume Delay Model

We use two different approaches to estimate near-term delays, each of which has certain advantages. The first approach uses an analytical model known as the Annual Service Volume (ASV). The ASVs used in our work are generated by the Capacity Analysis Group at the FAA's William J. Hughes Technical Center in Atlantic City, NJ. An ASV is the result of a detailed simulation of traffic at a particular airport, considering airfield configuration, aircraft fleet mix, and a host of other operational factors. After many iterations of the simulation, a curve is generated for an airport which relates the average delay at that airport to the annual number of operations at that same airport. An example ASV is shown in Figure 4.3.

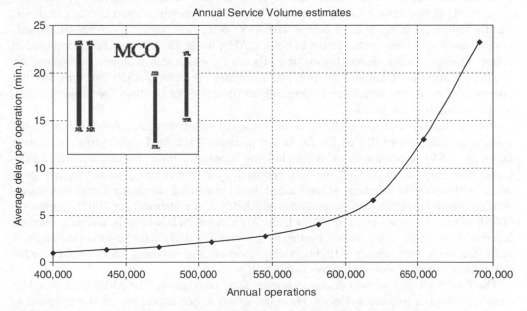

Figure 4.3 The Annual Service Volume (ASV) curve for Orlando International Airport (MCO), showing how modeled average delays increase as total traffic volumes increase

ASVs are easily interpretable. As the number of operations at an airport increases, the average delay at that airport will also increase. As seen in Figure 4.3, delays increase non-linearly as the number of operations increases. In this sense, the ultimate capacity of an airport is the point at which the ASV begins to increase very quickly with even small increases in demand.

The magnitude and other characteristics of delay depend strongly on the specifics of each airport, including the number, size, and layout of the runways, runway exit locations, fleet mix at the airport, and so on. Nevertheless, once the ASV is calculated, it is a straightforward exercise to estimate delays based on the annualized number of operations in the schedule. Annualization is necessary because the independent variable for an ASV is the annual number of operations, while the delays are projected on a monthly basis. We have chosen to annualize operations in the following way. The number of monthly operations for an airport is determined from the schedule for each month. From analysis of historical data, the fraction of annual operations which occurs in a given month is calculated. This fraction is based on a multiyear average of the historical scheduled traffic. The monthly operations count is then divided by the historical fraction in order to recover the corresponding annual total.

The ASV approach has the great virtue that it is simple to compute and is easy to interpret. It has several limitations, however. As mentioned previously, the monthly schedule data must be annualized to be used with ASVs. This assumes that the delay characteristics for an airport are approximately uniform throughout the year. This is a questionable assumption for most airports. As Figure 4.4 shows, delays are significantly higher in the summer and winter than in the spring and fall for the Core 30 airports. If airline schedules were to increase significantly in the spring, our annualization approach might significantly overestimate the impact. The size of this annualization effect is likely to be modest at most airports, but at a few airports, the size of the impact is likely to be significant. At San Francisco International Airport (SFO), delays are highly seasonal because persistent low-lying fog occurs only during certain times of the year. When fog is present, airport capacity is significantly reduced, leading to large delays. The ASV correctly captures this effect, but annualization can lead to substantial errors in delays. Also, while the ASV can be recomputed to reflect changes to the airport layout, the process by which this is done is complex and requires a significant amount of time and resources. Therefore, ASVs are typically not computed for airports which are undergoing construction, particularly for projects which are of relatively short duration.

Another limitation concerns change in the shapes of airline schedules. ASVs are calculated using a specific time-of-day profile for airport demand. While the latest information is used when the ASVs are computed, if the underlying schedule profile changes, without significantly changing overall demand, the ASV approach would not correctly predict the change in delay. Airlines adjust the shape of their schedules in response to changing economic circumstances in order to optimize their operations and their revenue (Petroccione, 2007; Burghouwt, 2009). Figure 4.5 shows an example for Delta Air Lines at ATL, where the airline's schedule became significantly more peaked, leading to an increase in delay at the airport despite a small reduction in the total number of flights. A corresponding analysis using ASVs would have led to the incorrect projection that delays would fall.

The final limitation we will discuss concerns delay propagation. The ASVs are focused on queuing delays at individual airports. However, delays at one airport are often propagated to other airports downstream. For example, Figure 4.6 shows that delays at St. Louis Lambert Field (STL) are better correlated with delays at JFK than they are with traffic at STL itself. While the

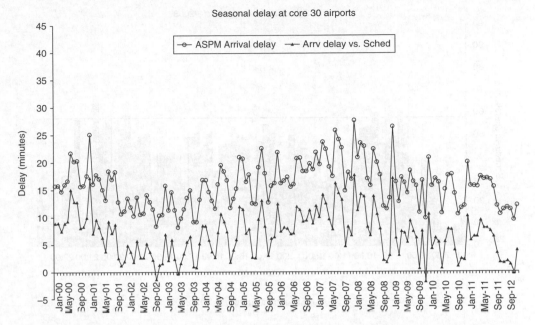

Figure 4.4 Arrival delay by month for the US's Core 30 airports. Two different arrival delays are shown: Aviation System Performance Metrics (ASPM) arrival delay and arrival delay versus schedule. ASPM arrival delay is measured against an ideal flight time, rather than against schedule. The data show periodic variations in arrival delay caused by seasonal variations in the weather

Airport Delay Forecasting initiative is focused on chronically delayed airports, it is also important to know what the *system-wide* impact of capacity constraints at any given airport is.

4.3.3 NAS-Wide Delay Model

We have attempted to address the shortcomings of the ASV approach by modeling delays using a NAS-wide model. The NAS-wide model we use, National Airspace System Performance Analysis Capability (NASPAC), models every individual flight, and therefore implicitly accounts for changes in airline schedules. NASPAC also links flights into aircraft "itineraries", so that delays generated at one airport are propagated to each subsequent flight in that aircraft's itinerary. We also simulate every day in the schedules, using seasonally appropriate weather, which avoids the issue of having to annualize the results. In addition, the detailed model enables the use of arbitrary airport capacities, which means that runway construction impacts can readily be modeled. Before describing the model in detail, we will discuss the manner in which the schedule input data is processed, and what information must be added to the airline schedule data in order to run the detailed simulation.

4.3.3.1 Flight Plan Data

One challenge we face with demand is that airline schedules provide insufficient information to model each flight. Schedules typically include information about gate departure and arrival

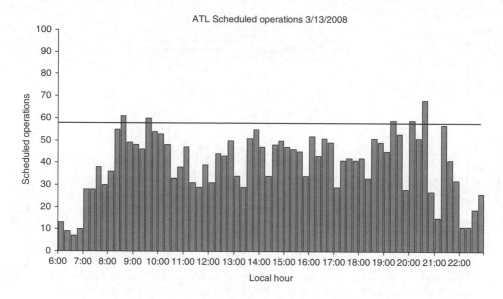

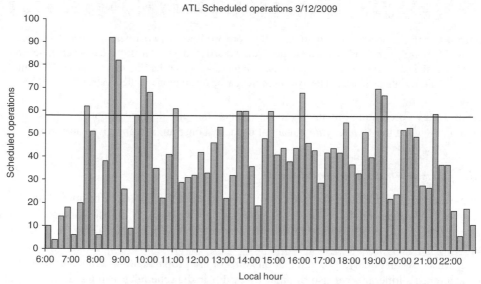

Figure 4.5   Scheduled operations at Hartsfield-Jackson Atlanta International Airport (ATL) for March 2008 and March 2009, showing how the schedule became significantly more banked despite a decrease in total operations

times, as well as the type of aircraft scheduled to operate each flight. However, critical information is missing, including the flight route that the aircraft is likely to take, the speed and altitude of the flight, and the gates the flight will use at the arrival and departure airports.

This missing operational information, which is needed to accurately model optimal flight time, as well as to determine which ATC resources each flight will use, is provided by analysis

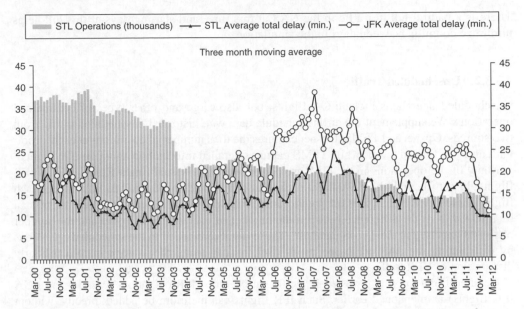

Figure 4.6 Monthly scheduled operations and delays at Lambert-St. Louis International Airport (STL) and delays at John F. Kennedy International Airport (JFK), showing that STL delays are correlated with JFK delays, and not STL operations

of historical operational data. The FAA's Traffic Flow Management System (TFMS) captures detailed information about the flight plans for Instrument Flight Rules (IFR) flights. Since all scheduled passenger flights are required to operate as IFR flights, we can use the TFMS database maintained by the FAA's Airspace Lab to fill in this missing data.

It might seem obvious that, to fill out the missing information, one should match each flight in the future summer schedule to a corresponding historical flight from the previous year. However, several problems arise in implementing this approach. First, it is not necessarily the case that all flights which are scheduled for the current year can be matched to a flight from a previous year. Second, we are interested in the optimal performance of a flight, but the filed flight plan may not correspond to what is optimal. For example, flights which occur later in the day are more prone to delays than those which depart early in the day, and it is possible that the operational data incorporate sub-optimal routes, speeds, or altitudes. Therefore, we have developed a hierarchical approach to matching historical operational data with current schedules which tries to balance providing as much detail about a given flight with these constraints on accuracy.

This hierarchical approach proceeds from the most specific match to the least specific. Where possible, a flight is matched to the most commonly filed route of flight, speed, and altitude for the same city-pair, airline, aircraft type, and day. If no matching flights were operated on that day for a given city pair and aircraft type, the flight is matched against a monthly table. Failing that, the flight is matched by city pair and aircraft type. Finally, if all else fails, the flight is assumed to fly a great circle route between origin and destination at the average speed and altitude for all aircraft in the ETMS dataset.

The input demand data is grouped into "simulation days". A simulation day begins and ends at 0900 UTC, which corresponds to 4:00 a.m. Eastern Standard Time (EST) in the US

There is a natural lull in traffic in the NAS at this time, and starting days at this lull helps minimize coupling between days in the simulation.

4.3.3.2   Unscheduled Traffic

Unscheduled operations include GA flights, but also cargo and other types of commercial operations. We supplement the airline schedule data with historical unscheduled traffic from two sources, Opsnet and TFMS. Opsnet captures the total number of operations at each airport with an FAA-funded tower, while TFMS captures detailed information about all IFR flights.

Many of the operations in the NAS are performed under Visual Flight Rules (VFR). VFR flights do not file with the FAA flight plans that contain sufficient detail to fully model each flight, nor are VFR flight plans retained for historical analysis. The number of VFR flights can be inferred from the difference between the number of operations in Opsnet and the number of flights captured in the TFMS database. These flights are added to the scheduled operations by distributing the daily VFR operations over the day using the average historical hourly distribution of VFR traffic.

Detailed flight information is available for unscheduled IFR operations (e.g., for most GA IFR flights) as mentioned earlier. Such IFR flights can therefore be added directly, without significant alterations, to the airline schedule data in preparing a demand schedule for NASPAC. However, certain commercial operators require additional considerations. For example, some of the cargo carriers operate a regular set of flights, but do not necessarily perform all of these flights. One example is Polar Air Cargo (PAC). While PAC publishes a schedule of flights, the actual number of flights operated is more variable than is typical for a scheduled passenger carrier. For example, the January 2010 schedule for PAC shows 22 departures from LAX, while only 14 were actually operated. As a result, PAC's schedule is only indicative of the true load the carrier imposes on the system. We therefore remove from the schedule data flights by freight and other commercial operators, which are not offering scheduled passenger service, and replace them with representative operations from the previous year.

In practice, we maintain a short list of carriers which offer schedule passenger service and filter them out of the TFMS data, then merge the resulting filtered data with schedule data for our list of passenger carriers to generate the final input demand for the simulations.

4.3.3.3   Model Description

As mentioned earlier, we have attempted to be parsimonious in simulating NAS-wide delays, in the sense that we have attempted to incorporate a minimal representation of the NAS that accurately captures delays. The current model uses capacities for airport runways, arrival and departure fixes, sector capacities, and airport gates. It also includes delay propagation through the construction of flightitineraries. The model does not include explicit representations of the airport surface in the form of taxiways, nor does it assign runways to arriving and departing flights. The model also does not consider passenger or crew delays and their interaction with propagated airframe delay.

While a full technical description of the simulation is beyond the scope of this chapter, it is worthwhile to review a few key points. (See Post et al., 2008; Thyagarajan and Murphy, 2011 for more details on the model.) The NAS-wide model used in this work, NASPAC, employs a traditional process-based simulation at its core. Each aircraft itinerary, consisting of one or

Table 4.1 Modeled resources

Resource	Level
Gate	By Airline and Aircraft Design Group
Departure Runway	By Airport
Departure Fix	By Fix
Sector	By Sector
Arrival Fix	By Fix
Arrival Runway	By Airport

more flights, is considered a single process, and all of the activities associated with a flight, such as gate pushback, taxi out, runway departure, and so on, are modeled in the simulation.

NASPAC contains a trajectory modeler which uses the detailed flight information in the demand input data to compute flight times, sector pierce and dwell times, fuel burn, and so on. The trajectory model in NASPAC implements the Base of Aircraft Data (BADA) performance model from Eurocontrol (Eurocontrol, 2010), which contains detailed models of a large number of aircraft types. The trajectory model explicitly accounts for winds, ingesting wind data for each simulation day and adjusting flight times accordingly.

Arrival and departure capacities are handled with separate single-server queues, even if multiple runways are used. Gate constraints are explicitly modeled, with separate gate queues for each airline and aircraft group at an airport (Thyagarajan et al., 2012). However, data on actual gate capacities and usage are available and have been analyzed for a subset of the airports in the NASPAC network.

Table 4.1 provides a summary list of NAS constraints explicitly modeled in NASPAC.

Our analysis shows that the dominant and systematic contributor to flight delays is queuing delay associated with limited airport capacity. Many simulation exercises rely upon modeled airport capacities when estimating flight delays. Capacities are modeled for a subset of runway configurations and weather conditions. Figure 4.7 shows a typical set of capacity curves for Newark Liberty International Airport (EWR). The lines on the graph, referred to as Pareto curves, indicate the optimal performance frontiers for the airport. The two curves shown are for Visual Meteorological Conditions (VMC, or good weather) and Instrument Meteorological Conditions (IMC, or poor weather). While the modeled rates show a trade-off between arrivals and departures for a given weather condition, the arrival rate does not change appreciably except under very large departure demand. In contrast, an analysis of historical data suggests that airport capacity is highly variable, even in good weather (Chung and Murphy, 2010; Houston and Murphy, 2012). Figure 4.8 shows the "called" arrival rates for EWR in VMC for the 10 most common runway configurations. (Some infrequent called rates have been omitted.) Called rates reflect what air traffic controllers feel is the best estimate of an airport's capacity at any particular instant in time. For what are nominally the same weather conditions (VMC), Newark's called rate varies between 38–52 arrivals per hour. Such large variations in airport capacity lead to very large differences in arrival delays, and if these variations are not captured in our modeling, our delay estimates will be inaccurate. Therefore, we have chosen to use the historical called rates as the airport capacities in the simulations, rather than the modeled rates.

Detailed analysis is also needed in generating aircraft itineraries. While we have access to airline flight schedules, these schedules do not include information on which specific aircraft will

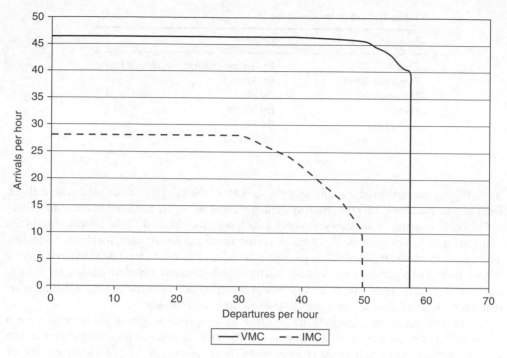

Figure 4.7   Modeled Pareto capacity curves for Newark Liberty International Airport (EWR), showing the difference in capacity between good weather (VMC) and bad weather (IMC)

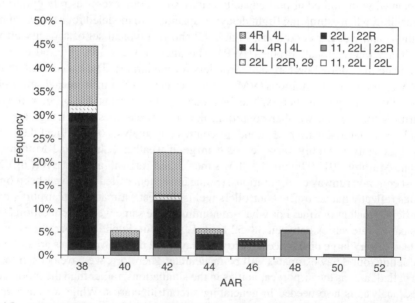

Figure 4.8   Airport VMC called rates at Newark Liberty International Airport (EWR) for different airport configurations, showing the variation in airport capacity in good weather

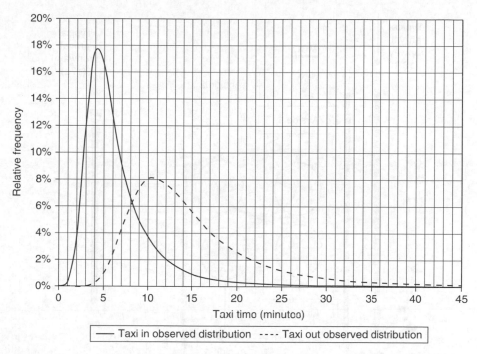

Figure 4.9 Observed distribution of taxi times for all domestic US flights reported in Airline Service Quality Performance (ASQP) data

operate a particular flight. In the absence of such data, scheduled flights must be linked together to form itineraries. Flights are added to an itinerary by a process which assumes that an arrival is linked to the first possible departure by the same aircraft type at an airport (Robinson et al., 2011; Robinson and Murphy, 2012). A departure is considered possible if it is scheduled to depart later than the arriving flight was scheduled to arrive, plus an extra amount of time referred to as the scheduled turn time. In order to determine the scheduled turn time, we examined historical data to determine how much time different airlines allow between flights operated by the same aircraft. In general, larger aircraft take longer to turn than smaller aircraft, because it takes longer to deplane and enplane the passengers, clean the aircraft, fuel the aircraft, load cargo, and so on. However, there are significant variations by airline. Most airlines schedule more time at the gate than is usually required, but Southwest Airlines chooses to absorb all delays in their block time, meaning that Southwest schedules less time at the gate than their flights typically require.

While we have chosen not to model all aspects of NAS performance, we have attempted, where possible, to capture the variability of some of the unmodeled aspects of NAS performance. We analyzed statistically historical data to characterize the underlying distribution of different operational quantities. For example, the US Department of Transportation (DOT) collects from most of the major American carriers operational data that include runway departure and arrival times, and gate departure and arrival times. From this information, the actual taxi in and taxi out times can be computed. The actual taxi times do not, of course, all have the same value. Instead, they provide a probability distribution that can be used to simulate taxi times in NASPAC. Figure 4.9 shows the distribution of the taxi-in and taxi-out times for all flights reported to DOT and included in the Bureau of

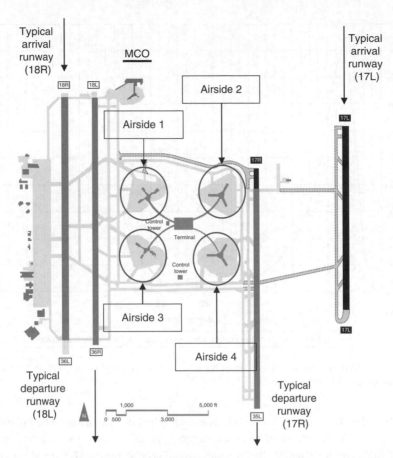

Figure 4.10 Airport diagram for Orlando International Airport (MCO), showing typical departure and arrival runways and highlighting the different terminals

Transportation Statistic's Airline Service Quality Performance (ASQP) data. Some of the variability comes from small differences in performance, such as slight differences in taxi speed, while others can be attributed to such effects as runway and gate assignments. Insofar as these effects are random and uncorrelated, the taxi times can be modeled using a normal distribution. In the simulation, the taxi time for a particular flight is drawn from this modeled distribution.

Of course, runway and gate assignments may be correlated with certain aspects of a flight. For example, in the United States, airlines typically own or lease a specific set of gates at an airport. Furthermore, not all airport gates are created equal; some gates may accommodate certain aircraft types, and not others. As a result, taxi times will depend upon the active runway configuration at the time of arrival or departure, which airline is operating a flight, and what type of aircraft is being flown. An example is shown in Figure 4.10. United Airlines controls gates at Terminal 3 at Orlando International Airport (MCO), while Southwest Airlines is housed at Terminal 2. Because of the runway layout, the unimpeded taxi-out time for Southwest is, on average, 4 min shorter than for United, as shown in Figure 4.11. We segregated

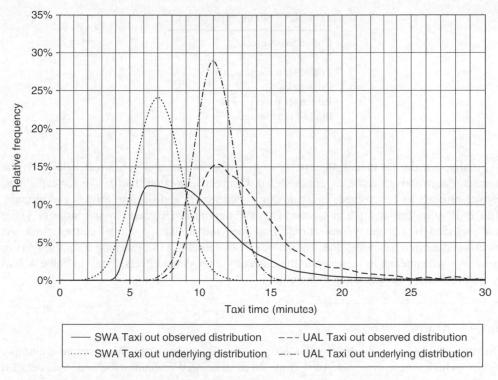

Figure 4.11 Taxi-out time distributions at Orlando International Airport (MCO) for Southwest Airlines (SWA) and United Airlines (UAL), showing the sensitivity of taxi times to airline terminal location

Table 4.2 Parameterized activities

Resource
Turn around
Carrier-caused delay
Pushback
Taxi-out
Taxi-in

the historical records during our data analysis to capture some of these effects. For taxi times, we grouped the data according to airline, airport, and equipment type before extracting the moments of the distribution. Examples of the fit distributions for taxi times are shown in Figure 4.11. (Note that the fit distributions look significantly different from the actual distributions. Our process attempts to remove the queuing effects evident in the long tails of the observed distribution to extract the underlying normal distribution. For more details, please refer to Robinson and Murphy, 2010.)

In addition to taxi times, similar analyses were conducted for actual aircraft turn times and gate pushback delays. The activities which have been parameterized in this fashion in the model are listed in Table 4.2.

Table 4.3 Future improvements

Convective weather
Rerouting
Traffic flow management
Crew delays
Passenger delays
Flight cancellations

Finally, we do not model all aspects of NAS performance which affect delays. At the moment, convective weather is not modeled. Convective weather reduces sector capacities, causes aircraft to be rerouted, and can shut down arrival and departure fixes and airports. Traffic flow initiatives, such as Ground Delay Programs (GDP) and Ground Stops (GS) are not included in our delay projections, nor does the simulation model cancellations. As mentioned earlier, the model does not include delays resulting from waiting for passengers or crew. We will attempt to incorporate the currently unmodeled effects, listed in Table 4.3, as we gain a better understanding of them.

4.3.4   Results

Our office uses both modeling methodologies described previously. We generate a monthly report which incorporates results from both the ASV calculations and the detailed simulation.

A typical page from our report is shown in Figure 4.12. This figure shows how the ASV approach is used to project airport-related delays for Chicago O'Hare International Airport (ORD) based on June 1, 2010 airline schedules. Starting at the top left of Figure 4.12, we see that approximately 10% more operations are scheduled at ORD in the summer of 2010 than were scheduled in the summer of 2009. The chart on the lower left of Figure 4.12 shows the graph that relates operations to delays at ORD. The symbols on this chart correspond to the different months of the year. The graph is used to compute average flight delays for each month, which are shown in the lower right of Figure 4.12.

The report also includes summary results for a cross-section of major US airports. Figure 4.13 shows an example of one such summary chart, taken from the monthly report based on June 2010 schedules. The chart shows that ORD is projected to have a 2.5 min increase in average delay per operation for August 2010 year over year.

As noted previously, the ASV methodology captures the impact of changes in the overall level of demand at an airport, but does not capture the effect of changes in the temporal distribution of that demand. The detailed simulation is used to capture these effects. Focusing again on ORD, Figure 4.14 shows the scheduled June 2010 operations by time of day for ORD as of April 2010, compared to the final June 2009 schedule. The 2010 schedule is much more peaked than the one for 2009. The increase in the peak is driven primarily by more aggressive departure banking, as shown in Figure 4.15.

While time plots of schedules convey a sense of changes in overall demand and temporal profile, they cannot show whether there will be a significant impact on operations. Figures 4.16 and 4.17 show results from the detailed simulation for four

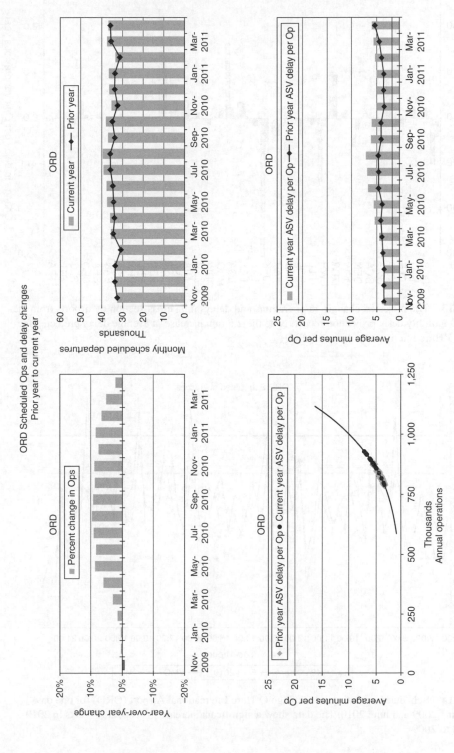

Figure 4.12 Representative page from the June 2010 monthly delay projection report. This page shows a significant increase in scheduled operations at Chicago O'Hare International Airport (ORD), and the projected increase in delays caused by the increase in operations

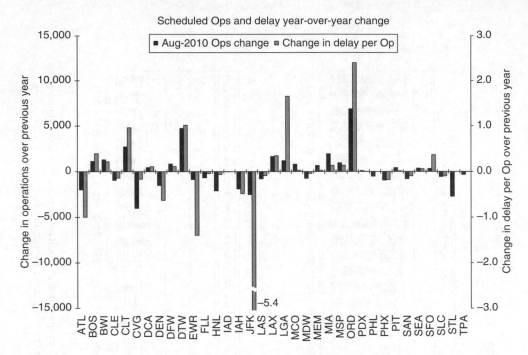

Figure 4.13 The projected changes in operations and delays for a cross-section of airports from the June 2010 monthly delay projection report. Note the 2.5 min increase in average delay projected for Chicago O'Hare International Airport (ORD)

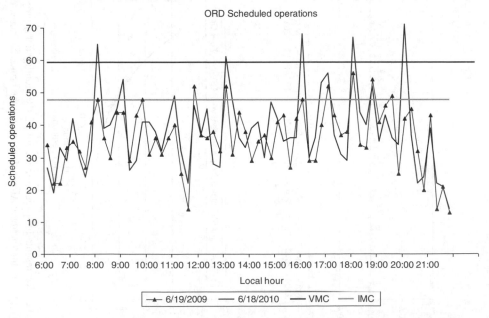

Figure 4.14 Scheduled operations at Chicago O'Hare International Airport (ORD) for one day each in June 2009 and June 2010. The data show a significant increase in scheduled banks in 2010 compared to 2009

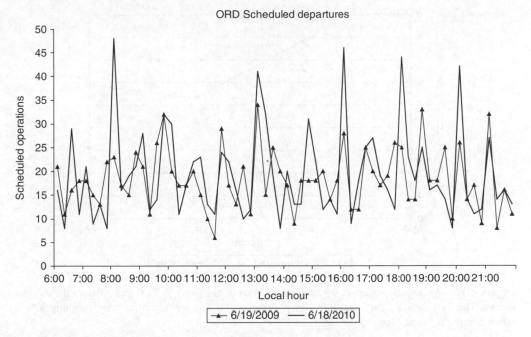

Figure 4.15 Scheduled departures at Chicago O'Hare International Airport (ORD) for one day each in June 2009 and June 2010. The data show that the increased in 2010 compared to 2009 is caused by the significant increase in the height of the departure banks

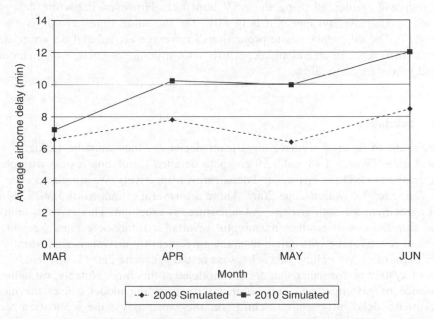

Figure 4.16 Average airborne delay from simulations of every day in 2009 and 2010 for the months shown. The data show a significant year-over-year increase in average airborne delay

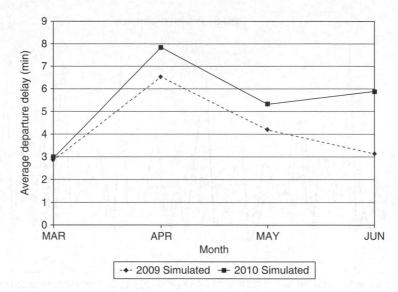

Figure 4.17   Average departure delay from simulations of every day in 2009 and 2010 for the months shown. The data show a significant year-over-year increase in average departure delay, reaching 80% when comparing June 2010 to June 2009

months in the first half of 2010. In Figure 4.16, we see that arrival delays were projected to increase by approximately 40% in June 2010, compared to June 2009. This is similar to the increases projected using the ASV approach. However, departure delays were projected to increase by approximately 80% for the same time period, as shown in Figure 4.17. The difference in the projections in average arrival and departure delay is due to the difference in arrival and departure bank structure at ORD, which can only be captured by detailed modeling.

4.3.4.1 Validation

If models are to be used to drive significant decisions, they must be validated against historical data. Figures 4.18 and 4.19 compare detailed simulation results with observed airport delays at ORD for arrivals and departures, respectively. We simulated every flight for the four months, March–June 2009. There are several points worth noting. First, the overall agreement for both arrivals and departures is excellent. This gives us confidence that the simulation will produce meaningful results for future schedules. Second, while both sets of results agree with actual data, the agreement for arrivals is significantly better than for departures. We believe this is because actual departure delay is affected by certain aspects of system performance that are not modeled at this time. Notably, we believe that the absence of passenger and crew connectivity from the model causes the model to underestimate delay propagation. Third, on the worst days, the simulation tends to overestimate arrival delay. This is because we do not allow flight cancellations or aircraft substitutions at this time. Overall, our validation gives us great confidence in using these models to project delays.

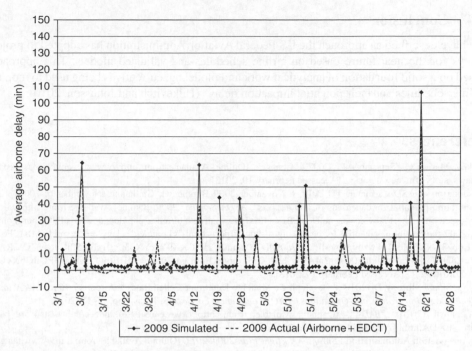

Figure 4.18   A comparison of simulated airborne delay to actual airborne delay for March through June 2009, showing excellent agreement

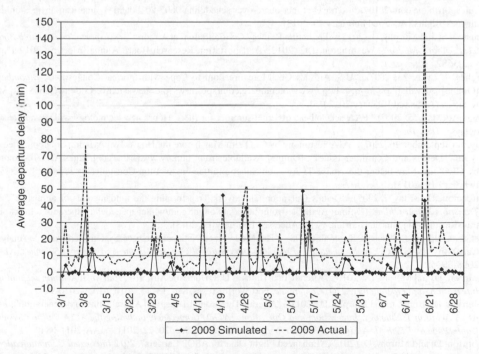

Figure 4.19   A comparison of simulated departure delay to actual departure delay for March through June 2009, showing excellent agreement

4.4   Conclusion

We have described an approach the US Federal Aviation Administration has adopted to project delays for the near future based on airline schedules and validated models. This approach, based on a solid foundation of analytical work and modeling, currently is being used to monitor airline schedules and their potential impact on delays (Hilkevitch and Johnsson, 2010).

References

Blakey, Marion C. (September 11, 2007). "Change", [Online] Available from: http://www.faa.gov/news/speeches/news_story.cfm?newsId=9532 [Accessed February 19, 2013].

Presentation at ASDA seminar, TU Delft, 6 November 2009 Burghouwt, Guillaume (November 6, 2009). "The economic potential for depeaking of hub airports", *ASDA Seminar 2009*, [Online] Available from: https://docs.google.com/viewer?a=v&q=cache:DomaBIYn4MsJ:www.airneth.nl/index.php/doc_download/984-burghouwt-g-2009-the-economic-potential-for-depeaking-of-hub-airports.html+&hl=en&gl=uk&pid=bl&srcid=ADGEEShm dL6bz3_-xV5Me11pEwznygahyML2Qz7q090PGbVAm0U25NCKX9YtxYXo3tP_JIxFpI5fbGvqs9CxWiRArX 9vL8HFF9KPmRyxfaIoz0Csaf2j4AWe_KG_GiS1qggzkE4pLgqC&sig=AHIEtbQ_KNIaWPOdGbXvhuB 26ZoRqdakmQ [Accessed February 19, 2013].

Chung, S. and Murphy, D. (2010). "Developing a Model to Determine Called Rates at Airports", *10th AIAA Aviation Technology, Integration, and Operations (ATIO) Conference*, Fort Worth, Texas, Sep. 13–15, 2010.

Eurocontrol (2010). "*BADA*", [Online] Available from: http://www.eurocontrol.int/eec/public/standard_page/proj_BADA.html [Accessed February 19, 2013].

Federal Aviation Administration (2009). *FAA Flight Plan 2009–2013*. [Online] Available from: http://www.faa.gov/about/plans_reports/media/flight_plan_2009-2013.pdf [Accessed February 19, 2013].

Hilkevitch, J. and Johnsson, J. (June 6, 2010). "Airlines overpack for summer travel; FAA says carriers can cut delays by reducing flights during peak airport hours", *Chicago Tribune*, [Online] Available from: http://articles.chicagotribune.com/2010-06-06/travel/ct-met-ohare-over-scheduling-0607-20100606_1_american-airlines-united-airlines-flights [Accessed February 19, 2013].

Houston, S. and Murphy, D. (2012). "Predicting Runway Configurations at Airports", *Transportation Research Board 91st Annual Meeting*, Washington, DC, 2012, Transportation Research Board Annual Meeting 2012 Paper #12-3682.

Hughes, J. (May 23, 2007). "U.S. Airlines Get More Freedom to Cope with Storms, Congestion", *Bloomberg*, [Online] Available from: http://www.bloomberg.com/apps/news?pid=newsarchive&sid=awQ25sSkGq.M& refer=us [Accessed February 20, 2013].

Levin, A. (May 25, 2007). "FAA acts to head off cruel summer for fliers; Hopes new technology will prevent record delays", *USA Today*.

May, J. (September 26, 2007). "ATA Testimony by CEO Jim May before the House T&I Aviation Subcommittee on Airline Delays and Consumer Issues", [Online] Available from: http://www.airlines.org/Pages/ATA-Testimony-by-CEO-Jim-May-before-the-House-TI-Aviation-Subcommittee-on-Aviation-Consumer-Issues.aspx [Accessed February 19, 2013].

Petroccione, L. (May 7, 2007). "Delta's Operation Clockwork Transforming the fundamentals of an airline", *Air Transport World Online*, [Online] Available from: http://atwonline.com/whitepapers/deltas-operation-clockwork-transforming-fundamentals-airline-0215 [Accessed February 19, 2013].

Post, J., Gulding, J., Noonan, K., et al. (2008). "The Modernized National Airspace System Performance Analysis Capability (NASPAC)", in *Proceedings of the 26th International Congress of the Aeronautical Sciences*, AIAA 2008-8945.

Robinson, D. and Murphy, D. (2010). "Aircraft Taxi Times at U.S. Domestic Airports", *10th AIAA Aviation Technology, Integration, and Operations (ATIO) Conference*, Fort Worth, Texas, Sep. 13–15, 2010, AIAA-2010-9147.

Robinson, D., Bonn, J., and Murphy, D. (2011). "Aircraft Turnaround Times and Aircraft Itinerary Generation", *11th AIAA Aviation Technology, Integration, and Operations (ATIO) Conference, including the AIAA Balloon Systems Conference and 19th AIAA Lighter-Than-Air*, Virginia Beach, VA, Sep. 20–22, 2011, AIAA-2011-6848.

Robinson, D. and Murphy, D. (2012). "Enhanced Flight Data for ASQP Carriers", *2012 Integrated Communications, Navigation, and Surveillance (ICNS) Conference*, Herndon, VA, April 24–26, 2012.

Thyagarajan, P. and Murphy, D. (2011). "Relative Impact of Various Resources in the US National Airspace System (NAS) On Total Flight Delays", 11th AIAA/ATIO Conference, Virginia Beach, Virginia, Sept. 20, 2011.

Thyagarajan, P., Shapiro, G., and Murphy, D. (2012). "Implementing Airport Terminal Gates in NAS-wide simulations", *12th AIAA/ATIO Conference Indianapolis, Indiana*, Sept. 17–19, 2012., [Online] Available from: https://www.google.co.uk/url?sa=t&rct=j&q=&esrc=s&source=web&cd=1&ved=0CDQQFjAA&url=https%3A %2F%2Fwww.aiaa.org%2FWorkArea%2FDownloadAsset.aspx%3Fid%3D13738&ei=8H0jUd-vHqSW0QXh3Y DQAw&usg=AFQjCNFHr5uynED9Vx-h1c6abg8-iveQ9Q&sig2=6W8smoC5CCq2r2nYIeXWCQ&bvm=bv.425 53238,d.d2k (accessed February 19, 2013).

5

Airport Operational Performance and Its Impact on Airline Cost

Mark Hansen[A] and Bo Zou[B]
[A]*National Center of Excellence for Aviation Operations Research, Department of Civil and Environmental Engineering, University of California, Berkeley, USA*
[B]*Civil and Materials Engineering, University of Illinois at Chicago, USA*

5.1 Introduction

Delays, many of them the result of limited or reduced airport capacity, are a fact of life in air transport. In 2007, nearly one in four US airline flights arrived at its destination over 15 min late (BTS, 2009). About a third of these late arrivals were a direct result of the inability of the aviation system to handle the traffic demands that were placed upon it, while another third resulted from airline internal problems. Most of the remainder was caused by an aircraft arriving late and thus having to depart late on its next flight (BTS, 2009). Between 2002 and 2007, as the air transport system recovered from the 9/11 attacks, scheduled airline flights increased about 22%, but the number of late-arriving flights more than doubled. Since 2007, traffic and delays have declined somewhat because of the recession, but the FAA expects growth to resume, with air carrier flight traffic reaching 2007 levels by 2013, and growing an additional 32% by 2025 (FAA, 2011). It is widely recognized that delay increases nonlinearly as demand approaches the capacity in the system (Figure 5.1).

Substantial investments are required in order to modernize and expand our aviation infrastructure so that it can accommodate anticipated growth without large increases in delay. In the US, the Next Generation Air Transportation System (NextGen) will deploy improved systems for communications, surveillance, navigation, and air traffic management and also require flight operators to invest in new on-board equipment. Substantial improvements in air transportation capacity also require airport infrastructure

Modelling and Managing Airport Performance, First Edition. Edited by Konstantinos G. Zografos, Giovanni Andreatta and Amedeo R. Odoni.

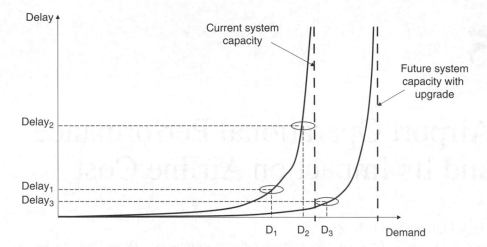

Figure 5.1   Illustration of the relationship between delay, demand and system capacity

enhancement. Estimates of these combined investments reach well into the tens of billions of dollars (GAO, 2008; ACI, 2009).

Much of the business case for these large expenditures rests on the value of reducing delay and its associated costs. The cost of delay includes many elements. Passengers are inconvenienced when their flight arrives late, or especially if they are forced to arrive on a different flight as a result of a cancellation or missed connection. Delays thus degrade the quality of air travel products, diminishing passenger willingness to pay for them and discouraging some from flying altogether. Adaptations to avoid or mitigate delay – such as leaving early for a business meeting to make sure it is not missed, or scheduling flights at less than ideal times to avoid congestion – also entail costs.

The delay cost element that receives the most attention, however, is that incurred directly by airlines through increased capital and operating expenses. This is the core element of most business cases for airport infrastructure investment to improve airport operational performance. Many in the community perceive that, since airlines must pay these costs with 'real money', they merit stronger consideration than the costs of passenger time loss. By the same token, there is a stronger basis for quantifying airline delay costs, and as a result a larger literature on methodologies for doing so.

This chapter reviews current knowledge and thinking about the costs of delay and related phenomena to airlines. We review – with a somewhat critical eye – various methods for determining the cost of delay for airlines. We also present results from the application of these methods. We focus on the US and Europe, where the flight delay problem gets the most attention and where the value of delay reduction is the most critical input to the cost-benefit analysis for airport and other aviation capital investments.

Section 5.2 reviews methods for measuring delay or – more broadly – operational performance in air transport. Section 5.3 discusses valuation methodologies. Section 5.4 identifies some unresolved issues regarding both measurement and monetization, while Section 5.5 summarizes and concludes this chapter.

5.2   Quantifying Operational Performance

5.2.1   Arrival Delay Against Schedule and Schedule Buffer

To quantify delay cost impact we must first quantify delay. This is not as simple as it may seem. To highlight the issues involved, we start with a stylized view of how airlines would operate in the absence of congestion and delays. An airline might start the process of scheduling a flight by determining an *ideal departure time (IDT)*. The ideal departure time would take into account not only preferred passenger travel times, but also internal airline constraints, such as those necessary to create efficient crew schedules and fleet plans. As part of this process, the airline would choose the most appropriate aircraft type from its fleet for the flight. Based on the characteristics of that aircraft and assuming it could fly the optimal, unimpeded origin-to-destination trajectory, an *ideal arrival time (IAT)* could be computed as illustrated in Figure 5.2. The interval between the ideal departure and arrival times, called *unimpeded flight time*, is a key quantity and will be discussed throughout this chapter.

Now let us consider how congestion and delays alter this situation. As illustrated in Figure 5.3, the airlines will typically increase scheduled flight times over unimpeded ones in order to account for delays resulting from flight restrictions imposed to organize traffic, congestion, and a variety of other factors. We call this added time the *schedule buffer*, which is the difference between the *scheduled arrival time (SAT)* and *IAT* Schedule buffers allow airlines to absorb some delays while preserving flight punctuality and predictability, and thus making the schedule more robust to flight time variations. Adding buffer also helps airlines keep good on-time performance, which has become an increasingly critical marketing tool. In principle, once the unimpeded flight time is determined, the schedule buffer can be obtained by comparing this with the flight schedule. This, of course, begs the question of how to define the unimpeded flight time, which will be discussed later.

The amount of buffer added depends in part upon the statistical knowledge of past delays (Cook et al., 2004), and is also subject to several constraints. First, as buffer is now part of

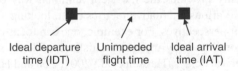

Ideal departure   Unimpeded   Ideal arrival
time (IDT)   flight time   time (IAT)

Figure 5.2   Unimpeded flight time

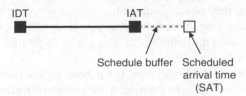

Schedule buffer   Scheduled
                  arrival time
                  (SAT)

Figure 5.3   Schedule buffer

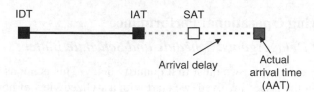

Figure 5.4   Flight delay against schedule

the schedule, it entails costs, for example by increasing crew pay hours and the travel time advertised to potential customers. A large buffer also decreases aircraft utilization and thereby increases capital cost per block hour. In short, while buffer time is valuable, particularly in a delay-prone system, it is also expensive. Setting the right amount of buffer requires trading off these costs with the potential cost savings from the buffer when delay occurs.

Of course, even after buffer is added, flight delays can still occur. Arrival delay, the type of delay most commonly considered, measures flight lateness against the *SAT* (Figure 5.4). Schedule buffer and arrival delay both represent excess travel times that would not exist in an operationally perfect air transportation system. While buffer is predetermined for a particular flight, arrival delay varies randomly from day to day and flight to flight, its magnitude shifted upward or downward according to the amount of buffer. Arrival delay is negative when the buffer exceeds the amount of time between the actual and ideal flight arrival times. Early arrivals are widely believed to have little cost impact since they save little in the way of resources as compared to an on-time flight. Thus, arrival delay truncated at zero, or positive arrival delay, is often used for the delay metric.

5.2.2   Alternative Metrics

The above description only provides the most conventional way of characterizing flights' operational performance. Various alternative metrics exist, looking at flight activities relative to schedule from different perspectives. For example, one can focus on the departure end, and develop a set of corresponding delay metrics. Statistical evidence shows strong correlation between departure and arrival delays (Hansen et al., 2001; Zou and Hansen, 2012a). However, it is more convenient to consider arrival delay due to its direct connection with schedule buffer. Mayer and Sinai (2003a) used excess travel time relative to the minimum feasible travel time observed on each flight segment as a single measure of flight operational performance. Essentially, this measure equals the sum of schedule buffer and arrival delay, with the minimum feasible travel time defined as the unimpeded flight time. As we shall see in the following section, since the definition of unimpeded flight time is not unique, neither is the excess travel time.

Another potential metric is delay variance. It has been argued that schedule disruption, an important delay impact, results mainly from delay variability rather than delay *per se*. At a hub airport, for instance, half of the flights arriving late by 20 min whereas the other half being on time would be more disruptive than all flights being delayed by 20 min. Hansen et al. (2001) used the Principal Component Analysis and found that, at the airline-quarter level, the variance of delay and the average delay have different patterns of variation. They identified the 'variability' factor that has much higher impact on the variance of delay than on the other delay metrics. Nonetheless, variance of delay has not been widely considered so far in delay cost analysis, in part because it is not clear how to assign it a cost factor.

It is also common to summarize delays in terms of the fraction of flights whose delay exceeds, or does not exceed, a certain value. We refer to these as on-time metrics. Airlines have developed a special nomenclature for this. The A-14 metric, for example, is the fraction of flights that arrive more than 14 min late. FAA and DOT have used A-14 – or, more precisely, its complement – as a key metric for airline on-time performance. Airlines attach greater significance to D-14, the fraction of flights that depart more than 14 min late, because this is a metric over which their own personnel, as opposed to the air navigation service provider, have more control. Other measures of this type that airlines follow include A/D-0, 30, 60, and 120. By tracking these different metrics, airlines develop a more complete picture of operational performance, and in particular the incidence of long, disruptive delays, than the average provides. As with delay variance, however, little has been done to relate on-time performance to airline costs.

In sum, while the alternative operational performance metrics discussed above have their uses, arrival delay and schedule buffer are the most widely used in delay cost analysis, because of their simplicity and mutual complementarity. For this reason the subsequent discussion of delay cost impact is based primarily upon these two metrics.

5.3   Estimating the Cost Impact of Imperfect Operational Performance

We consider two approaches to estimating the impact of imperfect operational performance on airline costs. The first, which we term the *cost factor approach*, is based on assigning unit costs to different categories of delay based on estimates of the resources consumed when a given category of delay occurs. The second, which we will call *the total cost approach*, is built on firm or industry level relationships between total operating cost and delay. The cost factor approach is more of a 'bottom-up' approach since, implicitly at least, it assumes that aggregate delay cost is the sum of the costs of individual delay events. Conversely, the total cost approach takes a 'top-down' perspective which allows for the possibility that delay can seep into every facet of air transport production cost.

5.3.1   Cost Factor Approach

The cost factor approach estimates delay cost as a linear function of one or more delay quantities, the coefficients of which are cost factors. In the simplest case with a single delay variable, the total cost impact of delay is:

$$C = P \cdot X \qquad (5.1)$$

where C is the total delay cost, P denotes the cost per unit time (e.g., measured in \$/min), and X represents the total delay minutes. X can be obtained by summing up delays experienced by individual flights.

Equation (5.1) is overly simplistic in that delay is not homogenous. One source of heterogeneity that affects cost is where the delay is taken, which may be at the gate, on the taxiway, or in the air. Other sources of heterogeneity with potentially significant cost implications include the length of delay, aircraft size, and whether the delay is caused by the flight itself or propagated from other ones. This suggests that delays should be disaggregated into categories before applying cost factors. The cost formula then becomes:

$$C = \sum_i P_i \cdot X_i \tag{5.2}$$

where subscript i denotes the delay category. Equation (5.2) represents the general formula of the cost factor approach, which admits many possibilities for classifying delay. How to determine X and P will be addressed in the ensuing two sub-sections. Further issues in applying (5.2) will be discussed in 5.3.1.3.

5.3.1.1 Categorizing Delay

Since delays in different categories can have different unit cost impact, how to appropriately categorize and quantify delays is a critical issue in applying the cost factor approach. There are two major bases for delay classification. The first is based on the phase of a flight in which the delay is taken. Gate, taxi, and airborne delays are the primary categories. Gate delay is measured as the difference between the actual and scheduled times a flight leaves the gate. Taxi and airborne delays are measured as differences between actual taxi and airborne times and unimpeded times. A flight delay can also be distinguished according to whether it is propagated from delay on an earlier flight, or is caused by some occurrence on the delayed flight itself. The latter are termed primary delays; they can occur at the gate, on the taxiway, or in the air. Propagated – sometimes termed reactionary – delays, on the other hand, are always manifested in late departure from the gate. Depending upon whether a propagated delay is caused by delay on previous legs flown by the same airframe, or delay on legs flown by a different aircraft, propagated delay can be further distinguished as rotational and non-rotational. Figure 5.5 illustrates these different concepts.

Also shown in Figure 5.5 is the role of schedule buffer. When a buffer is built into the schedule of a flight leg, it absorbs all or some of the delays on previous legs. In the above figure, we assume that the ith leg of aircraft 1 experienced no propagated delay. At the arrival end, an arrival delay equal to the sum of gate, taxi, and airborne delays minus buffer is incurred. Suppose no primary delay occurs on the successive two legs of the flight, but the primary delay on the ith leg propagates to the next two legs. As a result rotational propagated gate delays occur on the following two legs ($i+1$ and $i+2$). Due to the schedule buffer, as well as additional schedule layover time above that required to turn around the aircraft, such propagated delays are partly offset and diminish on legs farther downstream from the ith leg. Similarly, suppose some passengers from the delayed ith leg of aircraft 1 need to connect to the jth leg of aircraft 2, and/or the crew of the ith leg of aircraft 1 is assigned to the jth leg of aircraft 2. Thus, as a result of the delay on the ith leg of aircraft 1, aircraft 2 has to depart late on its jth leg, an example of non-rotational propagated delay. Given no other delays during the jth leg of aircraft 2, this flight will end up arriving at the destination airport with a smaller delay due to the schedule buffer and excess layover time. The propagated delay will be similarly further reduced on the next leg. Cook et al. (2004) proposed a delay categorization based on the above characterization, as shown in Figure 5.6. A schedule buffer is included and termed 'strategic delay'.

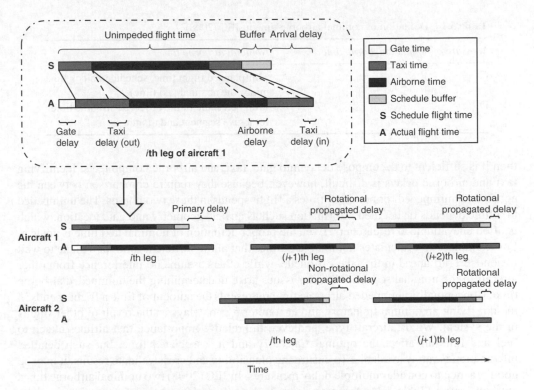

Figure 5.5 Illustration of different delay concepts

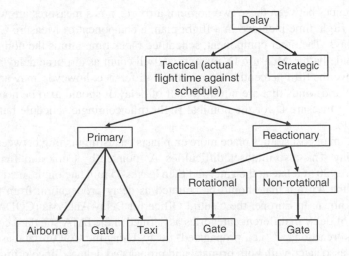

Figure 5.6 A comprehensive categorization of flight delay (*Source*: Cook et al., 2004)

While conceptually appealing, applying this decomposition to calculate delay cost is difficult in practice, given existing data and models. Instead existing cost studies employ parts of this framework. For example, the location of delay is widely considered. If one believes that one minute of gate delay costs the same irrespective of whether it is primary or propagated,

Table 5.1 Definition of nominal airborne time in JEC (2008)

Measure of nominal airborne time	Nominal airborne time
C	Min(planned flight time, scheduled block time minus the nominal taxi time)
D	Fifth percentile of the observed airborne time for a give segment and month

then it is sufficient to decompose delay into gate, taxi, and airborne components. Identifying taxi and airborne delays is difficult, however, because they require comparison between the actual and unimpeded (or nominal) times a flight spends in these two phases. The unimpeded taxi time depends on detailed information such as airfield geometry and gate location, which is often unavailable to researchers. Even the proper definition of nominal taxi time is unclear. For example, FAA calculates nominal taxi times that allow sufficient time for a plane to wait for one aircraft ahead in the take-off queue, while others assume no interference from other aircraft in the nominal scenario. Similar issues arise in determining the unimpeded airborne time. Conceptually, unimpeded airborne time represents the amount of time a flight spends in the air, flying an optimal trajectory and encountering no delays as the result of other flights in the system. Winds, aircraft type, and even the relative importance that airlines attach to fuel and time all affect the optimal trajectory and its associated time, but such detailed information is rarely available. To ensure the plausibility and cross validate results, it may be good practice to consider multiple delay measures. In JEC (2008) two nominal airborne times are considered (Table 5.1). Airborne delay is the actual elapsed time in the air minus the nominal airborne time.

Some differences between these two nominal airborne times measures are worth noticing. The estimated flight time included in a flight plan, a component of Measure C, may include anticipated delays. The other component, scheduled block time minus the nominal taxi time, includes schedule buffer. Measure D, in contrast, will count as airborne delay any additional flight time above the fifth percentile. Some of this difference, however, may arise as a result of aircraft type and winds that are not the result of delay. It should also be noted that delays calculated using measure D for the nominal flight time comingle schedule buffer and delay against schedule.

Another stream of cost studies place more emphasis on distinguishing between primary and propagated delay. This also imposes difficulties. As pointed by Cook and Tanner (2009), if an aircraft arrives 30 min late at the gate, and then leaves 45 min late on the next outbound leg, the portion in the 45 min that should be counted as delay propagation from the last leg is generally not known. In Europe, the Central Office for Delay Analysis (CODA) data allows the distinction of delay in different categories as defined in the IATA standard codes, in which one category is 'reactionary' (i.e., propagated) delay. Using the CODA database, ITA (2000) calculates cost associated with both primary and propagated delays. Since primary delays can take place at the gate, on the taxiway, or airborne, the cost factors the ITA applied are likely to be an average over those three possibilities.

On the other hand, if delays are measured against flight schedule, buffer needs to be identified separately. Again, because of the difficulty in measuring unimpeded flight times, there seems no widely acknowledged definition of buffer. ITA (2000) assumed an increased flight time of either 5 or 10% because of schedule padding. Inspired by the minimum feasible flight time introduced in Mayer and Sinai (2003a) and the nominal total flight time in JEC (2008), Zou and Hansen (2012a) proposed three buffer measures, defined as the difference between scheduled time and the fifth, tenth, and twentieth percentiles of the observed gate-to-gate time over all flights for each directional flight segment, airline and quarter. Since a consensus on defining buffer has not yet been achieved, trying different measures may be worthwhile in order to help a further understanding of buffer and related scheduling behavior.

In sum, the availability of delay data is a critical challenge for the cost factor approach. It is widely acknowledged that delays in different categories exhibit different cost impacts, and the categorization can be by location of delay or by the primary/propagated dichotomy. However, some difficulties exist in both ways of categorization; which one to choose largely depends on the specific data available to researchers. But of equal significance is the need to determine appropriate cost factors, to which we now turn.

5.3.1.2   Determining Cost Factors

Cost factors are based on the assumption that delay causes additional consumption of largely the same inputs as the airlines' normal line production process. These include fuel, labor (e.g., cabin crew, flight attendants, and ground personnel), capital (e.g., depreciation, rental, and lease), and others (e.g., airport charges, maintenance). These delay cost factors can be inferred from operating cost information that is regularly tracked – and in the US, publicly reported – by airlines on a per block-hour basis.

Delays occur on individual flights involving specific aircraft and circumstances. Ideally, therefore, cost data at the flight level would be used to develop flight-specific cost factors. Airlines themselves are unlikely to possess data at such fine detail; even if they did, this information would be highly sensitive and not publicly accessible. As a result, researchers often resort to two surrogates: one is more aggregate cost data; the other is information gathered from interviews. This is reflected in Table 5.2, which documents the data sources used by four major cost-of-delay studies.

Table 5.2   Data sources for determining cost factors

List of research	Data source
Cook et al. (2004)	Primary data from interviews with airlines, handling agents, aircraft operating lessors and other parties (e.g., Eurocontrol, research institutions, offices that set airport charges, and IATA). Supplementary data from ICAO, the Airline Monitor, and simulation results using Lido (for calculating fuel burn rate).
ITA (2000)	Interviews with individual airlines.
JEC (2008)	US DOT Form 41 (Schedule P5.2, T-2); FAA Aviation Environmental Design Tool System for fuel burn data; FAA rules for flight attendant hourly wage information.
ATA (2010)	US DOT Form 41 data for U.S. scheduled passenger airlines with annual revenues >= $100 million.

Aircraft operating cost data by carrier and aircraft type for US airlines are publicly available in the Form 41 database; US studies typically rely on this database as the primary resource for cost factors. Form 41 contains detailed airline financial and operating information, such as salary of pilots, direct expenses for maintenance of flight equipment, equipment depreciation cost, and the total number of flight hours for each combination of aircraft type, airline, and quarter. This allows construction of block hour cost for various cost categories – crew, maintenance, fuel, and so on. Cost factors for different types of delay are obtained by assuming which categories of cost are incurred for delays of a given type, as discussed further next.

In contrast, perhaps because cost data are less readily available publicly in Europe, cost factors are based on interviews in many European studies. Interview subjects include airline, airport, Eurocontrol, and research personnel. Such personal inquiries provide researchers with a richer picture of how delays affect airline operations and business methods. On the other hand, cost estimates obtained from interviews are inherently subjective. Interviewees usually tend to incorporate cost impacts that are obvious to them while omitting those that are not directly visible. In addition, even for the impacts that are manifest, estimates (like those based on block hour cost data) may be unduly influenced by accounting conventions that often have little empirical basis.

In developing cost factors from detailed cost data, as available in the US, judgments must be made about which cost categories to include for which forms of delay. Aircraft operating costs are categorized into various cost components such as fuel, crew, maintenance, depreciation, rental, airport charges, and so on. Decisions about which components to include in a delay cost factor are based upon knowledge and understanding of the researchers and/or subject matter experts. As is the case with delay categorization, this exercise can be carried out at varying levels of detail, ranging from a single cost factor representing the 'average' unit cost impact of all delays, to hundreds of different cost factors each describing the effect of one minute of delay for a given aircraft model, delay duration, delay location, cost scenario – even including consideration of network effects (e.g. Cook et al., 2004). Tables 5.3–5.6 reflect the cost components considered in several studies, and how they were used in constructing different cost factors. Table 5.3 provides a snapshot of the cost components that are included, as shown by checks, in different cost factors. In the first column, for example, the cost factor is defined by a combination of aircraft type (B733), delay duration (15-min), delay location (taxi), cost scenario (low), and network effect consideration (without network effect).

The common cost components appearing in the above tables are fuel, crew, and maintenance. Inclusion of other items varies from study to study. For example, airport charges can increase if there are long gate delays. Late arrivals induce additional ground and passenger handling costs. These are included in Cook et al. (2004) and ITA (2000). Probably due to more stringent delay compensation rules in Europe, all studies on the European side include passenger cost as an integral component. This could involve extra expenses on meals and hotel accommodation for passengers with long delays. ITA (2000) added a 'hub and connection' term, to capture the cost impact resulting from schedule disturbance and loss of operational efficiency at hub airports, for example, lengthening of connecting times, missed connections, and flight cancellations. Cook et al. (2004) further considered a potential drop of airlines' market share and revenue loss in the future due to a lack of punctuality. Some of these losses may be recouped by competing airlines, but this possibility is not typically considered.

Figure 5.7 illustrates the estimates of delay cost factor values and their composition from several studies, all converted to $US 2008.[1] In several delay factors, fuel, crew, and maintenance

Table 5.3 Cost components and cost factors in Cook et al. (2004)

Cost components	B733 15-min taxi delay: low cost scenario, w/o network effect	A319 65-min gate delay: high cost scenario, w/ network effect	......	B744 15-min gate buffer: used, low cost scenario	A321 65-min taxi buffer: unused, high cost scenario
Fuel	√	√	......	√	
Maintenance	√	√	......	√	√
Crew		√	......	√	√
Aircraft depreciation, rental, and lease			......	√	√
Airport charges	√	√	......		
Ground and passenger handling		√	......		
Aircraft operator passenger cost		√	......		

Table 5.4 Cost components and cost factors in ITA (2000)

Cost components	Primary delay	Propagated delay	Buffer
Aircraft operating cost	√		√
Operating staff cost	√	√	√
Structural cost	√	√	√
Passenger driven cost	√	√	
Hub and connection cost	√	√	√
Induced (long term) airline cost	√	√	

Table 5.5 Cost components and cost factors in CAA (2000)

Cost components	Long airborne delay	Short airborne delay	Long ground delay	Short ground delay
Flight crew	√		√	
Cabin crew	√		√	
Fuel and oil	√	√		
Flight equipment insurance	√		√	
Rental of flight equipment	√		√	
Maintenance and overhaul	√	√		
Flight equipment depreciation	√	√		
Handling charges & parking fees	√		√	√
Passenger services (e.g., meals)	√		√	

Table 5.6   Cost components and cost factors in JEC (2008)

Cost components	B733 gate delay	B744 taxi delay	......	A321 airborne delay
Taxiing fuel cost		√	......	
Airborne fuel cost			......	√
Pilot salary	√	√	......	√
Flight attendants expenses	√	√	......	√
Maintenance and depreciation			......	√

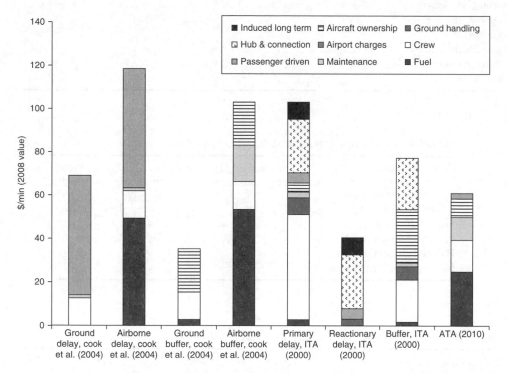

Figure 5.7   Cost factors and their composition

Notes: 1. All numbers updated to 2008 value in dollars, based on inflation and exchange rates provided by Eurocontrol (2010). 2. The numbers from Cook et al. (2004) are averaged across different aircraft models, as in Eurocontrol (2010). Passenger driven cost only includes 'hard' cost (passenger compensation and rebooking expenses). 'Soft' cost is counted as 'induced long term cost'.

cost account for the bulk in the total. When a buffer cost factor is measured, aircraft owner-ship, including airframe depreciation, rental, and lease, becomes an important cost compo-nent. The hubbing and connection cost factor, not estimated in other studies, accounts for most of the propagated delay cost factor and is generally estimated to be of the order of $15–20 per minute in ITA (2000). The passenger-related cost refers to compensation to passengers due to delay, and is substantial in Cook et al. (2004) compared to ITA (2000).

Table 5.7   Some characteristics of cost factors determined in existing studies

Source	Location where delay occurs	Aircraft size	Primary/ propagated
Cook et al. (2004)	√	√	√
ITA (2000)	√	×	√
CAA (2000)	√	×	×
JEC (2008)	√	√	×
ATA (2010)	×	×	×

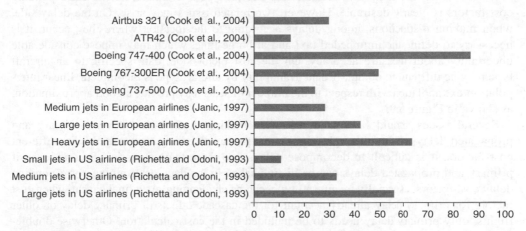

Figure 5.8   Unit ground delay cost in literature ($/min, updated to 2008 values)
Note: the values from Cook et al. (2004) are based on the high cost scenario for a typical delay
duration of 15 minutes, without considering the network effect.

The induced long term cost, referring to the loss of market share and revenue due to delay,
accounts for a major portion in Cook et al. (2004), as well. Overall, there is a significant variation of cost factors, with the high end reaching almost $175 per min of delay and the lower
at less than $40 per min.

The different dimensions of delay characteristics specified in each study, as shown in
Table 5.7, suggest the level of cost details the cost factors are able to provide if relevant
delay information is available. The location of delay significantly affects the magnitude of
delay impact; this is considered in most studies. It is also important to account for the
impact of aircraft size. Unless carefully aggregated, using cost factors not differentiated by
aircraft size would likely yield unreliable results. For example, aircraft size contributes to
the large differences among various ground delay cost factor estimates reported in the literature, as shown in Figure 5.8. Delay cost factors are also differentiated between primary
and propagated delays in two European studies,[2] where data permit such a distinction.

5.3.1.3   Shortcomings of the Cost Factor Approach

This section discusses several shortcomings of the cost factor approach. First, choosing the
appropriate delay categorization scheme is difficult. Second, the cost factor approach may
overlook potentially important aspects of the relationship between delay and airline cost.

Choosing the categorization scheme

Although multiple options for categorizing delays and associated cost factors may be available, choosing an appropriate one depends on the availability of necessary data from both the cost and delay sides. The cost side may be the less problematic, as long as all cost components are identified. In principle, one only needs to determine what cost components should be included for a specific type of delay. While this process involves a certain degree of judgment, the more significant questions concern the level of detail at which delay data are available in databases available to researchers. Sometimes applying a single cost factor to all delays may be necessary because delay cannot be further disaggregated with the data available.

If delay information by both location and aircraft type is available, having corresponding cost factors is clearly desirable. However, two related issues may arise. On the delay side, when making distinctions among delays according to the location where they occur, it is necessary to define the unimpeded taxi and airborne time, which may impose considerable uncertainty about measure accuracy. On the cost factor side, data specific to an aircraft type may be difficult to obtain. If data availability by aircraft type is incomplete, linear interpolation/extrapolation with respect to the number of seats may provide a good approximation, as shown in Figure 5.9.

Several issues should be considered when making distinctions between primary and propagated delays. First, although primary and propagated delays may entail quite different cost factors, it is difficult to decompose delays into these two categories. Second, even if primary and propagated delays can be distinguished in the data, determining corresponding delay cost factors, especially by aircraft type, can be challenging. As a final caveat, when cost factors for primary delay already account for the cascade effects of primary delay on other flights, only primary delay needs to be included in the cost calculation. Otherwise double-counting could occur.

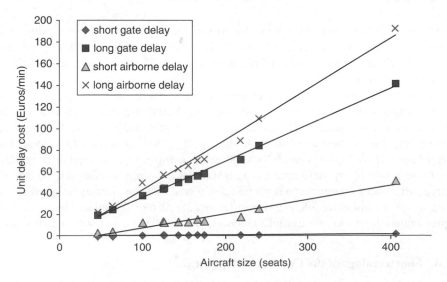

Figure 5.9   Cost of at-gate and airborne delays as a function of the number of aircraft seats (*Source*: Cook et al., 2004, without network effect)

Effects not captured

Recall that the cost factor approach implicitly assumes that cost increases linearly with delay, across all flights. In reality, however, a single four-hour delay will probably be more costly than 24 10-min delays, for two reasons. First, as delay becomes longer, additional cost items may be incurred. Figure 5.10 shows that handling surcharges, expenses for providing passengers with meals and accommodations, and other costs appear when delay exceeds a certain time threshold. In addition, the delay propagation effect increases non-linearly with the size of the initial delay, as demonstrated by Beatty et al. (1998). Longer delays propagate to more flights and are more likely to disrupt ground operations, gate assignments, crew schedules, and passenger itineraries (Figure 5.11).

A few cost factor studies, notably Cook et al. (2004) and CAA (2000), have attempted to take this effect into account. Cook et al. (2004) chose two specific delay durations (15 and 65 min) to typify 'short' and 'long' delays (Figure 5.9). Similarly, in CAA (2000), 20 min was selected as the threshold separating 'short' from 'long' delays. Both studies conclude that longer delays have higher cost factors. While such categorizations help address the issue of non-linearity, they are at best first steps toward representing the true relationship between length of delay and cost. Further efforts in this direction are certainly warranted.

Taking this one step further, the relationship between delay duration and cost can even be non-monotonic. Airlines sometimes add delays to flights in order, for example, to avoid having a flight arrive at a hub in the middle of a departure 'bank' (or 'wave'). In a hub-and-spoke network, the interaction of delays for many flights scheduled in a connecting bank plays an especially important role in maintaining the integrity of airport and airline operations. If all the flights in an inbound bank are delayed by the same amount, then the effect may be far less severe than if half the flights are delayed by a smaller amount of time than the others (Hansen et al., 2001). The cost factor approach cannot capture such effects, because of its underlying assumption that the cost of delay is an additively separable function of delay variables for individual flights.

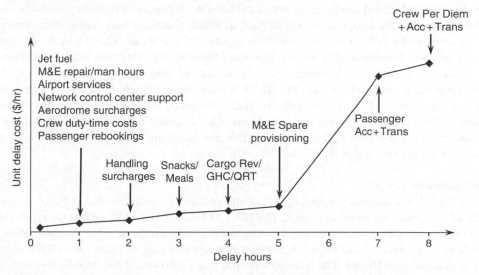

Figure 5.10 Delay cost increase as a function of delay duration with each arrow indicating the newly incurred cost (based on M2P Consulting, 2006)

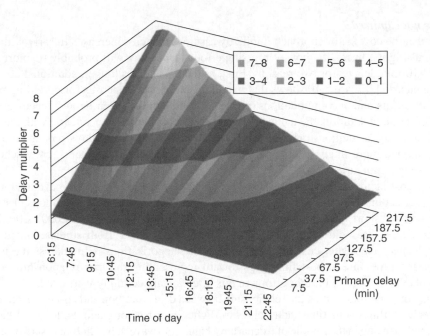

Figure 5.11   Delay propagation as a function of time of day and the amount of primary delays
(*Source*: Beatty et al., 1998)

Finally, the presence of delays may generate sizable indirect effects that may be difficult to capture with the cost factor approach. Carriers may take a variety of measures in flight scheduling to make their operations more robust to delay. One routine practice is building more padding into scheduled block times, which explains the necessity of including the buffer side when evaluating the whole delay cost impact. In addition, airlines may require extra aircraft, flight crew, and ground personnel, and load flights with more fuel. These adaptations entail overhead and capital costs that are not accounted for in determining cost factors. The cost of delay may thus permeate the entire cost structure of the airline in ways that are not tied to individual delay events (Hansen et al., 2001). It would be extremely difficult to capture such effects using cost factors. The majority of studies based on cost factors simply ignore these issues. (Only ITA (2000) tried to quantify such a 'structural cost' item, albeit in a vague manner.) As a consequence, it is likely that the cost factor approach may underestimate the true cost impact of flight delays.

Cost estimation results
Table 5.8 lists the estimated annual cost of delay obtained by several studies in Europe and the US, all using the cost factor approach. The ITA (2000) and Cook et al. (2004) are representative on the European side and widely cited by delay cost researchers. Since the publication of Cook et al. (2004), Eurocontrol has periodically reported cost estimates for ATFM delays using updated cost factors. The estimates for the years 2007 and 2009 from Eurocontrol are presented here, in order to compare with the estimates from two recent US studies, the JEC (2008) and ATA (2010).

Table 5.8 Existing estimates of delay (and buffer) cost to air carriers using the cost factor approach

Scope	Source	Evaluation year	Cost Estimate	(Averaged) cost factor	Note
Europe	ITA (2000)	1999	€3.0–5.1 billion (delay); €1.6–2.3 billion; buffer: €1.3–2.7 billion)	Delay: €35.5/min (using marginal cost), €50.9/min (using average cost); buffer: €45.2/min	Buffer is calculated as 5% to 10% of scheduled flight time.
	Cook (2004)	2002	€840–1200 million	€72/min	Focus on departures. Assume all delays are primary; classified into two types (< 15 min and >15 min). Buffer not considered.
	Eurocontrol (2010)	2007	€1.40 billion	N/A	Use updates of cost factors in Cook et al. (2004). Consider total ATFM delays. Buffer not considered.
		2009	€1.00 billion	N/A	Similar updates as for the 2007 estimate.
United States	Geisinger (1988) Odoni (1995) Citrenbaum and Juliano (1998)	1986 1993 1996	$1.2 billion $2–4 billion $1.8 billion		
	JEC (2008)	2007	$3.6/6.1 billion (corresponding to delay measures C and D in Table 5.1)		Delay is measured against 'unimpeded flight time'.
	ATA (2008)	2007	$7.8 billion	$60.46/min	Single cost factor; use block hour cost as a proxy. Buffer not considered.
	ATA (2010)	2009	$6.1 billion	$60.99/min	Same as for ATA (2008) estimate.

In addition to geographical and demand level differences, the seemingly large cost estimates for 2007 and 2009 in the US, compared to their European counterparts, may be explained by several reasons. First, the Eurocontrol estimates only considered delays due to ATFM, while the US estimates included all types of delays. Second, as discussed before, the delay measured in JEC (2008) is against some unimpeded, optimal flight time; schedule buffer is at least partly incorporated. This results in a greater amount of delay than one would obtain when only delay against schedule is considered. Block hour cost was also used for the ATA (2010) estimate as a proxy for a single delay cost factor. As has been discussed, many delays occur at the gate, where fuel consumption is marginal. Thus, using block hour cost, of which fuel expense constitutes a significant portion, will yield an overestimate of the cost of delays at the gate. On the other hand, neither the US nor the European estimates fully account for the non-linear, combinatorial, or indirect cost impacts described previously.

5.3.2   Aggregate Cost Approach

The weaknesses in the cost factor approach suggest that alternative methods for estimating the cost of delay to airlines should be considered. To a large extent, the weaknesses stem from allocating cost components to each cost factor, and assuming that cost bears a linear relationship with the delay duration. In order to address these two concerns, an 'aggregate cost' (or top-down) approach is proposed in this section. Rather than investigating explicitly the cost impact of each minute of delay, the top-down approach assesses the impact of delay impact in a more holistic manner. One way of doing this, the total time approach, is simple and intuitive, and uses readily available aggregate airline accounting information. Unfortunately, this approach is also almost certainly wrong. The second approach, an econometric one, is more rigorous, data intensive, and methodologically sophisticated, and hence probably more accurate.

5.3.2.1   Total Time Approach

This approach assumes that airline operating costs are proportional to total aircraft operating time, measured, for example, in plane-hours. If this were true, then delay cost could be estimated as the fraction of total aircraft operating time due to delay, multiplied by total airline operating costs.

The total time approach is based on the premise that if an airline operated zero aircraft hours, its costs would also be zero. Thus, the relationship between operating cost and aircraft hours includes the origin, and the point corresponding to the actual cost and hours. A straight-line interpolation between these two points can be used to predict how costs would change if delay hours were eliminated. This approach thus avoids the numerous details of determining cost factors, and only requires calculating the aggregated delay time. Since total operating cost, aircraft time, and delay time for an airline can be readily obtained, the total time approach is straightforward to implement. This was done in JEC (2008), and yielded much larger cost estimates than the cost factor approach. Using the same delay measures as employed under the cost factor approach (see Table 5.1), they conclude with considerably higher estimates using the total time approach ($19.1 vs $3.6 billion based on measure C; $23.4 vs $6.1 billion based on measure D).

While it is appealingly simple, the validity of this approach is very questionable. First of all, like the cost factor approach, it assumes a linear relationship between delay cost and time. Second, the approach ignores differences where delay and planned aircraft hours occur; for example, a much larger proportion of delay time occurs on the ground. Moreover, the total operating cost includes not only fuel, crew salaries, maintenance, and depreciation, but also advertising, ticket agents, landing fees, legal fees, and other items that may be relatively insensitive to delays. As a result, cost numbers generated through this approach will almost certainly be too high.

5.3.2.2 Econometric Approach

Basic concepts and cost model functional form
A second version of the aggregate cost approach rests on the idea that the provision of air transportation service to passengers constitutes a production process. Economic theory suggests that in the production process, each firm minimizes the cost C at which it can produce a given amount of output Y, given the prices *it* pays for inputs, $\overline{W}_{it}$. The airline cost function can be conceptually expressed as: $C = f(Y_{it}, \overline{W}_{it})$, where subscript i denotes a particular firm (airline), and t identifies the time period. A typical output measure is revenue ton-miles. Inputs include labor, fuel, capital, and materials. The function represents the cost of acquiring the optimal set of inputs, given the output and input prices (Hansen et al., 2001). In reality, however, capital inputs cannot be adjusted to the optimal level instantaneously (Caves et al., 1984; Gillen et al., 1990). We therefore relax the assumption of optimal capital stock by treating capital input, denoted by S, as quasi-fixed and employing a variable cost function to reflect the short-run cost minimization process. The airline variable cost function can then be written as a function of its output Y_{it}, the price of the three variable inputs (fuel, labor, and materials) $\overline{W}_{it}$, and capital input S_{it}, that is, $VC_{it} = f(Y_{it}, \overline{W}_{it}, S_{it})$.

In the airline cost literature, it has long been recognized that costs depend on the nature and quality of the airlines' output, as well as the quantity. Because the nature and quality of output also vary over time and across carriers, the specification of the airline cost function above needs to take these into account. A set of additional variables Z_{it} describing the nature of the output are introduced. Variables of this kind that often appear in the literature can be the size of the airline's network (often measured by the number of points served) and the average flight distance (stage length). We hypothesize that flight delays, or in a broader sense, imperfect operational performance also affects airline cost. Therefore we add a new variable or vector of variables $\overline{N}_{it}$ to the cost function. The cost function then becomes $VC_{it} = f(Y_{it}, \overline{W}_{it}, \overline{Z}_{it}, S_{it}, \overline{N}_{it})$.

Depending on the characterization of operational performance, different versions of airline cost models can be developed and estimated. Two most popular ones are Cobb-Douglas and Translog cost functions, both assuming nonlinear relationship between the cost variable and the functional arguments. In a Cobb–Douglas model, the nonlinear relationship is characterized by a log-log functional form, in which the coefficient associated with each explanatory variable is interpreted as the corresponding cost elasticity. Because of the log-log relationship, these elasticities are assumed constant. This constant elasticity assumption can be relaxed if a more flexible Translog functional form is adopted, which can be regarded as a second-order Taylor expansion of any general cost function. The Translog model imposes fewer a priori assumptions about the relationships between airline cost and the explanatory variables, and the cost elasticities vary with the variable values. In particular, because relationships are derived from observed co-variation between operational performance variables and cost, the results entail a minimum of assumptions about the delay-cost interaction mechanism involved. Interested readers can refer to Hansen et al. (2000; 2001) which demonstrated the employment of Cobb-Douglas and Translog cost functions to characterize the relationship between airline cost and operational performance.

Data, model estimation, and applications

Reliable airline financial and operational databases are critical to the estimation of airline cost models. In the US, the Bureau of Transportation Statistics (BTS) releases airline-level financial and operational information – which are reported as a mandate by major domestic airlines – on a quarterly basis. The BTS On-time Performance database, also publicly available, provides airline operational performance at the individual flight level. Aggregate airline financial and operational information outside the US can be accessed, for example, through the Digest of Statistics published by the International Civil Aviation Organization (ICAO) as well as other data sources including Monitor, IATA publications, and annual reports of airlines. In Europe, individual flight operation records that are comparable to the BTS On-time Performance data are collected and maintained by CODA. For the remaining parts of the world, however, access to detailed operational performance information may present a challenge.

The choice of estimation methods depends upon the nature of the dataset (e.g., cross section vs panel) and the specification of the cost model. To improve estimation efficiency, the Translog cost function is often jointly estimated with additional equations derived from production theory. Standard estimation techniques include Zellner's seemingly unrelated regression and the maximum likelihood method. Further details about model estimation can be found in Caves et al. (1984), Gillen et al. (1985; 1990), Oum and Yu (1998), and Hansen et al. (2000; 2001).

The estimated airline cost model can then be used to gauge the cost impact on airlines of delay as well as other dimensions of imperfect operational performance. One can, in principle, change operational performance variables to an improved level of interest and perform cost prediction. The cost savings then equal the difference between the new and original predicted cost. For example, if delay and buffer are included as two operational performance variables, one can reduce delay, or both and buffer, to zero, or some other value deemed to be the lowest attainable. A recent study on the total delay impact in the US (Ball et al., 2010), where multiple versions of airline Translog cost models have been estimated with different sets of operational performance variables, found the system-wide delay cost for US airlines in 2007 to range from $5–9 billion. Together with buffer, the total impact can reach $8–13 billion. These estimates lie between the results from the cost factor approach and total time approach for the same year estimated by JEC (2008). As discussed before, the other two approaches are likely to provide lower and upper bounds of the true values. It is therefore plausible that the econometric approach yields a closer estimate to the true cost impact.

5.3.2.3 Summary

This section has reviewed possible approaches for estimating the cost impact of delay, or more generally, of imperfect operational performance. The cost factor approach has gained popularity due to its intuitiveness and simplicity. It has significant shortcomings, however, as a result of which the estimates it generates may be inaccurate and biased toward underestimating true costs. In contrast, by treating total operating cost as proportional to hours of flight activity and ignoring important differences in distribution of time among phases of flight for delays and regular flight operations, the total time approach probably yields overestimates of delay cost. Building on production theory, the econometric approach investigates the statistical relationship between the operational performance of airlines and their costs. By employing flexible functional forms, the econometric approach avoids the strong assumptions, such as linearity and additive separability that are built into the cost factor approach. A recent system-wide delay cost study in the United States using airline cost models appears to yield plausible estimates (Ball et al., 2010). However, these econometric models also have limitations. They are data hungry and therefore can be applied only at a relatively aggregate level. If the focus is narrower, for example a specific airport, the cost factor approach may be the most appropriate way of gauging the cost impact of delays. Nevertheless, the existing estimates of cost impact of delay provide some assurance that, despite the intrinsic methodological differences, cost factor, total time, and econometric approaches all yield estimates of a similar magnitude when applied to the US system.

5.4 Further Issues

5.4.1 Cancellations

While the focus of the studies discussed previously is primarily on delay (and occasionally on the schedule buffer, as well), it is important to recognize that flight cancellations can also be an important cost driver. The decision to cancel a flight is intertwined with flight delays, and reflects the trade-off between avoiding excessive delays of subsequent flight legs and maintaining flight schedule integrity. Canceling a flight involves very complicated considerations in practice. Airlines have to weigh the cost incurred as a result of excessively long delays against the cost of cancelling a flight. The latter may include providing passengers on the canceled flight with meals and hotel accommodations, as well as transporting those passengers to their scheduled destinations. The second component often requires spare capacity, which incurs opportunity cost of capital and loss of potential revenue, and/or the amount one airline has to pay to another if such capacity is not available. Cancellation decisions are further compounded by the heterogeneity of airline behavior, which are subject to the operational preference and even corporate culture of individual airlines. Under similar circumstances, some airlines may choose to persevere through long delays while others may decide to cancel certain flights. This airline heterogeneity is also reflected at the schedule planning stage. As illustrated in Mayer and Sinai (2003b), risk-averse airlines may accept a low aircraft utilization rate to avoid cancellations. This involves another type of opportunity cost of capital. All these considerations make the quantification of airline cancellation cost a difficult task.

Partly as a result of those, but partly also because of the lack of data on airline cancellation decision-making, the body of literature on quantifying flight cancellation cost is rather limited. Among the few attempts, Eurocontrol (2009) suggested different cancellation cost factors by aircraft size, for which it considered service recovery cost (e.g., meals and hotel), interline cost, loss of future value (individual passenger delay expressed in value), and operational savings. Using post-operation ground delay program data, Xiong and Hansen (2009) modeled airline cancellation decision making in a discrete choice context. Their results revealed an upper bound on cancellation cost, at around $5000 per cancelled flight. An aggregate approach to quantify the cancellation effect on airline cost is seen in Hansen et al. (2000; 2001), where the authors employed the Principal Component Analysis and found the cancellation factor an important cost driver in airline operations. Since cancellation decisions are intertwined with flight delays (and possibly schedule buffer as well), it is difficult to precisely estimate the cost of cancellation purged of delay under an aggregate approach. Further research in this area is clearly warranted. A better understanding of the relative magnitude of the overall cancellation cost vis-à-vis that of delay and schedule buffer will enrich our knowledge base on airlines' cancellation decisions, and on the contribution of cancellations to the overall cost of imperfect operational performance.

5.4.2   Optimal Level of Operational Performance and System Response

While most of the focus so far has been given on the cost impact of imperfect operational performance relative to cases where delay (and buffer) was entirely eliminated, it is important to recognize that a delay-free world is a limiting – and unreachable – case. As long as winds and storms exist, aircraft parts fail, and people make mistakes, delay will remain part of operations in any aviation system. Similarly, to make schedules more robust to wind variability and other exigencies, airlines will continue padding extra time to make their schedules more robust to wind variability and other exigencies. As a consequence, the preceding estimates, which focus on the cost savings from reducing delay/buffer to zero, provide an upper bound of the achievable cost savings, and a high one at that.

In addition, a level of investment in aviation infrastructure and technologies that could achieve complete elimination of delay is neither realistic nor advisable from the efficiency standpoint. The inherent peaking nature of flight demand suggests that, if delay were reduced to zero at peak periods, the capacity of the aviation system would be significantly underutilized during off-peak hours. However, defining the attainable level of operational performance is both difficult and controversial. In effect, the question of the right level essentially reflects the trade-offs one has to make between throughput and operational performance.

Perhaps more importantly, the cost assessment methods outlined above assume that, except for operational performance improvement, everything else would remain unaffected in the system. This is certainly a simplification of the real world situations. Reduced delay – often through enhanced infrastructure supply – improves the quality of air service, attracting more people to use the air transportation system. This has been empirically observed by Hansen and Wei (2006) at Dallas-Fort Worth airport after the completion of a major capacity expansion project in 1996. The authors found that a large portion of the direct delay reduction benefits may have been offset by changes in airlines schedules. From a broader, longer-term perspective, both demand and supply in the air transportation system will respond to improved operational performance, leading to a shift of the system equilibrium. Adequately capturing this equilibrium shift should also account for the economies of density and the Mohring effect[3] that are inherent in an air transportation system when delay is absent. At the airport level, the impact of capacity change and flight delays on equilibrium shift has been examined in Morrison and Winston (1983; 1989; 2007), Jorge and de Rus (2004), Miller and Clarke (2007). Recently Zou and Hansen (2012b; 2011) have investigated this issue on a system-wide scale. Modeling the air transportation system respectively from the airline competitive and behavioral equilibrium perspectives, the latter two studies showed that realistic airport capacity increase will continuously reduce delay – at a diminishing rate. Zero delays can never be achieved. Facing reduced delay and increased passenger demand, airlines also adjust fare, frequency, aircraft size, and the number of passengers on each flight. Zou and Hansen (2012b) further showed that the net benefits of delay savings to airlines should be reflected by profit change rather than the operating cost alone. These studies present a first step to incorporate the equilibrium concept into airline delay cost analysis. Future research is warranted to advance the relevant methodologies, and make the equilibrium concept more practical, and easy to implement.

5.5 Conclusions

This chapter presents a comprehensive review of existing knowledge about the impacts of operational performance on airline costs. No single methodology exists that can serve as a panacea for translating changes in operational performance into changes in airline costs under all circumstances. The econometric approach makes the fewest assumptions about the mechanisms through which flight delays and related phenomena affect airline costs. This makes the approach a useful one for assessing the economic cost of air transport congestion and delay, when the necessary data are available. The much more commonly used cost factor approach is probably more appropriate for the day-to-day business of investment analysis. It cannot be overemphasized, however, that this approach rests on strong and untested assumptions about the cost-generating mechanisms involved. The cost factor method can be applied with varying levels of refinement, as time and data permit, but it rests on a shaky foundation regardless. It is somewhat re-assuring, however, that estimates of the total cost of delay in the US based on the econometric and cost factor approaches yield results of a similar magnitude.

Notes

1 The conversion to 2008 prices was made using the June value of the harmonized index of consumer prices (HICP) provided by EUROSTAT. All values were converted to dollars using the foreign exchange reference rate for Dec 1, 2008 published by European Central Bank. The above information is available in Eurocontrol (2009).
2 In Cook et al. (2004) multipliers were introduced in the cost factors to account for the 'knock-on' effect of primary delays.
3 The Mohring effect is an increasing return property that exists in many scheduled transport systems. It basically says that, as transport service frequency increases, passenger waiting time (or schedule delay) decreases, demand increases, which in turn results in further increase in service frequency.

References

Air Transport Association (ATA). (2008) *Cost of Delays: Direct Cost*. [Online] Available from: http://www.airlines. org/economics/specialtopics/ATC+Delay+Cost.htm [Accessed June 28, 2010].

Air Transport Association (ATA). (2010) *Annual and per-minute cost of delays to U.S. airlines*. [Online] Available from: http://www.airlines.org/Economics/DataAnalysis/Pages/CostofDelays.aspx [Accessed June 27, 2010].

Airport Council International (ACI, North America). (2009) *Airport capital development costs: 2009–2013*. [Online] Available from: http://aci-na.org/static/entransit/CapitalNeedsSurvey Report2009.pdf [Accessed June 28, 2011]

Ball, M., Barnhart, C., Dresner, M., et al. (2010) *Total delay impact study: a comprehensive assessment of the costs and impacts of flight delay in the United States*. NEXTOR Report prepared for the Federal Aviation Administration.

Beatty, R., Hsu, R., Berry, L., and Rome, J. (1998) *Preliminary evaluation of flight delay propagation through an airline schedule*. Presentation at the 2nd USA/Europe air traffic management R&D seminar. [Online] Available from: http://www.atmseminar.org/past-seminars/2nd-seminar-orlando-fl-usa-december-1998/papers/paper_038 [Accessed June 20, 2010].

Bureau of Transportation Statistics (BTS). (2009) *Airline On-Time Statistics and Delay Causes: On-Time Arrival Performance*. [Online] Available from: http://www.transtats.bts.gov/ot_delay/ot_delaycause1.asp[Accessed February 20, 2013].

Civil Aviation Authority (CAA). (2000) *Delay cost based on UK airline actual P&L figures*. SS4 working paper on delay cost methodology.

Caves, D. W., Christensen, L. R. and Tretheway, M. W. (1984) Economies of density versus economies of scale: why trunk and local airline costs differ. *Rand Journal of Economics*, 15, 471–489.

Citrenbaum, D. and Juliano, R. (1998) *A simplified approach to baselining delays and delay costs for the national airspace system*. Federal Aviation Administration, Operations Research and Analysis Branch, Preliminary Report 12.

Cook, A., Tanner, G., and Anderson, S. (2004) *Evaluating the true cost to airlines of one minute of airborne or ground delay*. Report prepared for the Eurocontrol Performance Review Unit. [Online] Available from: http://www.eurocontrol.int/prc/gallery/content/public/.../cost_of_delay.pdf [Accessed February 20, 2013].

Cook, A. and Tanner, G. (2009) *The challenges of managing airline delay costs*. Presentation at the Conference on air traffic management (ATM) economics, Belgrade. [Online] Available from: http://www.garsonline.de/Downloads/090910/Papers/Paper_Cook%20&%20Tanner.pdf [Accessed February 20, 2013].

Eurocontrol. (2009) *Standard Inputs for EUROCONTROL cost-benefit analyses. 4th edition*. [Online] Available from: http://www.eurocontrol.int/ecosoc/gallery/content/public/documents/CBA%20examples/Standard_inputs_for_CBA_2009.pdf [Accessed February 20, 2013].

Eurocontrol. (2010) *Performance review report: an assessment of air traffic management in Europe during the calendar year 2009*. Performance Review Commission. [Online] Available from: http://www.eurocontrol.int/prc/gallery/content/public/Docs/PRR_2009.pdf [Accessed February 20, 2013].

Federal Aviation Administration. (2011). *FAA Aerospace Forecast: Fiscal Years 2011–2031*. [Online] Available from: http://www.faa.gov/about/office_org/headquarters_offices/apl/aviation_forecasts/aerospace_forecasts/2011-2031/media/2011%20Forecast%20Doc.pdf [Accessed February 20, 2013].

Geisinger, K. (1988) *Airline Delay: 1976–1996*. U.S. Federal Aviation Administration, Office of Aviation Policy and Plans.

Gillen, D. W., Oum, T. H., and Tretheway, M. W. (1985) *Airline Cost and Performance: Implications for Public and Industry Policies*. Vancouver, Center for Transportation Studies, University of British Columbia.

Gillen, D. W., Oum, T. H., and Tretheway, M. W. (1990) Airline cost structure and policy implications: a multi-product approach for Canadian airlines. *Journal of Transport Economics and Policy*, 24 (1), 9–34.

Government Accountability Office (GAO). (2008) Next generation air transportation system: status of systems acquisition and the transition to the next generation air transportation system. Report prepared for the Congressional Requesters. [Online] Available from: http://www.gao.gov/assets/290/280621.pdf [Accessed February 20, 2013].

Hansen, M. M., Gillen, D. W., and Djafarian-Tehrani, R. (2000) Assessing the impact of aviation system performance using airline cost functions. *Transportation Research Record: Journal of the Transportation Research Board*, 1073, 16–23.

Hansen, M. M., Gillen, D. W., and Djafarian-Tehrani, R. (2001) Aviation infrastructure performance and airline cost: a statistical cost estimation approach. *Transportation Research Part E: Logistics and Transportation Review*, 37, 1–23.

Institute du Transport Aérien. (2000) *Costs of air transport delay in Europe*. Report prepared for the Eurocontrol Performance Review Unit. [Online] Available from: http://www.eurocontrol.int/prc/gallery/content/public/Docs/stu2.pdf [Accessed February 20, 2013].

Joint Economic Committee (JEC). (2008) *Your flight has been delayed again: flight delays cost passengers, airlines, and the U.S. economy billions*. [Online] Available from: http://jec.senate.gov/public/?a=Files.Serve&File_id=47e8d8a7-661d-4e6b-ae72-0f1831dd1207 [Accessed February 20, 2013].

Jorge, J. D. and de Rus, G. (2004) Cost–benefit analysis of investments in airport infrastructure: a practical approach. *Journal of Air Transport Management*, 10 (5), 311–326.

Mayer, C. and Sinai, T. (2003a) Network effects, congestion externalities, and air traffic delays: or why not all delays are evil. *American Economic Review*, 93, 1194–1215.

Mayer, C. and Sinai, T. (2003b) Why do airlines systematically schedule their flights to arrive late? *Working paper*, the Wharton School of Business, University of Pennsylvania. [Online] Available from: http://real.wharton.upenn.edu/~sinai/papers/Schedule-Mayer-Sinai-10-30-03-2.pdf [Accessed February 20, 2013].

Miller, B. and Clarke, J.-P. (2007) The hidden value of air transportation infrastructure. *Journal of Technological Forecasting and Social Change*, 44 (1) 18–35.

Morrison, S. and Winston, C. (1983) Estimation of long-run prices and investment levels for airport runways. *Research in Transportation Economics*, 1, 10–130.

Morrison, S. and Winston, C. (1989) Enhancing the performance of the deregulated air transportation systems. In: *Brookings Papers on Economic Activity: Microeconomics*, Washington D.C., Brookings Institution, pp. 61–112.

Morrison, S. and Winston, C. (2007) Another look at airport congestion pricing. *American Economic Review*, 97 (5), 1970–1977.

M2P Consulting. (2006) *The cost of delays and cancellations*. Presentation at the AGIFORS Annual Symposium 2006, Dubai.

Odoni, A. (1995) *Research directions for improving air traffic management efficiency*. Argo Research Corporation.

Oum, T. H. and Yu, C. (1998) Cost competitiveness of major airlines: an international comparison. *Transportation Research Part A: Policy and Practice*, 32(6), pp. 407–422.

Xiong, J. and Hansen, M. (2009) Value of flight cancellation and cancellation decision modeling: ground delay program postoperation study. *Transportation Research Record: Journal of Transportation Research Board*, 2016, pp. 83–89.

Zou, B. and Hansen, M. (2012a) Impact of operational performance on air carrier cost structure: evidence from U.S. airlines. *Transportation Research Part E: Logistics and Transportation Review*, 48(6), 1032–1048.

Zou, B. and Hansen, M. (2012b) Flight delays, capacity investment, and social welfare under air transport supply-demand equilibrium. *Transportation Research Part A: Policy and Practice*, 46(6), 965–980.

Zou, B. and Hansen, M. (2011) *Assessing Benefits from Aviation Capacity Investment: An Equilibrium Approach*. Presentation at the Institute for Operations Research and the Management Sciences (INFORMS) Transportation Science and Logistics Society Workshop, Asilomar, California.

6

New Methodologies for Airport Environmental Impact Analysis

Mark Hansen[A], Megan S. Ryerson[B] and Richard F. Marchi[C]

[A]National Center of Excellence for Aviation Operations Research, Department of Civil and Environmental Engineering, University of California, Berkeley, USA
[B]Department of Civil and Environmental Engineering, University of Tennessee, USA
[C]Senior Advisor, Policy and Regulatory Affairs, Airports Council International, North America, USA

6.1 Introduction

The relationship between airports and the environment has been a constant challenge with strong historical roots. This relationship is growing more complicated with the expanding umbrella of environmental concern. Considering pollutants airports have historically wrestled with, noise and criteria pollutants, along with the relatively recent concern of global Greenhouse Gas (GHG) emissions, the environmental impact of an airport is now spatially and temporally concentrated and distributed. This suggests the need to rethink longstanding policies governing airport environmental impact, and methods for environmental impact analysis.

In this chapter, we will review four key aviation pollutants that effect airports in different ways. Noise affects people who surround an airport; water run-off and criteria pollutants affect both the surrounding area and areas nearby; while Greenhouse Gas (GHG) emissions, a comparatively new concern for airports, have a global impact. This chapter reviews these pollutants and their impact on airports. It also focuses on the methodological challenge in assessing the environmental impact of noise and GHG emissions, and their trade-offs, as well as new methodologies to handle the policy and operational challenges of these pollutants.

Modelling and Managing Airport Performance, First Edition. Edited by Konstantinos G. Zografos, Giovanni Andreatta and Amedeo R. Odoni.
© 2013 John Wiley & Sons, Ltd. Published 2013 by John Wiley & Sons, Ltd.

6.2   Pollutant Overview

6.2.1   Noise

Since the introduction of turbojet airplanes into commercial service in the late 1950s, aircraft noise has been generally recognized as the most significant environmental impact at airports. Airport managers and government regulators consistently rank noise mitigation among their highest environmental priorities. Public controversies relating to aircraft noise have impeded airport capital development programs at airports in all regions of the world, sometimes adding decades to the approval of airport infrastructure projects. As an example, the case of a new runway first proposed for Seattle's SeaTac International airport in 1988 is illustrative:

SEA-TAC INTERNATIONAL AIRPORT'S THIRD RUNWAY OPENS ON NOVEMBER 20, 2008.

On November 20, 2008, the new third runway at Seattle-Tacoma (Sea-Tac) International Airport opens to scheduled air traffic … The 8,500-foot-long runway is the culmination of more than 20 years of planning, construction, and controversy. Planners at the Port of Seattle, which owns and operates Sea-Tac, first recognized the need for a third runway in 1988. … Construction work began in 1997 after the FAA approved the runway, but stopped several times as the result of the ongoing litigation. (Klass, 2010)

Likewise, a proposal to construct a third runway at London's Heathrow airport is another highly controversial action of long standing. As part of the noise mitigation proposed by the airport operator, BAA, modifications to a series of runway restrictions called the Cranford Agreement, originally instituted in the 1950s, were recently proposed: "The Cranford Agreement is a Government agreement preventing easterly take-offs from the northern runway over the village of Cranford. It was introduced in the 1950s when departure noise was considered more disruptive to residents than noise from arriving aircraft." (London Heathrow, 2011).

The duration of the London Heathrow noise controversy spans 60 years. Serious noise controversies, while perhaps not as enduring as the LHR example, are, nevertheless, widespread, having been experienced at many major airports in Europe, North America, Asia, and Australia. One serious impact of the aviation industry's inability to resolve these protracted public noise controversies is the failure to add needed airport infrastructure in a timely manner and the protracted absorption of flight delays during the interim years.

Despite decades of efforts, effective noise mitigation remains elusive and public opposition to airplane noise has not abated significantly, despite reductions in the number of people exposed to noise as measured by traditional indices by 90% from 1975 to 2000 (FAA AEE, 2010). This paradox arises because both ICAO, through its Committee on Aviation Environmental Protection (CAEP), and national aviation regulators have adopted energy averaging indices to describe aviation noise impacts. These are usually expressed as 24-hour averages showing contours of equal noise energy resulting from airplane flight activity, experienced at locations near the airport. The average noise levels are typically time weighted to reflect a presumed increase in community annoyance during nighttime hours (DNL) or a combination of evening and night time hours (CENEL, in California, LAeq in the United Kingdom and Lden in the European Union). The basis for these average metrics stems from seminal work by Schultz (1978). In his original work, Schultz establishes a relationship between the percentage of people highly annoyed by various types of transportation noise and the A-weighted day-night average noise level. However, later researchers have raised questions about the comparability of annoyance resulting from noise from different sources, such as rail, aircraft, and road, as well as the parameters used to establish the statistical relationships between the noise levels and annoyance (Schomer, 2005) and Fidell (Fidell et al., 2004).

Aside from questions about the proper fitting of annoyance data to average sound levels, questions are often raised by community representatives about the basic validity of using long term average sound levels to describe community annoyance. A significant controversy over a new runway at Sydney Kingsford airport in Australia focused on the inadequacy of published predictions of DNL impacts to realistically describe the actual noise impacts experienced by the community of the proposed new runway, which was opened in 1994. Although the airport had faithfully used the approved metrics (DNL) to predict the effect of the new runway, the actual community reaction once the runway was placed into service was substantially more negative than was predicted by the accepted noise-annoyance relationships. In response to those concerns, the Australian government developed a package of alternative metrics to describe the projected impacts of flights using the new runway (*Transparent Noise Information Package* which can be downloaded via the Internet (search Transparent Noise Information Package, Australian Department of Infrastructure and Transport). This suite of alternative metrics considers such factors as the individual movements or numbers of movements over a given time period, the number of hours with no jet movements, expressed as a percentage of the total number of hours during the period of interest, the number of events louder than 70 dB(A), and the total number of instances where an individual is exposed to an aircraft event above a specified noise level over a given time period. The use of similar supplemental metrics is now becoming more widespread.

6.2.1.1 Pollutant Harm and Impact

The generally acknowledged negative physiological effects of noise exposure include hearing loss, physiological effects such as hypertension and irritability, community annoyance, speech interference, sleep interference and effects on learning. However, the levels of airplane noise in communities surrounding airports are seldom high enough to threaten acute health damage, such as hearing loss. For example, industrial health guidelines normally regard noise exposure above 85–90 dBA averaged over an eight-hour work day to produce an unacceptable risk of long term hearing loss (OSHA, 2002).

Likewise, single event exposure to noise levels above 110 dBA can produce permanent shifts in hearing acuity. By contrast, inspection of actual DNL contours at many airports reveals that community noise exposure from airplanes in excess of 75 dBA averaged over a day or single event exposures above 100 dBA are extremely rare. However, the more subtle effects of long term exposure to airplane noise are less well established due, in part, to the difficulty of eliminating socio economic or other factors that play a role in assessing community health effects near airports. Lack of acceptance within the aviation industry and regulatory bodies of clear evidence of negative health effects has impeded more concentrated action to reduce noise impacts around airports. Nevertheless, the negative effects of exposure to aircraft noise and overflight are clearly manifested in community objections and vigorous opposition to airport expansion projects. These controversies are based largely on fears of increased noise exposure, in many cases magnified by the difficulty of conveying the effects of the projected noise environment to community residents.

6.2.1.2   Aircraft Noise Policy

The dramatic increases in airplane noise that accompanied the widespread introduction of jet airplanes was initially manifested as a local problem in the political jurisdictions surrounding airports, and many communities adopted local ordinances in attempts to control the impacts. However, national governments soon realized that the proliferation of local regulations affecting national transportation were counterproductive and assumed a more active role in establishing national standards for airplane noise emissions and reducing the scope of local regulation.

Aircraft noise regulation takes two primary forms: regulation of noise at the source and regulation of aircraft operating conditions to minimize community noise exposure. Source noise is regulated by national governments, following recommendations set by the International Civil Aviation Organization and its Committee on Aviation Environmental Protection (ICAO, 1944; FAA AEE, 2003). Source noise regulation applies to individual airplane and engine combinations, setting maximum allowable limits for the noise at the maximum gross certificated weights (takeoff and landing) and for specific climb/power profiles. Separate limits apply to propeller and turbojet aircraft. ICAO periodically reviews these certification noise limits, assessing the current and expected future airplane technology characteristics and the economic reasonableness of more stringent noise limits to revise the certification limits.

Regulation of airplane operating conditions in the form of night time curfews, restricted flight paths, prohibition of late night engine tests or limits on the type of aircraft allowed to operate at an airport have variously been used in attempts to minimize noise. Recently, the integration of new navigation technologies, such as the FAA's Next Generation Air Transportation System, NextGen in the United States and the Single European Sky ATM Research Programme in Europe have presented new opportunities to mitigate noise through modified operational procedures (FAA, 2011; SESAR/Eurocontrol, 2008). Area navigation capabilities of modern airplanes (RNAV) allows precise adherence to flight tracks through the use of the airplane's flight management system and autopilot. These procedures can be helpful in avoiding areas with sensitive populations. However, they present the risk of aggravating noise complaints if used to establish new flight tracks over sensitive areas. Continuous Descent Arrivals (CDA: uses advanced navigation and flight management capability to allow arriving airplanes to descend from cruise altitude to the airport with a constant, low power setting, rather than the traditional use of multiple level flight segments during the descent, with accompanying increases in engine power (Clarke et al., 2004). However, much of the noise reduction from CDA approaches occurs outside the area normally considered as highly impacted by airplane noise (65 DNL in the US) and may not contribute to substantial reductions in community annoyance.

6.2.1.3   Noise and Community Reaction

Contributing to the difficulty of explaining the noise effects of a proposed project is a significant gap between the formal and technical language of acoustics and the intuitive experience of non-technical community residents. The human ear can perceive changes in sound pressure level over a range of 11 or 12 orders of magnitude. In order to simplify the description of such a large range of intensity, acoustics has adopted a logarithmic measurement scale using decibel notation. The reference level of 20 μPa RMS is equivalent to the lower threshold of detectability for a young, healthy ear, roughly equivalent to the sound of a leaf falling onto a forest floor. Other commonly use reference sounds include the sound of gentle wind in a forest (approximately 40 dB), a normal conversation at a distance of a few feet (approximately 65 dB), the sound of an individual airplane overflight near an airport (80–100 dB, depending on the airplane type and distance from the observer) and sound of a rock band or an overflight by a Concorde supersonic jet takeoff near the airport (approximately 120 dB). The logarithmic scale of sound level has proven very difficult to explain to the non-technical public: decibels do not add arithmetically, a doubling of intensity is 3 dB, 10 over-flights by an airplane producing an 80 dB sound level is equal to one overflight producing a 90 dB level, and so on.

Complicating the difficulty of conveying acoustic information to the public are the spectral weighting scales used to characterize the human hearing response and perception of a given sound based on the presence or absence of low frequencies or pure tones. Further compli-cating factors include the difference between measurement of the peak sound level of a single event (Lmax), the sound level of that event integrated over its duration (SEL) and the level of sound integrated over an hour (Leq) or a day (Lday) with, or without, consideration of addi-tional weighting factors reflecting increased annoyance during evening or night time hours (Ldn, CNEL, etc.). All of the technical metrics mentioned can be well characterized by mea-surements or models with a high level of accuracy. None of them can be effectively used to describe the acoustic effects of a change in flight track or a runway extension in a community setting. Noise regulators have almost universally adopted cumulative energy averaged metrics with adjustments for increased annoyance during sensitive night or evening hours as the preferred means of expressing community noise impacts and the resulting annoyance. However, the difficulty of devising metrics that accurately characterize, and allow prediction of, community response to airplane noise is complicated by underlying uncertainties over the effect of socio economic effects, the effect of exposure to new flight tracks or new noise, all of which seem to provoke more vigorous community opposition than current modeling would suggest (Fidell, 2003).

Finally, there are tradeoffs between noise emissions and other pollutants, which are not well understood. As an example, much of the noise from jet engines is the result of high velocity, turbulent airflow from the hot jet exhaust. Modern turbofan engine noise has been steadily reduced as new designs increase the ratio of air flowing through the engine fan relative to the pure jet exhaust volume. This increase in bypass ratio has allowed the cooler fan airflow to envelop and shield the hot, turbulent jet exhaust, resulting in reduction in noise. The higher combustion efficiency of these engines has also reduced CO_2 emissions. However, these increased bypass ratios are achieved, in part, through the use of higher combustion tempera-tures, which results in increases in the production of nitrogen oxide (NO_x) emissions. Understanding of the relationship between various pollutants and noise emissions is still evolving and the implications for future regulation remain uncertain.

6.2.1.4   Models for Noise at Airports

Given that regulators have chosen cumulative measures of noise exposure, various models have been developed to represent those exposures, usually as contours of equal value covering the geographic area around an individual airport. Depending on the regulatory regime, these contours might represent DNL (day night sound level averaged over 24 h with nighttime noise weighted by a factor of 10), or a similar metric with different weighting factors for evening and night noise (CENEL in California, DEN in the European Union). These models calculate the ground level noise of each airplane landing or taking off at an airport, using the known certification noise levels at standard distances and classical noise propagation behavior. Input to such models require average daily flight schedules, airplane type, series, and engine, takeoff and arrival thrust management practices and other detailed information in a format compatible with the model needs. Early models poorly represented such effects as excess attenuation due to atmospheric conditions, topographical variations of the ground elevation or the effects of ground cover in either reducing or increasing low angle propagation. These deficiencies have largely been resolved in new versions of the various noise exposure models approved for use by various national regulatory agencies. In this regard, the US FAA is developing a new suite of interlaced models called the Aviation Environmental Design Tool that uses common inputs to drive noise and air quality modeling. This model is being proposed to ICAO CAEP as a new tool for the calculation of global aviation noise and air quality impacts.

Given community concerns that time averaged metrics like DNL do not properly describe the effect of noise on activities, such as speech interference, sleep interference or interference with learning, many of these models also produce supplemental metrics such as values for the amount of time a given point, or contour, will experience instantaneous sound levels above a given value. An example might provide the number of minutes per day that sound levels exceed a recognized speech interference level, or a threshold for sleep awakening. Although formal regulatory actions rely on the use of cumulative measures, like DNL, regulators encourage analysts to employ relevant supplemental metrics to help convey a full understanding of aircraft activity.

6.2.2   Greenhouse Gas Emissions

It is well known that the operation of transport vehicles is a major component of anthropogenic climate change – the warming of the Earth's temperatures due to human activities. The production, delivery, and combustion of transportation fuels increase levels of greenhouse gases (GHG) in the atmosphere (EPA, 2008). In this section, we will explore characteristics of these emissions globally and at airports, current and future policies surrounding GHG emissions, and models for GHG emissions at airports.

The transportation sector is responsible for 13% of global GHG emissions and 28% of United States domestic GHG emissions, making it the fifth and second largest contributor respectively (Pew Center on Global Climate Change, 2004). The CO_2 contribution of both international and domestic aviation in the EU-27 (the 27 member countries of the European Union) was 3% in 2007; transportation's contribution was 24% (EEA, 2009).

While aviation continually seeks improvements in fuel efficiency, this share is expected to increase as other transport modes shift away from carbon-based fuels. Such an action will further increase the pressure on the aviation sector to reduce GHG emissions (Yang et al., 2009). Rapid growth is also increasing pressure to reduce GHG emissions. Both the Federal Aviation Administration (FAA) and the European Organization for the Safety of Air Navigation are planning for system growth, further increasing the pressure on the aviation sector to reduce GHG emissions (Ryerson, 2010a,b; Ky and Miaillier, 2006).

6.2.2.1   Pollutant Characterization and Challenges

According to the United States Environmental Protection Agency (EPA), the principal anthropogenic greenhouse gases are Carbon Dioxide (CO_2); Methane (CH_4); Nitrous Oxide (N_2O); and Fluorinated Gases. As related to transportation, CO_2 is produced through the burning of fossil fuels; CH_4 is produced during the production and transport of oil; N_2O is produced during the combustion of fossil fuels and fluorinated gases are emitted from industrial processes (EPA, 2010). These pollutants all trap heat in the atmosphere resulting in a warming of the Earth, yet at different levels of warming. For this reason, measurement of greenhouse gas emissions is often reported in units of carbon dioxide equivalent, or CO_2e, to consider the warming potential of each pollutant compared with the warming potential of CO_2. A discussion of how to account for global warming potentials of pollutants is well described in Kim et al. (2009).

Regulating and reducing the impacts of greenhouse gas emissions is more challenging than doing so with other pollutants, as GHGs are spatially and temporally distributed rather than concentrated. Unlike noise emissions from an aircraft overflight, GHG emission impacts are long term rather than immediate. Unlike criteria pollutants, GHG emissions are felt world-wide through the warming of the Earth, rather than localized. Greenhouse gas emissions also have varying warming potential in the atmosphere that can vary depending on the spatial distribution of the emission. These challenges motivated a common metric in GHG policy: the emission of the most abundant greenhouse gas – Carbon Dioxide equivalents (CO_2e) that is directly correlated with the burning of fossil fuels (Environmental Protection Agency, 2010).

As global rather than local pollutants that are not experienced at the point of emission (compared with noise), the pressure to reduce their emission is diminished and the locus of responsibility for doing so is unclear. The spatial distribution of GHG emissions is matched by the spatial distribution of aviation, a mode with local (airports), regional, national, and international components. Aviation organizations are grappling with unclear roles to regulate and reduce the emission of a pollutant with impacts outside their jurisdiction along with the need to remain competitive.

6.2.2.2   GHG Policies

While greenhouse gases are global pollutants, policy organizations are more local in nature. Policy has followed the structure of policymaking organizations: from local airport authorities to the International Civil Aviation Organization (ICAO), many organizations are grappling with GHG policy. However, all organizations face the challenge of placing themselves at a competitive disadvantage. If airport A, for example, institutes a greenhouse gas policy, carriers or passengers may take their business to a competing airport B in response to higher costs. While GHG emissions may drop at airport A, both revenues and GHG emissions could rise at airport B. The result is a negligible change in GHG emissions system-wide and a strongly disadvantaged airport A. This same scenario holds at the state level (for example, if the state of California institutes a policy, perhaps a share of air traffic shifts to the state of Washington) and the international level. In this section we will briefly review GHG policies related to aviation at the multi-national, national, and local level.

According to the Kyoto Protocol, international aviation emissions are under the auspices of ICAO. Countries that signed the Kyoto Protocol include domestic air travel emissions in their emissions allocation, yet exclude international aviation. ICAO is currently developing a GHG emission program at a global level. In 2010, there was consensus on high level principles for engine standards on manufacturers but the consensus document stops short of recommending global policy impacting operations. In contrast and on a more local scale, the European Union under Directive 2008/101/EC, aviation will be included in the European Union Emissions Trading Scheme (EU–ETS) starting January 2012. EU-ETS is the largest cap-and-trade emission program, which has been operational since 2005. Emissions allocations will be distributed to the airline operators through both a benchmarking and auctioning method (Vespermann and Wald, 2010). All CO_2 emissions from flights touching a European airport will be considered.

A challenge is the relationship between these two GHG policies. There is current no level of coordination or synchronization between ICAO's GHG plans and EU-ETS. The steps taken by ICAO are mild, such that there are no global market based measures planned (ICAO, 2007). ICAO faces a complex set of institutional challenges compared with a smaller body such as the EU. Many members of ICAO oppose market based measures; different philosophies lead to a weakened, or less definitive, policy. This has led to EU-ETS as the only international climate change initiative in place. This may not always be the case going forward, as under EU-ETS Directive 2008/101/EC, third country measures to reduce the climate change impact of aviation could lead to an EU-ETS exemption. The interpretation is that an ICAO based measure could nullify EU-ETS. However, since no third country has introduced such measures, there is no practical experience as to how measures can be considered as comparable.

Policies at the US national level are currently being debated. Different Federal policies recommended different measures to reduce GHG emissions; however none have been successfully implemented. However, examples of more local regulation at the state and airport level are available from the United States. Many US states have adopted climate action plans (Arizona Climate Change Advisory Group, 2006; New Mexico Climate Change Advisory Group, 2006) that defer to the Federal Government on aviation GHG emissions. GHG considerations are a fairly new issue for airports. In California, the state government can use the 2007 People of California versus San Bernardino County ruling to encourage and potentially compel local governments to incorporate GHG emissions forecasting into their regional planning (People of California v San Bernardino County, 2007). Other successful state-level settlements are discussed in Anders et al. (2009). The San Diego County Regional Airport Authority (2008) includes a GHG inventory in the Environmental Impact Report of their Master Plan, and plans to replacing airport shuttle buses with alternative-fuel vehicles, encourage the use of gate-electricity; and promote greener building practices.

The spatially and temporally distributed nature of GHG emissions is challenging traditional policy-making organizations. In a similar way, the nature of GHG emissions challenges the understanding and inventory of GHG emissions.

6.2.2.3 Inventories and Models

The practice of GHG emissions inventorying has led to both an understanding of GHG emissions sources and has raised questions about the methods used for inventorying. The Airport Cooperative Research Program recently published the *Guidebook on Preparing Airport Greenhouse Gas Emissions Inventories*, which clearly presents the state of the art in airport GHG inventorying (Kim et al., 2009). To summarize, in developing an inventory, there are two key questions that must be addressed. The first is: *What should be included?* and the second is: *How should the inventory be performed?* The first question addresses the boundaries for the inventory. The second involves the methodology for accounting for any emissions producing activities that fall within that boundary.

To answer *What should be included?* a key to inventories is that they are "policy relevant," such that the emissions sources considered are able to be effected by the organization performing the inventory. An example is the GHG emissions inventory performed by the Port of Seattle (2008). The Port of Seattle divides GHG emissions from the airport it operates, Seattle Tacoma International Airport, into ownership categories. Emissions owned by the port include those produced from hotel and parking lot shuttles (while on airport grounds) and facility power. Airline or tenant emissions are those from aircraft. Emissions owned by the public include passenger vehicles and hotel and parking lot shuttles (while not on airport grounds).

In response to *How should the inventory be performed?* inventories are either top-down or bottom-up. Bottom-up methods develop estimates directly from spatially resolved detail and aggregate these activities to represent the inventory domain; top-down methods apply non-spatial estimates to a spatial domain using some proportionality assumptions. Top-down inventories often involve a calculation of the amount of aviation fuel consumed scaled by emission coefficients. These emissions coefficients can incorporate the emission from actual combustion of that aviation fuel in operation, or they can incorporate an entire emission from the life cycle (and are well documented by Chester, 2008). The Federal Aviation Administration, the European Environmental Agency, and the German Research Labs have independently developed bottom-up methods to calculate GHG emissions from an individual flight moving through the airspace (EEA, 2006; Kim et al., 2007; Scheelhaase and Grimme, 2007). These models are of higher fidelity than top-down models and involve a spatial understanding of aircraft movements.

A key component of understanding aircraft GHG emissions, whether it be through a top-down or bottom-down inventories, is the ability to model how aircraft operations and technology will change in response to policy. Models that capture inventories at a snapshot of time provide an understanding of GHG emissions sources and levels. However, under a wide range of environmental policy scenarios, carriers and other stakeholders might find aircraft costs variable. Models that capture how aircraft costs compare over a range of environmental policies will help understand the demands airports can expect due to environmental policies, and are explored in the future models section.

6.2.3   Water Runoff

Although noise pollution has long been the primary environmental issue defining objection to airport operations and capital expansion, water pollution has become a significant source of objection, both from regulators and from communities concerned with the airport's operation. Environmental controversies based on water pollution have been used to successfully slow down the acquisition of regulatorily required discharge permits. Community opposition to stream contamination by airport pollutants, or to odors from holding ponds used to store contaminated water has been observed at airports in Europe, North America, and elsewhere. Future increases in regulatory scrutiny are expected and opposition to airport expansion based on concerns over water pollution is likely to continue.

6.2.3.1   Pollutant Harm and Impact

Water pollution at airports shares some of the same impacts as water pollution from other industrial sites: potential for contamination of drinking water supplies or aquatic life in receiving waters with toxic compounds, potential for depleting dissolved oxygen levels in receiving waters to the point that fish kills occur, potential siltification of adjacent streams from construction runoff and the potential for acute exposure to irritants of persons exposed to contaminated water. There are several predominant sources of water pollution at airports. Although not all are unique to airports, together they have a high likelihood of occurrence at airports. The principal pollutants and sources are: residual contaminants from military or industrial manufacturing of munitions, contaminants from aircraft maintenance, runoff of silt from construction sites, chronic leakage or acute spills of aviation fuels and aircraft or runway deicing agents.

- Contaminants from military or industrial manufacturing of munitions have arisen at airports with a history of military use or prior weapons manufacture. These contaminants often show up decades after the manufacturing sites have been abandoned as leachates into the airport ground water, possibly escaping outside the airport's boundaries to raise drinking water contamination issues. Compounds of chromium, arsenic, mercury, magnesium, antimony, chlorine, chlorinated hydrocarbons, benzenes and others are often associated with long abandoned industrial or military sites at airports.
- Improperly controlled current or past aircraft maintenance practices often result in ground-water contamination by hydraulic fluids, lubricants, cleaning chemicals, etchants, paint strippers and related maintenance supplies.
- Earth moving at airport construction sites can result in the transport of silt to nearby streams or ponds if not properly contained.
- Chronic leakage of aviation fuels from underground storage facilities or distribution pipelines has often polluted significant portions of airports, in some cases reaching the airport perimeter and fouling adjacent waters. Likewise, acute spills, both large and small, have occurred from catastrophic failure of tanks, of pipeline delivery systems supplying on-airport fuel storage facilities or uncontained refueling spills that reach unprotected storm water drains in terminal aprons.
- As the use of increased quantities of aircraft and/or runway deicing fluids increases to address safety issues during winter operations, the quantities of spend fluids has received increased regulatory scrutiny.

Existing policies at major airports have successfully addressed most, but not all, of the water pollution issues at airports. Legacy pollutant sites associated with former military munitions or industrial manufacturing activities have largely been identified and remediated through ground-water extraction and purification. The installation of approved oil – water separators on all apron drains, or in major storm drains serving terminal and maintenance aprons is a nearly universal and highly effective method of eliminating liquid hydrocarbon releases. Tenant leases require periodic testing of effluents and mandate approved containment and recovery of spent fluids. New fuel storage and distribution facilities are equipped with double containment and/or advanced leak detection capabilities. Increasingly, spent aircraft deicing fluids are recovered using vacuum devices, containment booms or dedicated deicing aprons having fluid recovery and remediation capability. At many airports, significant volumes of containment ponds have been built to modulate the rate at which contaminated deicing runoff is discharged into receiving waters, or to permit on site remediation through aeration, distillation, reverse osmosis, biological or chemical means. The quantity of runway deicing fluids used is increasingly being scrutinized with the goal of reducing the amount of fluid used to the minimum quantity that will adequately protect against snow or ice contamination. In addition some regulators have prohibited the use of deicing fluids with extremely high biological or chemical oxygen demand, like urea, in favor of less biologically damaging compounds, like potassium acetate or formic acid.

Despite progress having been made in controlling many water pollutants at airports, the risk of acute spillage or chronic leakage from fuel facilities is still a concern. In addition, the quantities of and disposal of both aircraft and runway deicing runoff continues to be the subject of regulatory activity and of concern.

6.2.3.2   Models for Water Runoff at Airports

Standard hydrologic modeling techniques have been used to predict water pollutant concentrations in airport discharges. These models are heavily dependent on the availability of robust measurements of discharge concentrations for calibration. Transport models for subsurface migration of contaminants in groundwater flows are also available from hydrologic engineering practices. New techniques to use real time mass accumulation measurement of the liquid water content of frozen or freezing contaminants are being developed to allow both airlines and airports to measure, then model, the potential accumulation of winter contaminants on airplane lift surfaces and on runway surfaces. These new measuring and modeling techniques will permit accurate determination of hold over times following aircraft or runway deicing by estimating the rate at which recontamination will occur following initial application of deicing agents . Models like SWMM (storm water management model) are used to measure the dispersion of these pollutants (Bloomington Airport South Drainage and Water Quality Modeling update, Barr Engineering, 2008).

Recommendations for improved modeling capability should include sensing and modeling systems to more accurately allow the management of water pollutants. For airports with a regulatory limit of discharges into receiving streams, automated monitoring and modeling systems are needed that warn airport staff of conditions that will lead to exceedances of those regulated discharge limits. For aircraft and runway deicing, systems are needed that use in pavement or on wing sensors to measure the type and depth of contaminants, the concentration and residual freeze suppression capability of applied deicing fluids and models that correlate those measurements of surface condition with the aforementioned mass accumulation monitors to predict when additional chemical application is required.

6.2.4   Criteria Air Pollutants

6.2.4.1   Pollutant Harm and Impact

Airports harm air quality through emissions of several pollutants. These include nitrogen oxides (NO_x), volatile organize gasses (VOC), carbon monoxide (CO), particular matter (PM), sulfur oxides (SO_x), and so called hazardous air pollutants (HAP) also known as air toxics. Sources of airport emissions include aircraft, ground service equipment, airport-related surface vehicles, and stationary sources.

Table 6.1   US Large Hub Communities, by Air Quality Standard Attainment Status

Attainment Status	**Air Quality Standard**	
	OZ (8-hour)	*PM (PM-2.5 1997, PM-2.5 2006, or PM-10)**
Non-attainment	17	16
Attainment	9	10
Total	26	26

*PM-2.5 is particles less than 2.5 µm in diameter. PM-10 is particles less than 10 µm in diameter.
Sources: EPA (2010); BTS (2007)

Emissions from airports affect regional air quality, and in many cases contribute to non-attainment of government air quality standards. The standards typically pertain to five forms of air pollution: sulfur dioxide, nitrogen dioxide, carbon monoxide, ozone (a product of reactions between NO_x and VOC), and particulate matter. These are referred to as criteria pollutants, because they are specifically addressed by air quality standards, even though in some cases it is recognized that they also are indicators for the presence of other pollutants. Standards are based primarily on the goal of protecting human health. Human health consequences of the criteria pollutants and those related to them include:

- Respiratory problems, including bronchoconstriction, airway irritation, breathing difficulties during exercise, aggravation of asthma symptoms, and permanent lung damage.
- Coronary problems, including chest pain in persons with heart disease, irregular heartbeat, and non-fatal heart attacks.
- Skin inflammation.
- Premature death in people with heart and lung disease.

Air quality standards are premised on the theory that below some threshold concentration of a given pollutant, its human health effects are very minor or non-existent. The air quality impacts of airports are thus of greatest concern in areas that do not meet the standards. Unfortunately, large airports are typically located in large urban areas, which are more likely to be in non-attainment status. Table 6.1 shows that, as of 2010, over 60% of large hub communities in the United States, those with 0.5% or more of US passenger traffic, are non-attainment areas for ozone or PM.

Nonetheless, airports are comparatively small contributors to regional air pollution. Table 6.2 presents estimates of these contributions for various airports and pollutants. The values are not strictly comparable because some estimates are for specific airports while others consider all airports in a region, and the different studies do not all address the same set of airport sources. Nonetheless, the figures suggest that, aside from NO_x, contributions are usually in the 1% or less and well below 1% for particulates. Given that NO_x is a precursor for ozone, and that non-attainment for ozone is widespread, it is fair to say that control of NO_x emissions is the biggest air quality challenge airports face.

Figure 6.1 presents the contributions of various airport sources to overall airport emissions, for Denver International Airport. Airport and airport equipment are the predominant sources

Table 6.2   Airport contributions to regional emissions

Region	Year	PM10	PM2.5	VOC/ROG	NO_x	SO_2	CO
San Francisco Bay Area	2008	0.09%	0.23%	1.21%	4.54%	0.43%	2.39%
Los Angeles Area	2006	0.29%	0.77%	0.98%	1.49%	2.41%	1.26%
Atlanta (ATL)	2000	0.05%	0.10%	0.25%	2.84%	0.69%	0.50%
Boston (BOS)	2000				0.7%		
Seattle (SEA)	2000				2.0%		
St. Louis (STL)	2000				1.4%		
Phoenix (PHX, WGA, DVT)	2005	0.1%	0.2%		7.8%		
Chicago (ORD, MDW)	2001				2.0%		
New York (JFK, LGA, EWR)	2000				4.0%		

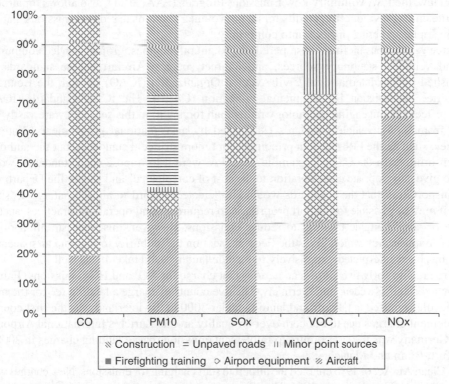

Figure 6.1   Emissions contribution by source, Denver International Airport (2001). *Source*: Ricondo and Associates (2005)

for NO_x and CO, while unpaved roads and airport construction are the primary sources for particulate emissions. The latter are not considered in the data for Table 6.2, which therefore understates the airport particulate contribution to some degree, although not enough the change the conclusion that airports are a minor source of particulates.

6.2.4.2   Policies

Airport emissions are subject to two forms of government policy. The aforementioned air quality standards set maximum ambient concentrations of various pollutants. In addition, source regulations set allowable emissions from various fixed and mobile sources.

In the US, the Clean Air Act [42 U.S.C. §7401 et seq. (1970)] requires that national ambient air quality standards (NAAQS) are established by the US Environmental Protection Agency. The US states must then prepare State Implementation Plans (SIPs) that identify specific measures for improving local air quality (or maintaining it) so that the NAAQS are attained. The SIPS consider emissions from all sources, including airports. The CAA also prohibits Federal agencies, including the Federal Aviation Administration, from taking actions that interfere with a state's ability to attain NAAQS. Thus, in taking any action affecting airport development, the FAA must assure that it conforms to the applicable SIP. This can be done by ensuring that the emissions targets included in the SIP contain the additional emissions expected to result from the project, or offsetting the project emissions increases with reductions of the same size. Federal law, the FAA Voluntary Low Emissions Program (FAA, 2011), also allows for airports to accrue emissions reduction credits for taking voluntary actions to reduce emissions that can later be applied to bring projects into conformity.

Source regulations include those pertaining to aircraft engines, ground service equipment, on-road vehicles, stationary sources, and indirect sources. Aircraft engine standards are established by the International Civil Aviation Organization (ICAO) through the Council's Committee on Aviation Environmental Protection (CAEP). The ICAO standards (ICAO, 1944) reflect current engine technology rather than forcing it; in this sense they are vastly different from motor vehicle standards established by many national governments. National agencies, such as the EPA, can, in principle adopt more stringent standards, but the authority is often limited. In the US, for example, the EPA, were it to impose such standards, would have to give "appropriate consideration to the cost of compliance" and satisfy the Department of Transportation that the standards would not "create a hazard to aircraft safety." In some cases, it may be possible for airport proprietors to require ground operations practices, such as towing or single engine taxiing, to reduce emissions. However, this may run afoul of the Federal Aviation Act, which gives the Federal Aviation the authority to regulate airspace, and which has been interpreted expansively to include landings and takeoffs as well.

There are currently five European nations that charge for NO_x and HCs emissions; France, UK, Switzerland, Sweden, and Germany. The five countries charge a landing fee for the mass of molecules emitted at takeoff and landing under 3000 Ft. Switzerland was the first country to implement charges due to concern over air quality around Zurich's International Airport in 1986. Germany was the last country to implement emissions based landing charges in 2008 in Frankfurt, Bonn and Munich.

The Clean Air Act of 1986 enabled the airport to start charging for emissions. New charges were announced in 1993 and began in 1996. The effect at Zurich's airport has been a significant reduction of high NO_x/HC emitting aircraft at the airport. The law that allows the airport to do this is the Swiss Air Quality Act (1986). This law states that the government is allowed to pass laws to get levels of several pollutants that they monitor to appropriate standards. The Swiss Government set a national model for emission charges in 1995 (Aircraft Emission Charges at Zurich Airport, 2010).

The Swiss model has been applied by several airports, Stockholm in 1998 and Geneva the same year. These laws charging for ton of emissions had three positive effects. One, Enough revenue for Zurich's airport to pay for monitoring the air quality, a 10% reduction in emissions of NO_x and HC in 1997, and the introduction of better engines from aircraft manufacturers. Zurich's airport earns approximately €2–3.2 million a year as of 2009 and has used funds collected in the past to pay for air quality monitoring and improvements to the airport for air quality like fixed power for parked aircraft.

There have been questions to the effectiveness of these emission charges. A German study from 2008 through 2009 monitored emissions of both NO_x and HCs and found that there was a very small addition to the total operating cost and suggested that emission charges may only raise awareness about the problems that come with those two pollutants and that they don't put enough pressure on airlines to force change. The study concluded that the charges did bring awareness to the problem, but couldn't force any change (Heinitz and Doerig, 2010). Another ICAO study had a similar conclusion when monitoring Stockholm and Zurich from 1998 to 2007 showed a very small change in airline fleets and suggests that the current policies that have been adopted do not affect the Total Operating cost to bring significant change in airline fleets.

Ground service equipment emissions are regulated by EPA under the non-road engine category. The regulations were phased in between 2004 and 2007 and will eventually reduce emissions by 70–90%. Equipment turnover is gradual, however, so it will take some time before the full effect of these regulations is realized. While local authorities are pre-empted from imposing such standards, they may take certain actions to reduce emissions from GSE, such as regulating their operation, requiring purchase of certain models, or even or even equipment that uses alternative fuels. Exactly what localities can and cannot do in this regard remains fairly murky, however.

The situation for motor vehicles operating on the airport property is similar. The EPA sets emission standards that cannot be locally pre-empted, but restrictions can be placed on vehicles accessing the airport property. For example, the state of California, under a waiver granted to the FAA Administrator based on a provision of the CAA, has defined a series of categories for motor vehicles with low emissions, such as Low Emissions Vehicle, Super Low Emissions Vehicle, and Ultra Low Emissions Vehicle. It is probable that any state could require that vehicles operating on the airport be in one of these categories.

States have much more authority to regulate emissions from stationary sources. Although airports contain certain stationary sources (boilers, fuel tanks, etc.), these are small contributors to overall emissions, and airports as a whole cannot be considered stationary sources under the CAA. It has, however, recently been proposed that "Airport Bubbles" be subject to state regulation under the authority of states to regulate "indirect sources". The CAA allows states to have "indirect source" review programs to assure that "a new or modified indirect source will not attract mobile sources of air pollution" that would jeopardize NAAQS attainment. Thus, through indirect source review, states could make approval of airport development projects contingent SIP emissions targets. This concept remains untested in practice, at least in the United States.

6.2.4.3 Models

Air quality modeling is a highly evolved science, and the basis for models that focus on airports. There are two main categories of modeling: emissions modeling and dispersion modeling. Emissions models quantify emissions from various airport sources. Dispersion models, predict concentrations of pollutants throughout the airport region, taking into account both airport and non-airport emissions. The EDMS is an example of a model designed to handle both emissions and dispersion. This is the model the model the FAA provides to all airports to approximate

Table 6.3 Approaches to modeling aircraft emissions

Key Parameters	Simple Approach	Advanced Approach	Sophisticated Approach
Fleet (aircraft/engines combinations)	Identification of aircraft types	Identification of aircraft and corresponding engine types	Identification of aircraft, corresponding engine types, and age
Time in Mode	N/A (indirectly accounted for via United Nations Framework Convention for Climate Change (UNFCC) LTO Emission Factors)	ICAO Databank Certification Values adjusted if possible to reflect airport specific information	Refined values (e.g., with consideration of performance, number of engines in use)
Emission Indices and Fuel Flow	UNFCC LTO Emission Factors by Aircraft Type	ICAO Databank Certification Values	Refined values using aircraft specific performance and operational data
Movements	Number of aircraft movements by aircraft type	Number of aircraft movements by aircraft-engine combination	Same as advanced

Source: Modified from ICAO (2007)

emissions. It's also used to predict the emissions of airport expansion and is a both an emission and a dispersion model (Emissions and Dispersion Modeling System (EDMS), FAA OEE, 2011).

Emissions Models
Aircraft emissions are based on a landing-take-off (LTO) cycle that typically considers aircraft operations below 3000 feet above ground level, the typical mixing height below which emissions affect local air quality. The simplest emissions models are based on LTO emissions factors by aircraft type; programs like Piano X (see Table 6.3) (Lissys Limited-Piano X, 2010). In more sophisticated models, five operational modes are considered: approach, taxi in and ground idle, taxi out and ground idle, take-off, and climb. Time and thrust setting in each mode for each engine type (or aircraft-engine type combination) are estimated, and engine performance data are then used to obtain emissions. The engine performance data can be based on the ICAO engine certification values obtained in a laboratory setting, or more refined values that take into account real-world influences on engine performance. If the goal is an emissions inventory, total emissions for an LTO cycle is sufficient, while emissions by location and time are required if the results are to be input to a dispersion model.

GSE emissions are also based on an aircraft LTO. The amount and duration of GSE activity required for a cycle depends upon the aircraft type, in particular whether it is a narrow or wide body, and GSE emissions factors are accordingly disaggregated. More sophisticated models like the EDMS take into account the specific aircraft type and parking location, which in turn require specific types of GSEs for specific lengths of time, and at specified engine loads (Emissions and Dispersion Modelling System (EDMS), FAA OEE, 2011). Emission factors may also depend on the level of deterioration on the equipment.

Airside and landside vehicle traffic emissions are based on methods that apply to motor vehicle traffic more generally. Emission factors are defined on a vehicle kilometer basis, and are differentiated by vehicle class, age, and speed. More advanced methods like MOVES and SIP calculate emissions from different road segments and times of day, each with characteristics such as traffic volume, speed, weather, and vehicle fleet composition (Motor Vehicle Emission Simulator, 2011).

Other activities that must be considered for comprehensive modeling of airport emissions include:–

- Aircraft refueling – emission factor based on the amount of fuel delivered per year, and differentiated by type of fuel and delivery mode (truck or pipeline). Additional emissions from refueling facilities based on the amount of fuel stored.
- Deicing – emission factor based on the amount of deicing fluid used.
- Boilers – emission factor based fuel used or power generated, and differentiated by type fuel. Emissions are adjusted for the efficiency of control equipment.
- Incinerators – emission factor based on amount of refuse incinerated, with adjustment for control equipment efficiency.
- Maintenance facilities – emission factors based on the quantities of evaporated substances (paints and solvents) used.
- Fire training – emission based on the quantity for fuel used in such training, differentiated by fuel type.

Dispersion Models
Dispersion models link emissions to actual air quality, as measured by pollutant concentration. In addition to emissions input, these models require detailed meteorological data, including wind velocity, turbulence, temperature, and temperature gradient. Terrain data is also required. Many working dispersion models are based on the concepts of a Gaussian steady state plume. The models assume that winds and emissions rates vary slowly enough so that the atmosphere reaches steady state. Other dispersion models like PUFF-PLUME consider time variation in weather and other factors explicitly (Turner, 1994). Most modern dispersion models account for deposition – pollutants falling out of the atmosphere.

Dispersion modeling is very complex, particularly when non-steady state conditions and chemical reactions are taken into account. To correctly assess impact, model runs must include a baseline, no airport, case in which only non-airport emissions sources are included. The models can yield important insights, however. For example, a recent study of Atlanta Hartsfield (Unal et al., 2005), a non-attainment area for the 75 ppb 8-hour ozone standard, found the airport impact to be as much as 56 ppb just north of the airport, and to average 5 ppb over the whole non-attainment area. These differences are clearly large enough to affect the attainment status of a region.

6.3   The Future of Airport Environmental Impact Analysis

This is an era of unprecedented environmental sensitivity in the airport and aviation community. Substantial efforts are being made to improve the fidelity, usability, and level of integration of environmental models. Progress in these aspects will likely continue. However, there is another need that has received less attention. If, as many expect, environmental policies place stronger demands on airports and aircraft operators, we will need models that predict the impacts of these policies on the aviation system. In the future, therefore, environmental impact models will be complemented by environmental *policy* impact models.

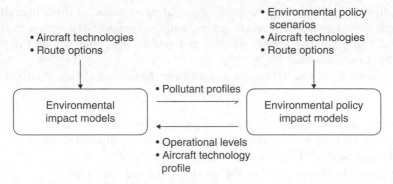

Figure 6.2   Framework for airport environmental impact assessment

The need for an integrated modeling framework that includes environmental impact and environmental policy impact models is clear. Environmental impact models characterize the level of emissions given a set of aviation system inputs, while environmental policy impact models estimate what the aviation system will look like – what and how aircraft will be operated over what network – after an environmental policy is instituted. As pollutants are both constraints on operations and the output of operations, airports both impact the environment and are impacted by environmental policy. A future framework for airport environmental impact assessment must reflect the dual role of pollutants as model inputs and outputs.

A proposed framework for aviation system environmental impact assessment includes environmental impact models and environmental *policy* impact models. Environmental impact models characterize the level of emissions and resulting ecological and welfare impact from a given transportation system, while environmental policy impact models estimate what the system would look like – what and how vehicles will be operated over what network – after an environmental policy is instituted. The environmental impact models, generally computer-based models that take intercity transportation system characteristics as inputs, create a baseline picture of system pollutants and their effects. These inputs may inform the environmental policy scenarios which become the inputs for the environmental *policy* impact models. This is a group of models that capture how an environmental policy scenario impacts system characteristics, such as vehicles in the system, operational levels and network structures. The output of the environmental policy impact models becomes an input for a second run of the environmental impact models to calculate a level of pollutants after an environmental policy. This framework, depicted in Figure 6.2, shows the relationship between environmental impact models and environmental policy impact models and that environmental considerations can be incorporated as both model inputs (as a policy) and outputs (as resulting pollutant profiles).

Environmental impact models and environmental policy impact models are explored in greater detail.

6.3.1   Environmental Impact Models

It can be expected that future models will continue the decades long trend toward increased modeling capability using more accessible, powerful and less expensive computer resources with better user interfaces and output graphics. Furthermore, the trend in computer-based environmental models is integration of functionalities to present a complete profile of pollutants movement. Future models to integrate tools for the assessment of multiple pollutant impacts are needed. Current practices involve constructing separate models for future traffic demand, runway and taxiway capacity, airspace procedures, noise, air pollution and other environmental impacts. These different models are often developed at different times during a project, by different organizations using different data sources and assumptions.

Data collection and formatting for the various models are often laborious, time consuming and expensive. As a result of using different input data, the overall environmental analysis is susceptible to disagreement in results among the several models being used, with a concomitant risk of legal challenge. Work is underway in Europe, the United States and at ICAO/ CAEP to develop integrated models with common assumptions and data sources. In some models being developed, operational data for multiple pollutant analysis is fully integrated, with some data being derived from real time aircraft flight tracks extracted from air navigation service providers' surveillance radars (Roof et al., 2007).

Additionally, combining multiple pollutants in a single analysis tool is also proving challenging. Researchers are devoting considerable effort to reexamining the basic relationships between noise exposure and annoyance, exploring the effects of income, age, previous exposure to airplane noise and other characteristics thought to affect community reaction to airplane noise (Stallen and Smit, 1999). Researchers are also greatly improving the fidelity of GHG and fuel consumption modeling, including the FAA's System for assessing Aviation's Global Emissions (SAGE) (Kim et al., 2007). However, weaving emissions with different profiles and histories into a single analysis tool is proving a challenge. The negative health impacts of noise and many local pollutants are spatial and temporally concentrated, such that exposure to these pollutants is a very important factor. Greenhouse gas emissions, however, have impacts that are spatial and temporally distributed, and therefore the actual quantity of emission rather than the intensity of the emission in a particular location is the component to be captured by models. Development and refinement of tools that allow for the joint estimation of pollutants are underway, such as the FAA's Aviation Environmental Design Tool (AEDT) and Aviation environmental Portfolio Management Tool (APMT). Performing sensitivity analysis is a goal of these and other tools, and a critical component of planning aviation systems concerning these multiple pollutants. There are examples of existing literature that considers pollutant trades, including Sridhar et al. (2011).

Another component of the future of environmental modeling is the integration of a broader definition of pollutants and viewing outputs as communication tools. Evidence indicates that adverse reaction to airplane noise often occurs well outside areas where traditional annoyance criteria based on standard physical metrics predict negative community reaction (Guski, 1999). As an example, a major redesign of airspace in the United States was undertaken by the Federal Aviation Administration from February 1999 through September 2007. The New York, New Jersey, Philadelphia Metropolitan Area Airspace Redesign affected flight paths over an area of 31 000 square miles covering five states and affecting operations at 21 airports. The noise analysis (Federal Aviation Administration, 2007) was based on models that predicted DNL levels and changes to those levels resulting from the proposed modifications to flight tracks. Although the analysis complied fully with the existing legal requirements under the US National Environmental Policy Act (GAO, 2008), the project was vigorously opposed by citizens and national and local politicians in the five states, resulting in years of litigation and adverse publicity. Although the final analysis was upheld on technical grounds by the courts, the significant adverse citizen reaction was not predicted by the modeled analysis due, in part, to the models' inability to factor in intangible moderating factors. Similar adverse reactions have occurred at airports around the world demonstrating that strict adherence to current legal requirements to assess impacts prior to undertaking actions, such as flight track changes or new runway construction, may not be sufficient to avoid significant public and political adverse reaction. Future models will need to incorporate these findings and provide analysts with estimates of community reaction to noise that include consideration of psychological or socio-economic moderating factors that contribute to adverse community reaction to airplane noise.

In addition to the need to properly model intangible moderating factors, future models should explore more effective means of conveying the resulting information in community settings. Mere presentation of contour maps or data tables, which has been the norm to date, must be supplemented by additional graphic, audio or video tools that faithfully simulate the impact of proposed airport or airspace development in ways that are easily understood by laypersons. For example, through creative use of interactive audio, it should be easy for someone to experience directly what it "sounds like" inside a 65 dB contour.

6.3.2   Environmental Impact Policy Models

Environmental impact policy models take an environmental policy as an input and output how the aviation system behaves. In many scenarios, especially in the United States, the ultimate environmental policy is unknown. Therefore, environmental policy is most conveniently modeled as a variable rather than an input. This can be done as an explicit constraint on pollutants that varies with pollutant levels. An example of considering an environmental policy as a variable and performing scenarios analysis is found in Ryerson and Hansen (2010a).

The environmental policy will dictate the scope of pollutants to be considered. If a policy is considering constraining emissions concentrated around an airport, then emissions from vehicle and airport operation in a confined area are calculated. A policy can similarly be constrained to focus on operational emissions, manufacturing emissions, or another subset of the system. However, if a policy is focusing on emissions from the entire system, then the scope of emissions considered should be from the entire life-cycle. Using the metrics developed by Chester (2008) that include vehicle components (manufacturing, operation, maintenance, and insurance), infrastructure components (construction & maintenance, operating, and insurance), and fuel production, one can capture emissions from the entire system lifecycle.

The aviation system changes due to an environmental policy differ in the short-run or long-run. A short-run impact is one in which the aviation system actors – airports, airlines, and others – have limited time to respond to the policy. Because of this, the components related to capital inputs, such as aircraft and network structures, will remain relatively constant. However, in the long-run these same actors are able to adjust to the policy. Operations may change, along with the airline network structure and the aircraft vehicle technologies that serve them. The change in fleets and the resulting impact to airports is an important component of capturing the long-run impact of an environmental policy at airports. How carriers may adopt or use existing vehicle technology differently due to an environmental policy is uncertain. The GAO (2009) reports that in the US there was a shift to larger jets due to the fuel price increases of 2008. Ryerson and Hansen (2010a) find that turboprops could provide lower operating costs per seat compared with regional jets and narrow body jets at high fuel prices.

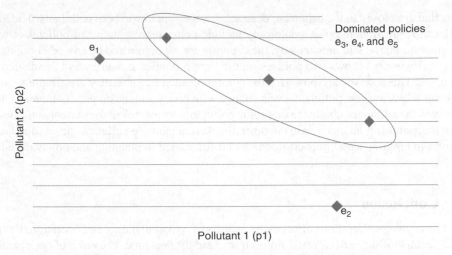

Figure 6.3   Pareto frontier of pollutants

In light of these and other findings, it is possible that an environmental policy could encourage larger aircraft sizes among the same technology (for example, larger jets compared with smaller jets) or a shift in technology (for example, a shift from jets to turboprops).

The methodology that underlies the environmental policy impact model itself can take many forms. To capture the short-term impact of a policy, economic models that capture the short-term economic behavior of carriers such as those described in Scheelhaase and Grimme (2007) can be used. For example, Scheelhaase and Grimme (2007) develop operating cost models for the existing fleet and network, and consider what the increase in costs would be under an environmental policy. To capture the long-term impact, however, capturing system details through optimization or performing detailed empirical modeling through large-scale data collection is both challenging and of questionable value, due to high levels of uncertainty. Daganzo (1999) describes how analytic modeling moves away from the need for large amounts of data inputs and also complicated optimization models, while providing key insights through variable interactions. Performing scenario analysis of long-run aviation system impacts using analytic system-wide planning models can capture the trends and potential large-scale system shifts due to environmental policy. These analytic system-wide planning models are well explored in the logistics research and less explored in the aviation research. Generally, logistics scheduling literature considers how a central planning body would schedule operations and choose vehicle types to minimize costs (see Smilowitz et al., 2002). Examples in aviation are both presented and discussed in Ryerson and Hansen (2010a,b). Such a perspective captures the ideal solution in terms of technology and cost to meet each policy scenario; this provides the basis for setting a policy.

Upon determining the changes in the air transportation system due to an environmental constraint or policy, the computer-based simulation models are used to calculate actual pollutant output (emission and exposure). These models provide calculations that can be translated into Pareto Curves which communicate pollutants trade-offs. For example, for environmental policy level e_i there is an output level of pollutant 1 ($p1$) and pollutant 2 ($p2$) (Figure 6.3). Such a plot would include one pollutant on each axis and plot the coordinates ($p1$, $p2$) at each e_i.

Points that are closest to the origin are desirable, as they are low in both pollutants. The Pareto frontier to result from the plot would only include points representative of policies of which we cannot do better; said another way, these points are not dominated by any other points. For example, Figure 6.3 shows two policies on the Pareto frontier, e_1 and e_2, and dominated policies e_3–e_5. These curves are powerful for communication because they are simple to understand and showcase how pollutant levels change with policy and the output of other pollutants. As there is the additional challenge that some pollutants "compete" in the sense that a reduction of one translates into an increase in the other, displaying multiple pollutant curves illustrate the tradeoffs in environmental impacts that must be considered in planning and policy decisions.

6.4   Conclusion

Environmental considerations affect many aspects of airport planning and operation. Environmental factors influence the level of airport traffic and the fleet mix. As a result of environmental review policies and political processes, these factors also constrain the ability of airports to provide facilities that are commensurate with traffic demands. Many of the responsibilities of airport planners and managers flow from environmental imperatives.

The challenges of reconciling environmental concerns with the role of airports in providing mobility and enabling commerce have become more complex over time and will continue to do so in the future. Technologies to reduce populations exposed to high levels of time-averaged noise impact have succeeded on their own terms, but the limited public reaction to that success has demonstrated that this was only a small part of the problem. Airport noise impact is more a psycho-social phenomenon than an acoustic one, as the emerging problems of "new noise" and "rich noise" demonstrate. Meanwhile there has been growing concern about the contributions of airports to greenhouse gas emissions, water pollution, and air pollution. While airports are not the primary sources of any of these, it has proven uniquely difficult to mitigate their impacts due to the overriding importance of safety, and the need avoid measures that add weight to aircraft.

Perhaps for this reason, much of the effort to address airport environmental impacts goes toward models that can accurately assess the environmental harm done by airports. With continuing advances in computational capability, air and noise models are increasingly focusing on understanding detailed spatio-temporal variations that play a pivotal role in determining human reactions and conformity with clean air standards. Regarding GHG, the focus continues to be on emissions inventories, but ones that focus on emissions sources that are "policy relevant" and consider the whole product life cycle rather than just the "tailpipe". More fundamentally, the challenges of airport environmental modeling have changed because with the recognition that airports have limited jurisdiction over pollutants, while their environmental impact strongly depends on fleet mix and schedules, and thus on the private business and public policy decisions that shape them. This creates a need for a framework for airport environmental impact assessment to capture short-term and long-term system changes in fleet selection and operations; the outputs of these models are then integrated with models that estimate local environmental impacts. Further, because of the broader concept of airport environmental impact along with the interrelated nature of pollutants, new methods for communication are also needed. In this chapter we have attempted to explore the full range of airport environmental impacts and propose a new framework for performing and communicating airport environmental impact analysis.

Acknowledgements

The authors would like to thank the following people for their comments, contributions, and invaluable assistance in revising the manuscript for this chapter: Jack Lee, Summer intern at NEXTOR; Wolfgang Grimme, Research Associate at German Aerospace Center; Mary Ellen Eagan, President at Harris Miller Miller & Hanson Inc.; and John E. Putnam, Partner at Kaplan Kirsch & Rockwell, LLP.

References

Aircraft Emission Charges Zurich Airport, (2010) Zurich Airport [Online] Available from: http://www.zurichairport. com/Portaldata/2/Resources/documents_unternehmen/umwelt_und_laerm/Broschuere_Emission_Charges.pdf [Accessed February 19, 2013].

Anders, S. J., De Haan, D. O., Silva-Send, N., et al. (2009) Applying California's AB 32 targets to the regional level: A study of San Diego County greenhouse gases and reduction strategies. *Energy Policy*, 37 (7), 2831–2835.

Ang-Olson, J. (2009) What Does Climate Change Mean for Airport Planning? Presented at the 88th TRB Annual Meeting, Washington DC: Transportation Research Board, January 13, 2009. Available from: http://www.trbav030. org/pdf2009/TRB09_Ang-olson.pdf [Accessed 18th June 2010].

Arizona Climate Change Advisory Group, (2006) *Climate Change Action Plan*. [Online] Available from: http://www. azclimatechange.gov [Accessed February 19, 2013].

Barr Engineering co., Bloomington Airport South Drainage and Water Quality Modeling update, (2008) [Online] available from: http://www.ci.bloomington.mn.us/cityhall/dept/commdev/planning/longrang/enreview/auar/ appendixa.pdf [Accessed February 12, 2013].

Bureau of Transportation Statistics (BTS). (2007) *BTS Hub Status 2007*. [Online] Available from: http://www.rita.dot. gov/bts/ [Accessed February 19, 2013].

Cal Climate Action Partnership and the Office of Sustainability, (2009) *Climate Action Plan, University of California, Berkeley*. [Online] Available from: http://sustainability.berkeley.edu/calcap/docs/2009_UCB_Climate_Action_ Plan.pdf [Accessed 15th June 2010].

Chester, M. V. (2008) *Life-cycle Environmental Inventory of Passenger Transportation in the United States*. UC Berkeley: Institute of Transportation Studies. [Online] Available from: http://escholarship.org/uc/item/7n29n303 [Accessed February 19, 2013].

Commonwealth Scientific and Industrial Research Organisation (CSIRO) – Australia *The Generic Reaction SET (GRS) Model for Ozone Formation. (2006)* Division of Energy Technology [Online] Available from: http://www. engr.ucr.edu/~carter/Mechanism_Conference/15%20B%20Azzi.pdf

Clarke, J. P., Ho, N. T., Liling, R. et al. (2004). Continuous descent approach: design and flight test for Louisville international airport. *Journal of Aircraft*, 41(5), 1054–1066.

Daganzo, C. (1999) *Logistics Systems Analysis*, 3rd *Edition*. Berlin: Springer.

Department of Transportation (DOT), (2003) *Revised Departmental Guidance: Valuation of Travel Time in Economic Analysis*. [Online] Available from: http://regs.dot.gov/docs/VOT_Guidance_Revision_1.pdf [Accessed 18th June 2010].

Environmental Protection Agency, (2008) *Inventory of US Greenhouse Gas Emissions and Sinks: 1990–2006*. USEPA #430-R-08-005. [Online] Available from: http://www.epa.gov/climatechange/emissions/usinventoryreport.html [Accessed February 19, 2013].

Environmental Protection Agency, (2010) *Currently Designated Nonattainment Areas For All Criteria Pollutants*. [Online] Available from: http://epa.gov/airquality/greenbk/ancl3.html [Accessed February 19, 2013].

European Environment Agency, (2006). *EMEP/CORINAIR Emission Inventory Guidebook – 2007: Group 8: Other mobile sources and machinery*. [Online] Available from: http://reports.eea.europa.eu/EMEPCORINAIR4/en/ page017.html [Accessed February 19, 2013].

European Environment Agency, (2009). *Greenhouse gas emission trends and projections in Europe 2009: Tracking progress towards Kyoto targets*. [Online] Available from: http://www.eea.europa.eu/publications/eea_ report_2009_9 [Accessed February 19, 2013].

Federal Aviation Administration Office of Energy and Environment, (2010) *FY 2010 Methodology Report, FAA Flight Plan Performance Measures*. [Online] Available from: http://www.faa.gov/about/plans_reports/media/ FY10%20Portfolio%20of%20Goals.pdf, p. 3 [Accessed February 19, 2013].

Federal Aviation Administration Office of Energy and Environment, (2011) *Emissions and Dispersion Modeling System (EDMS)*. [Online] available from: http://www.faa.gov/about/office_org/headquarters_offices/apl/research/models/edms_model/ [Accessed February 19, 2013].

Federal Aviation Administration Office of Energy and Environment (FAA AEE), (2003) *Noise Standards: Aircraft Type and Airworthiness Certification*. Federal Aviation Administration Office of Energy and Environment Advisory Circular 36-4C.

Federal Aviation Administration, (2007) *Final Environmental Impact Statement New York/New Jersey/Philadelphia Metropolitan Area Airspace Redesign*. United States Department of Transportation Environmental Impact Statement.

Federal Aviation Administration, (2011) *Voluntary Airport Low Emissions Program (VALE)*. [Online] Available from: http://www.faa.gov/airports/environmental/vale/ [Accessed February 19, 2013].

Federal Aviation Administration, (2011) *Next Generation Air Transportation System (NextGen)*. [Online] Available from: http://www.faa.gov/nextgen/ [Accessed February 19, 2013].

Heinitz F. M. and Doerig, N. (2010) *Economic Effects Of Local Aircraft Emissions Charges At German Airports*, Erfurt University of Applied Sciences.

Fidell, S. (2003) The Schultz Curve 25 Years Later: A Research Perspective. *Journal of the Acoustical Society of America*, [Online] 114 (6), 3007–3015, Pt. 1. Available from: http://asadl.org/jasa/resource/1/jasman/v114/i6/p3007_s1?isAuthorized=no [AccessedFebruary 19, 2013].

Fidell, S., Sanford and Silvati, L. (2004) Parsimonious Alternative To Regression Analysis For Characterizing Prevalence Rates Of Aircraft Noise Annoyance, *Noise Control Eng. J.*, 52 (2), 56–68, March–April.

Government Accountability Office (GAO). (2008) *FAA Airspace Redesign – an Analysis of the New York/New Jersey/Philadelphia Project*. Government Accountability Office report GAO-08-786.

Government Accountability Office (GAO). (2009) *Commercial Aviation: Airline Industry Contraction Due to Volatile Fuel Prices and Falling Demand Affects Airports, Passengers, and Federal Government Revenues*. Government Accountability Office report GAO-09-393.

Guski, R. (1999) Personal and social variables as co-determinants of noise annoyance. *Noise Health*, 1 (3), 45–56.

Lissys Limited (2010) *Piano-X*, [Online] Available from: http://www.piano.aero/ [Accessed February 12, 2013].

London Heathrow Airport. (2011) Heathrow launches community consultation on noise schemes. [Online] Available from: http://www.heathrowairport.com/portal/page/Heathrow%5EHeathrow+press+releases/943a1a99f94df210VgnVCM10000036821c0a____/a22889d8759a0010VgnVCM200000357e120a____/ [Accessed February 12, 2013].

Klass, T. (2010) Sea-Tac International Airport's third runway opens on November 20, 2008, *Seattle Post-Intelligencer*. [Online] Available from: http://www.historylink.org/index.cfm?DisplayPage=output.cfm&file_id=8855 [Accessed February 12, 2013].

Intergovernmental Panel on Climate Change (IPCC). (2007) *Climate Change 2007: The Physical Science Basis*. Contribution of Working Group I to the Fourth Assessment Report of the Intergovernmental Panel on Climate Change. Solomon, S., D. Qin, M. Manning (eds.).

Intergovernmental Panel on Climate Change (IPCC). (2001) *Aviation and the Global Atmosphere: Executive Summary*. [Online] Available from: http://www.grida.no/climate/ipcc/aviation/064.htm [Accessed 12th October 2007].

International Civil Aviation Organization (ICAO) Environmental Report 2007, (2007) [Online] Available from:http://legacy.icao.int/icao/en/env2010/pubs/env_report_07.pdf [Accessed July 19, 2011].

International Civil Aviation Organization (ICAO). (2007) *Airport Air Quality Guidance Manual*. [Online] Available from: http://www.icao.int/environmental-protection/Documents/Publications/FINAL.Doc%209889.1st%20Edition.alltext.en.pdf [Accessed June 21, 2010].

International Civil Aviation Organization (1944). *Convention on International Civil Aviation*. [Online] Available at: http://www.icao.int/publications/pages/doc7300.aspx

Kim, B. Y., Fleming, G. G., Lee, J. J., et al. (2007) System for assessing Aviation's Global Emissions (SAGE), Part 1: Model description and inventory results. *Transportation Research Part D*, 12, 325–346.

Kim, B., Waitz I. A., Vigilante M. and Bassarab, R. (2009) *Guidebook on Preparing Airport Greenhouse Gas Emissions Inventories*. [Online] Airport Cooperative Research Program Report 11, Available from: http://onlinepubs.trb.org/onlinepubs/acrp/acrp_rpt_011.pdf [Accessed February 19, 2013].

Ky, P. and Miaillier, B. (2006) SESAR: Towards the new generation of air traffic management systems in Europe. *ATCA Journal of Air Traffic Control*, January–March.

Morrison, S. A., and Winston, C. (2007) Another Look at Airport Congestion Pricing. The American Economic Review, 97(5), 1970–1977.

New Mexico Climate Change Advisory Group, (2006) *Final Report*. [Online] Available from: http://www. nmclimatechange.us [Accessed February 19, 2013].

Occupational Safety and Health Administration (OSHA) (2002) *Hearing Conservation* (revised) Occupational Safety and Health Administration, U.S. Department of Labor, Document 3074.

People of California versus San Bernardino County, (2007) [Online] Available from: http://ag.ca.gov/globalwarming/ pdf/SanBernardino_complaint.pdf [Accessed February 19, 2013].

Pew Center on Global Climate Change, (2004) *Global Warming Facts and Figures*. [Online] Available from: http:// www.c2es.org/facts-figures [Accessed February 19, 2013].

Port of Seattle, (2008) *Greenhouse Gas Emissions Inventory*. [Online] Available from: http://www.airportattorneys.com/ files/greenhousegas06.pdf [Accessed July 28, 2008].

Plaut, P.O. (1998) The comparison and ranking of policies for abating mobile-source emissions. *Transportation Research Part D* 3(4), 193–205.

Roof, C., Hansen, A., and Fleming, G. (2007) *Aviation Environmental Design Tool (AEDT) System Architecture*. USDOT Volpe Center Doc #AEDT-AD-01.

Ricondo and Associates (2005), *Airport Emissions Inventory*, Denver International Airport.

Ryerson, M. S. and Hansen, M. (2010a) Capturing the Impact of Fuel Price on Jet Aircraft Operating Costs with Engineering and Econometric Models. In: *Fourth International Conference on Research in Air Transportation*, ICRAT 2010 June 2010, Budapest, Hungary.

Ryerson, M. S. and Hansen, M. (2010b) The potential of turboprops for reducing aviation fuel consumption. *Transportation Research Part D: Transport and Environment*, 15(6), 305–314.

San Diego County Regional Airport Authority, (2008). *Final Environmental Impact Report* (EIR), *Airport Master Plan, Appendix E*. [Online] Available from: http://www.san.org/sdcraa/airport_initiatives/master_plan/eir.aspx [Accessed February 19, 2013].

Scheelhaase, J. D. and Grimme, W. G. (2007) Emissions Trading for international aviation – an estimation of the economic impact on selected European airlines. *Journal of Air Transport Management*, 13, 253–263.

Schomer, P. (2005) *"Biases Introduced by the Fitting of Functions to Attitudinal Survey Data"*, ASA/NOISE-CON 2005 Meeting, Minneapolis, MN, [Online] available at: http://www.acoustics.org/press/150th/Schomer.html [Accessed February 12, 2013].

Schultz, T.J. (1978) Synthesis of social surveys on noise annoyance, *J. Acoust. Soc. Am.*, 64(2), 377–405.

Smirti, M. and Hansen M. (2007) Achieving a Higher Capacity National Airspace System: An Analysis of the Virtual Airspace Modeling and Simulation Project. Presented at the *AIAA Modeling and Simulation Technologies Conference*, Hilton Head, SC, USA: American Institute of Aeronautics and Astronautics.

Smilowitz, K., Atamtürk, A. and Daganzo, C. (2002) Deferred Item and Vehicle Routing within Integrated Networks. *Transportation Research E: Logistics & Transportation Review*, 39, 305–323.

Sridhar, B., Chen, Y., Ng, H. and Linke, F. (2011) Design of Aircraft Trajectories based on Trade-offs between Emission Sources. Proceedings of the Ninth Annual FAA/EUROCONTROL Air Traffic Management Research and Development Seminar, Berlin, Germany, June 2011. Available at: http://www.atmseminar.org/seminarContent/ seminar9/papers/20-Sridhar-Final-Paper-4-14-11.pdf [Accessed February 12, 2013].

Stallen, P. J. and Smit, P. (1999) A Theoretical Framework for Environmental Noise Annoyance. *Noise Health*, 1(3), 69–80.

Transparent Noise Information Package (2010) Australian Department of Infrastructure and Transport. [Online] Available at: http://www.infrastructure.gov.au/aviation/environmental/transparent_noise/tnip.aspx [Accessed February 12, 2013].

Turner, D. B. (1994). *Workbook of Atmospheric Dispersion Estimates: An Introduction to Dispersion Modeling*, 2nd edn. CRC Press.

Vespermann, J. and Wald, A. (2010) Much Ado about Nothing? – An analysis of economic impacts and ecologic effects of the EU-emission trading scheme in the aviation industry. *Transportation Research Part A: Policy and Practice*, doi:10.1016/j.tra.2010.03.005.

SESAR. (2008). *Eurocontrol*. [Online] Available at: http://www.eurocontrol.int/sesar/public/standard_page/ overview.html [Accessed February 12, 2013].

Yang, C., McCollum, D., McCarthy, R. and Leighty, W. (2009) Meeting an 80% reduction in greenhouse gas emissions from transportation by 2050: A case study in California. *Transportation Research Part D*, 14 (3), 147–156.

7

Airport Safety Performance

Alfred Roelen[A] and Henk A.P. Blom[A, B]

[A]*Air Transport Safety Institute, National Aerospace Laboratory (NLR), The Netherlands*
[B]*Aerospace Engineering Department, Delft University of Technology, The Netherlands*

7.1 Introduction

Statistics of worldwide air transport accidents indicate that flight safety has continuously increased since aviation became a means of mass transportation in the 1960s (Boeing, 2011). Thanks to this historical development, society has become accustomed to the fact that flying with commercial airlines is a very safe means of transportation. Nevertheless, each time a rare catastrophic accident happens in commercial aviation, this typically makes worldwide headlines in the news, by which society is now and then reminded of the potentially dramatic consequences of accidents in commercial aviation. Because these headlines have such a large impact, there is a widely supported objective that the frequency of these headlines in the news should not increase. This means that the accident risk per flight should at least continue to improve with the future growth in commercial aviation flights. In view of this widely supported objective, it makes sense to evaluate in more detail how accident risk has evolved during the recent past.

The aim of this chapter is to analyse how safety performance has evolved for the ground segment of a flight relative to the airborne segment over the period 1990–2008. In order to have sufficient data, this study makes use of statistical data of worldwide accidents of scheduled commercial flights by fixed-wing aircraft with a maximum take-off weight of more than 5700 kg. This means that this paper does not differentiate between fatal accidents and non-fatal accidents, and neither takes the number of fatalities into account. Though, there are important differences in the severity of various accident types. For example, the average number of fatalities for an airborne accident is significantly higher than for an accident on the ground (Roelen and Smeltink, 2008).

Modelling and Managing Airport Performance, First Edition. Edited by Konstantinos G. Zografos, Giovanni Andreatta and Amedeo R. Odoni.
© 2013 John Wiley & Sons, Ltd. Published 2013 by John Wiley & Sons, Ltd.

Section 7.2 discusses the main data sources and taxonomies of accidents and incidents. Subsequently, it is explained how the information from these sources is used to estimate yearly rates for various accident types in commercial aviation. Following this approach, a systematic comparison of accident rate development over the period 1990–2008 has been made for *Take-off, Landing and Ground Operations* versus developments of rates of other groups of accident categories such as *Airborne* and *Weather*. This comparison shows that for Take-off, Landing and Ground Operations the accident rate is not really improving over the period 1990–2008, which is in contrast with the overall improvements during this period.

In Section 7.3, the accidents related to Take-off, Landing and Ground Operations are further analysed at the level of the main accident categories, such as runway excursion, abnormal runway contact, ground handling, ground collision and runway incursion. This analysis shows that the non-decreasing accident rate applies to each of the main accident categories of the Take-off, Landing and Ground Operations group.

In Section 7.4, the accident data base is further analysed at the level of the main airborne and aircraft related accidents, such as accidents due to Controlled flight into terrain, In-flight loss of control, Mid-air Collision, Turbulence and Aircraft system failure. This analysis shows that for most of these categories there is a systematic decreasing accident rate over the period 1990–2008.

In Section 7.5, an analysis of the safety driving mechanisms that apply to commercial aviation is performed. This reveals significant differences for the air and ground segments of commercial aviation.

In Section 7.6, an overview is provided of various safety initiatives, addressing both the airborne and ground segments of a flight. Subsequently, it is verified whether these safety initiatives have a clear correlation with the trends in accident rates analysed in Sections 7.3–7.4. This shows that there are differences in the way safety improvement mechanisms (identified in Section 7.5) work for air and ground segments of commercial aviation.

Finally, Section 7.7 presents the concluding remarks of the chapter and provides recommendations for ensuring the future continuation of the extremely good safety records of commercial aviation. Occurrence rates of accidents related to take-off, landing and ground operations do not show a clear positive or negative trend over the period 1990–2008. If nothing is done this could result in an increasing number of accidents per year in commercial aviation due to the growth of air traffic. To prevent increasing accident numbers, past safety initiatives should be continued, current safety initiatives should be strengthened and new safety initiatives should be started.

7.2 Accident Rates in Commercial Aviation

This section explains the data sources and accident taxonomy used for the estimation of yearly rates for various accident types. First subsection 7.2.1 explains the statistical data sources and their specific use. Subsection 7.2.2 introduces the internationally accepted occurrence categories in aviation safety. Subsequently, subsection 7.2.3 uses the data sources and accident taxonomy to show how occurrence rates for Taking-off, Landing and Ground Operation compare to occurrence rates for all other categories.

7.2.1 From Accident Statistics to Accident Rates

Worldwide registration of accident events in commercial aviation is maintained by the International Civil Aviation Organization (ICAO) in its Accident and Incident Data Reporting (ADREP) data base. It is precisely defined when an occurrence is an accident, which can be summarized as follows:

an occurrence associated with the operation of an aircraft which takes place between the time any person boards the aircraft with the intention of flight until such time as all such persons have disembarked, and in which i) any person suffers death or serious injury, or ii) in which the aircraft receives substantial damage, or iii) aircraft gets missing or inaccessible". (ICAO, 2001)

An incident is defined as 'an occurrence, other than an accident, associated with the operation of an aircraft which affects or could affect the safety of the operation' (ICAO, 2001).

For our analysis in this chapter, ICAO's ADREP database has been searched on accidents of fixed-wing aircraft in scheduled commercial air transport with a maximum take-off weight of more than 5700 kg. There are no restrictions regarding the location of the occurrence. One should be aware that this selection excludes the following flights:

- Non fixed-wing aircraft;
- Non-commercial air transport;
- Non-scheduled flights;
- Aircraft with a maximum take-off weight below 5700 kg.

The main reason(s) for making these choices are:

- Non fixed-wing aircraft form a negligible fraction of all flights;
- Non-commercial air transport falls outside the study objective;
- Including non-scheduled flights would require an out of proportion effort in counting the number of non-scheduled flights for each year;
- Aircraft with a maximum take-off weight above 5700 kg covers the largest collection of flights for which the accident data in ADREP is reliable.

The resulting curve of number of accidents per year is given in Figure 7.1. This curve shows that the total number of accidents exhibits a negative trend suggesting a decrease from about 140 per year in 1990 to about 100 per year in 2008.

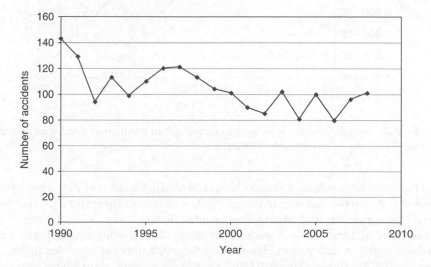

Figure 7.1 Number of accidents per year (1990–2008); scheduled commercial air transport by fixed-wing aircraft with a maximum take-off weight of more than 5700 kg (*Source*: ICAO's ADREP data base)

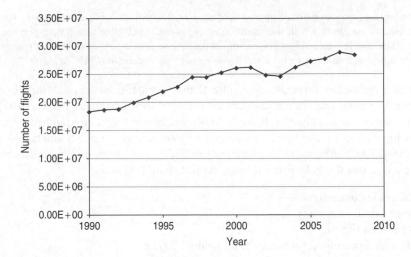

Figure 7.2   Number of flights per year; scheduled commercial air transport by fixed-wing aircraft with a maximum take-off weight of more than 5700 kg (*Source*: Official Airline Guides Timetable Database: Database is maintained by OAG Aviation, Luton, UK)

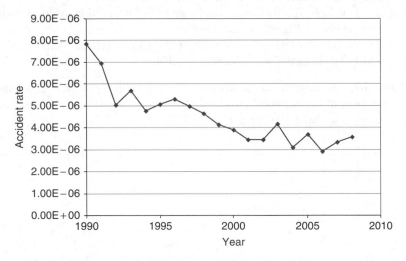

Figure 7.3   Calculated accident rates; scheduled commercial air transport by fixed-wing aircraft with a maximum take-off weight of more than 5700 kg

Using the same flight selection criteria, the corresponding number of flights per year in the same period (1990–2008) is shown in Figure 7.2. The number of flights for each year has been established through analysis of worldwide scheduled flights.

The calculated accident rates in Figure 7.3 are obtained by dividing the number of accidents for scheduled flights in each year in Figure 7.1 by the total number of scheduled flights for that year in Figure 7.2. The curve in Figure 7.3 indicates that the accident rate exhibits a clear tendency to decrease over the last two decades, although the rate of improvement seems to be levelling off.

7.2.2   CICTT categories

Since 1999, ICAO and a Commercial Aviation Safety Team (CAST), have jointly set-up a CAST/ICAO Common Taxonomy Team (CICTT). CICTT has been in charge of the development of common taxonomies and definitions for accident and incident reporting systems. One of their products (CAST/ICAO, 2011) defines specific occurrence categories that apply both to incidents and to accidents, and clusters them in six groupings (see Table 7.1).

An element of the CICTT category design is that it sometimes associates one accident to more than one incident/accident category. Examples of airport related accident types that are associated with two CICTT categories are:

1. *High speed aborted take-off and subsequent runway overrun.* In CICTT terms, this kind of accident is typically allocated to the category 'runway excursion' (RE) and also to one of the following two categories: 'system component failure – powerplant' (SCF-PP) or 'system component failure – non-powerplant' (SCF-NP).
2. *Runway veer-off during landing.* In CICTT terms, this kind of accident is typically allocated to the categories 'runway excursion' (RE) and 'loss of control – ground' (LOC-G).

In line with the CAST/ICAO taxonomy, each of these types of accidents should be allocated to two CICTT categories. Similarly, the first example is associated to two CICTT groupings.

In order to better understand the characteristics of the accident data that lie behind the curve in Figure 7.3, in the subsequent sections of this chapter, the overall accident rate of Figure 7.3 is decomposed according to the six occurrence category groupings of (CAST/ICAO, 2011) in Table 7.1, which are:

1. Take-off, Landing and Ground Operations
2. Airborne
3. Weather
4. Aircraft
5. Miscellaneous
6. Non-Aircraft Related

7.2.3   Take-off, Landing and Ground Operation versus Other Categories

The ADREP database provides for each accident event a category classification according to the CICTT taxonomy. This categorization of each accident allows to 'split' the accident rate curve in Figure 7.3 into two curves, that is, one curve for accidents falling in the first CICTT group (Take-off, Landing and Ground Operation), and one curve for accidents falling in the other five groups. The results are presented in Figure 7.4. Because the CICTT taxonomy sometimes classifies a single accident under more categories, the sum of the curves in Figure 7.4 is slightly higher than the accident rate curve shown in Figure 7.3. In order to make this explicit, from now on we will refer to occurrence rates rather than accident rates.

The two curves in Figure 7.4 show that the occurrence rate for Take-off, Landing and Ground Operations has not improved during 1990–2008, while the *Other* occurrence rate clearly has. Figure 7.4 also shows that the Take-off/Landing and Ground Operations occurrence rate seems to have taken the lead over the rate of all Other occurrences. This finding coincides well with

Table 7.1 CICTT incident/accident categories and their grouping (In (CAST/ICAO, 2011) six categories have not been allocated to any of the six groupings. For two of these six categories the ADREP database contains accident events that have some relevance for the current study. These two categories are (1) Collisions with obstacle(s) during take-off and landing; and (2) (Near) Collision with bird(s)/wildlife (CAST/ICAO, 2011)

Take-off, landing and ground operations	
Ground handling	RAMP
Ground collision	GCOL
Loss of control – ground	LOC-G
Runway excursion	RE
Runway incursion – vehicle, aircraft or person	RI-VAP
Runway incursion – animal	RI-A
Undershoot/overshoot	USOS
Abnormal runway contact	ARC
Fire/smoke (post impact)	F-POST
Evacuation	EVAC
Airborne	
Airprox/TCAS alert/loss of separation/near mid-air collision/mid-air collision	MAC
Controlled flight into/toward terrain	CFIT
Loss of control – in flight	LOC-I
Fuel related	FUEL
Low altitude operations	LALT
Abrupt manoeuvre	AMAN
Weather	
Windshear or thunderstorm	WSTRW
Turbulence encounter	TURB
Icing	ICE
Aircraft	
System/component failure or malfunction (powerplant)	SCF-PP
System/component failure or malfunction (non-powerplant)	SCF-NP
Fire/smoke (non-impact)	F-NI
Miscellaneous	
Security related	SEC
Cabin safety events	CABIN
Other	OTHR
Unknown or undetermined	UNK
Non-aircraft-related	
ATM/CNS	ATM
Aerodrome	ADRM

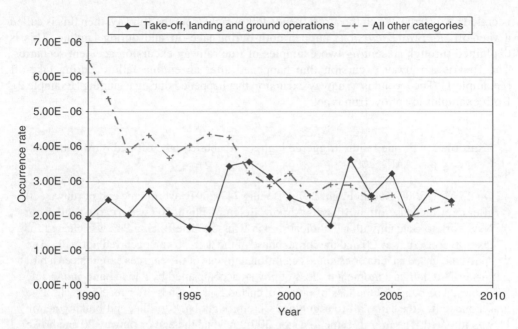

Figure 7.4 Calculated occurrence rates for Take-off, Landing and Ground Operations (connected line) versus occurrence rate for the other five CICTT accident groups (dashed line); scheduled commercial air transport by fixed-wing aircraft with a maximum take-off weight of more than 5700 kg

the earlier finding by (Bloem and Blom, 2010) that the ground related accident rate seems to have taken over the air related accident rate since the beginning of the twenty-first century.

In order to further improve our understanding of the two curves in Figure 7.4, in the next two Sections (7.3 and 7.4) we analyse the occurrence rate and look at the nature of the accidents for some key occurrence categories of these two curves.

7.3 Analysis of Take-Off, Landing and Ground Operation Accidents

In this section we analyse the trend and nature of the accidents for the following occurrences in the Take-Off, Landing, Ground Operations group:

- Runway Excursions (RE);
- Take-Off and Landing occurrences other than runway excursion (i.e. LOC-G, USOS, ARC);
- Ground Operations occurrences (i.e. RAMP, GCOL, RI-VAP, RI-A, F-POST, EVAC).

These three are covered in subsections 7.3.1, 7.3.2 and 7.3.3 respectively. Subsequently, a summary of the main findings for the Take-off, Landing and Ground Operations group is given in subsection 7.3.4.

7.3.1 Runway Excursions

A runway excursion is an event where an aircraft ends outside the runway during take-off or landing. When the aircraft does not stop prior to reaching the end of a runway, then this is called an overrun. When the aircraft ends left or right from the runway, then this is called a veer-off. *Runway Excursions* happen both during take-off and during landing. This is explained through presenting two examples of true runway excursion accident scenarios. The first is a runway excursion that happened after an engine failure during take-off (Example 1). The second is a runway excursion that happened during a landing (Example 2). Both examples are of overrun type.

Example 1   Engine failure triggered high speed aborted take-off and runway overrun (AAUI, 2009)

On 25 May, 2008, Kalitta Flight 207, a Boeing 747-209 F, overran the end of runway 20 after a rejected take-off at Brussels Airport, Belgium. Flight 207 was a cargo flight from New York to Bahrain with a technical stop at Brussels. The Boeing 747 was cleared for take-off from runway 20 and the initial phase of the take-off run occurred normally. The speed increased under a constant acceleration until one of the engines experienced a bird strike. This caused a momentary loss of power, accompanied by a loud bang and visible flames. The bang and the loss of power occurred four seconds after reaching V1[1]. Two seconds after the bang, all four engines were brought back to idle, and braking action was initiated. The aircraft came to a stop 300 m beyond the end of runway 20 and broke in three parts. The crew of four, and a passenger suffered only minor injuries.

Example 2   Landing overrun (Transportation Safety Board of Canada, 2007)

On 2 August, 2005, Air France Flight 358, an Airbus A340-313, crashed after it overran the end of runway 24 L during landing at Toronto Airport, Ontario, Canada. Flight 358 departed Paris, France, on a scheduled flight to Toronto with 297 passengers and 12 crew members on board. On final approach to Toronto, the flight crew were advised that the crew of an aircraft landing ahead of them had reported poor braking action, and the aircraft's weather radar was displaying heavy precipitation encroaching on the runway from the northwest. At about 200 feet above the runway threshold, while on the instrument landing system approach to Runway 24 L with autopilot and autothrust disconnected, the aircraft deviated above the glideslope and the groundspeed began to increase. The aircraft crossed the runway threshold about 40 feet above the glideslope. During the flare, the aircraft travelled through an area of heavy rain, and visual contact with the runway environment was significantly reduced. There were numerous lightning strikes occurring, particularly at the far end of the runway. The aircraft touched down about 3800 feet down the runway, reverse thrust was selected about 12.8 seconds after landing, and full reverse was selected 16.4 seconds after touchdown. The aircraft was not able to stop on the 9000-foot runway and started its overrun at a groundspeed of about 80 knots and caught fire. All passengers and crew members were able to evacuate the aircraft before the fire reached the escape routes. A total of two crew members and 10 passengers were seriously injured during the crash and the ensuing evacuation.

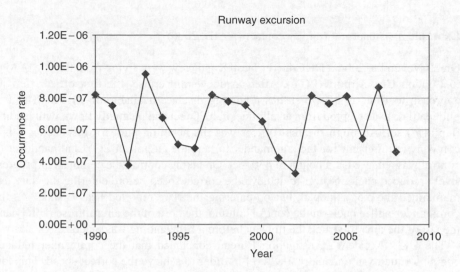

Figure 7.5 Calculated occurrence rates for Runway Excursions

Figure 7.5 presents the occurrence rates for Runway Excursions. It shows that the calculated occurrence rates for Runway Excursions do not show a clear positive or negative trend over the period 1990–2008. Moreover, the occurrence rate is rather high, that is, on average it happens around 0.6 times per million flights.

7.3.2   Take-Off and Landing Categories other than Runway Excursion

There are three Take-Off and Landing categories other than runway excursions:

1. Abnormal Runway Contact
2. Undershoot/overshoot
3. Loss of Control – ground

Each of these categories is explained through presenting an illustrative example of a true accident scenario. The first is a landing with *Abnormal Runway Contact* (Example 3). The second is an *Undershoot landing* (Example 4). The third is *Loss of Control* – ground, during a landing (Example 5).

The calculated occurrence rates for each of the three occurrence categories are given in Figure 7.6. It shows that the occurrence rates for 'Abnormal Runway Contact', for 'Undershoot/overshoot' and for 'Loss of Control – ground' do not show a clear positive or negative trend over the period 1990–2008. The total rate of Take-Off and Landing accidents other than runway excursion is as high as it is for runway excursions, that is, on average it happens around 0.6 times per million flights, and this rate is non-decreasing.

Example 3 Abnormal runway contact (AAIB, 2006)

On 29 November 2004 a DHC-8-311 Dash 8 and crew were flying from the Isle of Man to London City Airport (LCY). After an uneventful cruise, the first officer flew an approach to Runway 10 which has a landing distance available of 1319 m. The 5.5° glidepath was intercepted from an altitude of 2000 feet and manually flown with landing flap (15°) set. Although the first officer was the handling pilot for this sector, it is a company requirement for the commander to land the aircraft at LCY and handover of control was achieved at a height of 430 feet. On taking control, the commander progressively reduced engine power to achieve the correct speed before entering the flare. He maintained the 5.5° glidepath but on entering the flare reported heavy 'sink'. As the commander pulled back on the control column, the nose of the aircraft rose rapidly and the tail of the aircraft struck the runway before a firm landing was made.

The UK's Accident Investigation Bureau concluded that the commander reduced engine torque to an unusually low level in order to achieve the correct speed. This low power setting provided the aircraft with less energy than normal approaching the flare. Without significant power increase and with the onset of sink, the pitch angle had to be increased rapidly to reduce the rate of descent, resulting in the tail strike.

Example 4 Undershoot (NTSB, 2003)

On 26 July, 2002, FedEx flight 1478, a Boeing 727-232 F, struck trees on short final approach and crashed short of runway 09 at the Tallahassee Regional Airport, Tallahassee, Florida. The flight was operating a scheduled cargo flight from Memphis International Airport, in Memphis, Tennessee, to Tallahassee. The captain, first officer, and flight engineer were seriously injured, and the airplane was destroyed by impact and resulting fire.

The approach to runway 09 in Tallahassee was a visual approach. The runway was equipped with high intensity runway lights, in-pavement runway centreline lights, and runway end identifier lights. A four-box Precision Approach Path Indicator (PAPI) light system was located on the left side of runway 09 to provide lighted signal glidepath guidance relative to the published 3° glidepath to the runway's touchdown zone.

As the aircraft descended through 500 feet above ground level the approach was not stabilised because its rate of descent was greater than FedEx's recommended 1000 feet per minute, the engines' power settings were less than expected, and its glidepath was low as indicated by the PAPI light guidance. According to FedEx procedures at the time of the accident, if a visual approach was not stabilized when the airplane descended through 500 feet above ground level the pilots were to perform a go-around. However, the pilots continued the approach, remained significantly below the proper glidepath, collided with trees then impacted the ground, coming to about 1556 feet west-southwest of the runway.

The National Transportation Safety Board (NTSB) determined that the probable cause of the accident was the captain's and first officer's failure to establish and maintain a proper glidepath during the night visual approach to landing.

Example 5 Loss of Control – ground (DGCA, 2010)

On 10 November, 2009, Kingfisher Flight 4124, an ATR-72, skidded off runway 27A during landing at Mumbai Airport, India. The flight was operated as a scheduled service between Bhavnager, India and Mumbai. There were 38 passengers, including two infants and four crew members on board the aircraft.

Runway 27A was available for visual approach only and landing distance available was reduced to 1703 m due to maintenance activities. At the time of accident there were water patches on the runway. Prior to the Kingfisher aircraft, an Air India Airbus 319 had landed and reported to Air Traffic Control (ATC) that it had aquaplaned and damaged two runway edge lights. The ATC person was not familiar with the terminology of 'aquaplaning' and not realizing the seriousness of it, cleared the Kingfisher aircraft for landing and did not mention that the previous aircraft had aquaplaned. The Kingfisher aircraft's approach was high and fast. The aircraft landed late on the runway with a remaining runway length of around 1000 m. During landing the Kingfisher aircraft aquaplaned and did not decelerate even though reversers and full manual braking was applied by the cockpit crew. The aircraft skidded off the runway finally came to a stop. All the passengers safely deplaned after the accident.

According to (DGCA, 2010) the accident occurred due to an unstabilized approach and decision of crew not to carry out a go-around. Contributing factors were water patches on runway 27A, the failure of ATC to communicate about aquaplaning of the previous aircraft, and the lack of input from the co-pilot to abort the marginal approach and make a go-around.

7.3.3   Ground Operation Categories

There are six categories where ground operations play a key role. These are:

1. Ground handling
2. Ground collision
3. Fire/smoke (post impact)
4. Evacuation
5. Runway Incursion – vehicle, aircraft or person
6. Runway incursion – animal

Of these six, the Runway Incursion – vehicle, aircraft, or person, typically receives a lot of attention because the consequences of an accident event in this category can be very large. This is illustrated through Example 6. It should be also taken into account that according to the CICTT definitions, the Ground Collision category covers collisions which happen during taxiing to or from a runway in use. It is explicitly stated that events categorized under Runway Incursion or Ground Handling are excluded from the Ground Collision category (CAST/ICAO, 2011). Hence, Example 6 is not categorized as a Ground Collision.

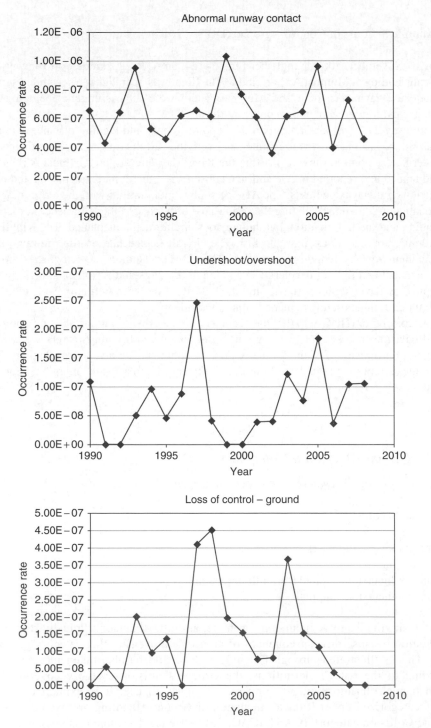

Figure 7.6   Calculated occurrence rates for abnormal runway contact, undershoot/overshoot and loss of control – ground

Example 6   Runway incursion – aircraft (ANSV, 2004), (Nativi and Learmount, 2001)

On 8 October 2001, a Boeing MD-87, operated by SAS, while on take-off from runway 36R of Milano Linate airport, collided with a Cessna 525-A, registration D-IEVX, which taxied into the active runway. The Cessna had been cleared to taxi from the general aviation apron to the runway along the north side R5 taxiway but it instead took the R6 taxiway to the south-east. It then called for clearance to proceed from the R5 hold while it was in fact at the R6 stop point. ATC cleared the aircraft to continue within seconds of clearing the MD-87 to take-off. After the collision the MD-87 continued travelling down the runway, the aircraft was airborne for a short while and came to a stop impacting a baggage handling building. The Cessna remained on the runway and was destroyed by post-impact fire. All occupants of the two aircraft and four ground staff working inside the building suffered fatal injuries. Four more ground staff suffered injuries and burns.

Italy's accident investigation board concluded that the immediate cause of the accident was the runway incursion on the active runway by the Cessna. Immediate and systemic causes that led to the accident include the low visibility, the absence of proper airport lights, markings, signs and publications, and radio communications which were performed in the Italian and English language and were not performed using standard phraseology.

The occurrence rates for ground handling and ground collisions are presented in Figure 7.7. The Ground Handling curve covers accident occurrences during or as a result of ground handling operations such as servicing, boarding, loading and deplaning. The Ground Collision curve covers accident occurrences while taxiing to or from a runway in use. The curves in Figure 7.7 show that these occurrence rates for ground handling and ground collision do not show a clear positive or negative trend over the period 1990–2008. Moreover, the combined rate of ground handling and ground collision is rather high compared to the overall accident rate, that is, on average it happens around 0.8 times per million flights.

The occurrence rates for fire/smoke (post impact) and evacuation are shown in Figure 7.8. These curves show that the occurrence rates for fire/smoke (post impact) and for evacuation do not show a clear positive or negative trend over the period 1990–2008. Moreover, the combined rate of fire/smoke (post impact) and evacuation is rather high compared to the overall accident rate, that is, on average it happens around 0.4 times per million flights.

The occurrence rates for the two runway incursion categories are shown in Figure 7.9. These curves show that the occurrence rates for runway incursion do not show a clear positive or negative trend over the recent period. Fortunately, the combined rate of the two runway incursion categories is quite low compared to the overall accident rate, that is, on average it happens around 0.05 times per million flights.

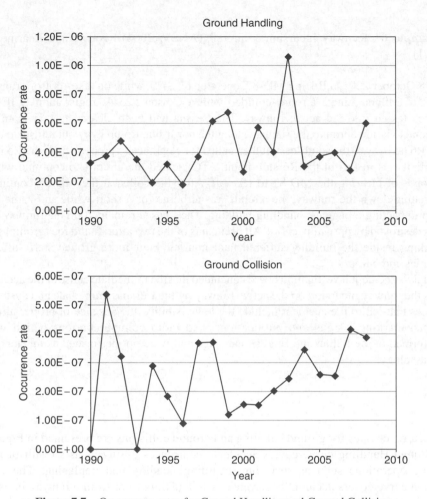

Figure 7.7   Occurrence rates for Ground Handling and Ground Collision

7.3.4   Summary of Take-Off, Landing and Ground Operation Analysis

Subsection 7.3.1 has shown that the calculated occurrence rates for runway excursions do not show a clear positive or negative trend over the period 1990–2008. Moreover, the occurrence rate is at a rather high level compared to the average overall accident rate, that is, it happens around 0.6 times per million flights.

Subsection 7.3.2 has shown that the calculated occurrence rates for 'Abnormal Runway Contact' for 'undershoot/overshoot' and for 'Loss of Control – ground' do not show a clear positive or negative trend over the period 1990–2008. The total rate of Landing accidents other than runway excursion is as high as it is for runway excursions, that is, it happens around 0.6 times per million flights.

Subsection 7.3.3 has shown that the rates for all Ground Operation occurrences do not show a clear positive or negative trend. Moreover, the total rate of these occurrences is rather high

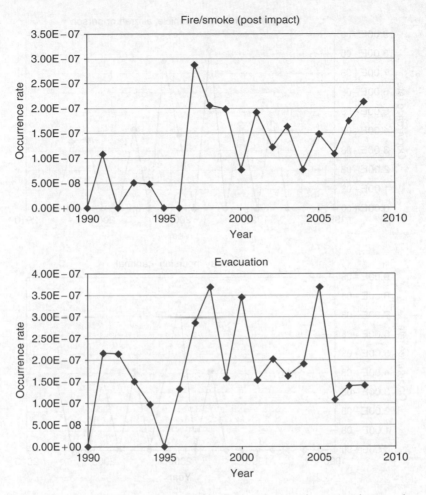

Figure 7.8   Calculated occurrence rates for fire/smoke (post impact) and evacuation

compared to the average overall accident rate, that is, overall it happens around once per million flights.

In conclusion, over the period 1990–2008, the occurrence rate for categories in the group Take-Off, Landing and Ground Operation does not exhibit a tendency to decrease. Moreover, these non-decreasing rates apply to all occurrence categories that are associated with the airport, that is:

- Runway Excursion,
- Abnormal Runway Contact, Undershoot/overshoot, Loss of Control – ground,
- Ground Handling, Ground Collision, Fire/smoke (post impact), Evacuation, Runway Incursion.

Because all of them seem to lack a significant downward trend, this could imply that, without an effective change in the trend, the number of airport related Take-Off, Landing and Ground Operations accidents is expected to continue to grow with worldwide traffic growth.

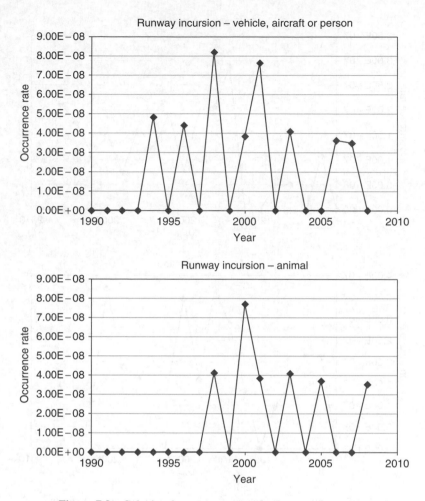

Figure 7.9   Calculated occurrence rates for Runway Incursions

7.4   Analysis of Other CICTT Categories

This section analyses rates and nature of accidents for the categories in the five groups other than the Take-off, Landing and Ground Operation group. First, in subsection 7.4.1, the occurrence rates in each of the other five category groupings are evaluated. Next, in subsection 7.4.2, the categories in the *Airborne* grouping are analysed. Then in subsections 7.4.3–7.4.6 the main categories in the other four groups are analysed. Finally, subsection 7.4.7 summarizes the main findings.

7.4.1   Occurrence Rate per Category Grouping

To further investigate the trend of other main accidents in aviation, separate occurrence rates are calculated for each of the five other groups of CICTT categories, where the associations of accident scenarios to categories come directly from the ADREP database. The majority of

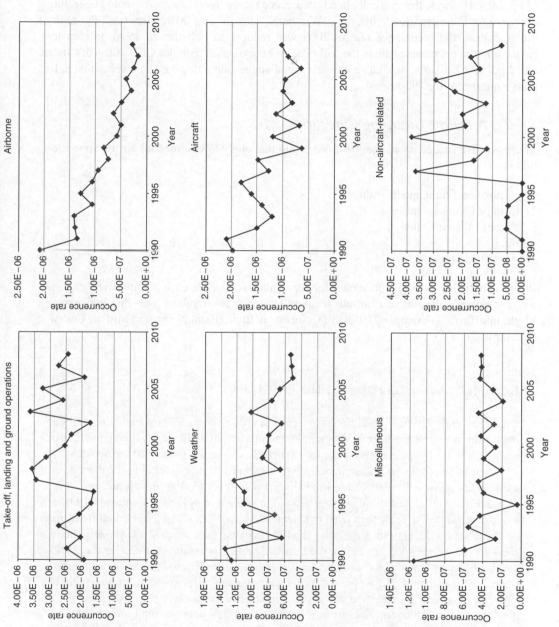

Figure 7.10 Calculated occurrence rates for CICTT category groups; scheduled commercial air transport

accident scenarios are associated to one accident category only, but if an accident scenario is associated to multiple categories, all categories were included in the analysis as separate counts. The results are presented in Figure 7.10.

Figure 7.10 shows that the calculated occurrence rates exhibit a negative trend suggesting a decrease for the categories 'Airborne', 'Weather', 'Aircraft' and 'Miscellaneous'. For each of those groups, the occurrence rate in 2008 was reduced to half that of 1990 or even less. Especially the occurrence rate in the Airborne category, which includes CFIT, Loss of Control and Mid-air Collisions, exhibits a negative trend suggesting a decrease to a value that is less than a quarter of the 1990 rate.

7.4.2   Airborne Grouping Categories

There are five categories in the *Airborne* group that apply to commercial air transportation[2], that is:

1. Controlled Flight into Terrain
2. Loss of Control – Inflight
3. (Near) Mid-air Collision
4. Fuel
5. Abrupt Manoeuvre

Prior to analysing the occurrence rates for accidents associated to each of these five categories, we first provide illustrative example accident scenarios for numbers 1–3, that is Controlled Flight into Terrain (Example 7), Loss of Control – in flight (Example 8), and Mid-air Collision (Example 9).

Example 7   Controlled Flight into Terrain (FSF, 1998)

On 20 December, 1995, American Airlines Flight 965, a Boeing 757 on a scheduled flight from Miami to Cali, Columbia, struck mountainous terrain while descending from cruise altitude. Of the 163 passengers and crew on board, 4 passengers survived the accident.

When the aircraft was approximately 60 nm from Cali and descending the crew contacted Cali Approach. Cali cleared the flight for a direct Cali approach and report at Tulua. Followed 1 min later by a clearance for a straight in approach to runway 19 (the Rozo 1 arrival). The crew then tried to select the Rozo NDB (Non Directional Beacon) on the Flight Management Computer (FMC). Because their Jeppesen approach plates showed 'R' as the code for Rozo, the crew selected this option. But 'R' in the FMC database meant Romeo. Romeo is a navaid 150 nm from Rozo, but has the same frequency. The aircraft had just passed Tulua when the autopilot started a turn to the left (towards Romeo). This turn caused some confusion in the cockpit since Rozo 1 was to be a straight in approach. The left turn brought the B757 over mountainous terrain, so a Ground Proximity warning sounded. With increased engine power and nose-up the crew tried to climb. The spoilers were still activated however so the aircraft did not achieve a sufficient rate of climb and crashed into a mountain at about 8900 feet.

Example 8   Loss of Control – Inflight (NTSB, 2010)

On 12 February, 2009, a Colgan Air, Inc., Bombardier DHC-8-400, operating as Continental Connection flight 3407, was on an instrument approach to Buffalo-Niagara International Airport, Buffalo, New York, when it crashed into a residence about 5 nautical miles (NM) northeast of the airport. The two pilots, two flight attendants, and 45 passengers aboard the airplane were killed, one person on the ground was killed, and the airplane was destroyed by impact forces and a post-crash fire.

The flight had been normal until the last part of the approach as the aircraft reached 2300 feet altitude at a speed of about 180 knots. The captain then began to slow the airplane less than 3 miles from the outer marker to establish the appropriate airspeed before landing while maintaining the same altitude. The engine power levers were reduced, the landing gear was deployed, and flaps were selected to 10°. The airspeed dropped to about 131 knots and the stall warning system (the stick shaker) activated. The pilots increased engine power but moved the control column aft instead of forward. This caused the aircraft to stall, it rolled left and right, pitched down and hit the ground.

NTSB (2010) concludes that the probable cause of this accident was the captain's inappropriate response to the activation of the stick shaker, which led to an aerodynamic stall from which the airplane did not recover.

Example 9   Mid-air Collision (Aviation Safety Network, 2011)

On 29 September, 2006, a mid-air collision occurred over the Brazilian Amazon jungle, between a Boeing 737–800 operated by Gol Airlines of Brazil, and an Embraer Legacy 600 business jet owned and operated by Excelaire of Long Island, New York. The Embraer Legacy pilots experienced control difficulties after the collision but were able to land the aircraft. The 737 descended out of control after losing the outer portion of its left wing and was destroyed by the in-flight breakup and impact forces; all 154 occupants were fatally injured.

The accident investigation concluded the following on the cause of the accident:

The evidence collected during this investigation strongly supports the conclusion that this accident was caused by ATC clearances given to and followed by the two aircraft. This directed them to operate in opposite directions on the same airway at the same altitude resulting in a mid-air collision. The loss of effective air traffic control was not the result of a single error, but of a combination of numerous individual and institutional ATC factors, which reflected systemic shortcomings in emphasis on positive air traffic control concepts. Contributing to this accident was the undetected loss of functionality of the airborne collision avoidance system technology as a result of the inadvertent inactivation of the transponder on board the Embraer Legacy. Further contributing to the accident was inadequate communication between ATC and the Embraer Legacy flight crew.

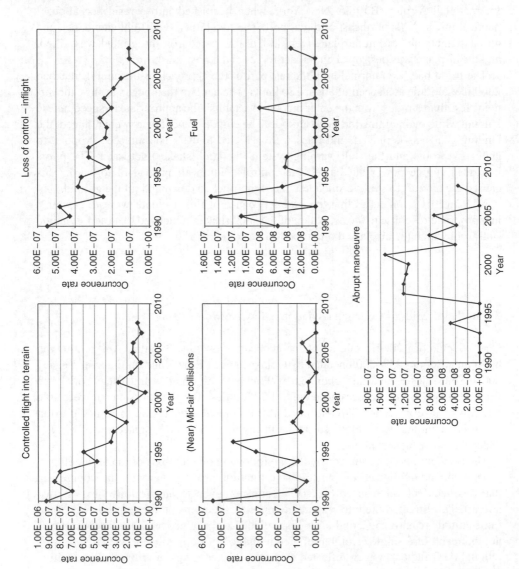

Figure 7.11 Calculated occurrence rates for the categories of the 'Airborne' group

Figure 7.11 presents the occurrence rates for accidents associated to the five accident categories in the group Airborne. The curves in Figure 7.11 show that with the exception of the 'Abrupt Manoeuvre' category, the calculated airborne occurrence rates exhibit a negative trend suggesting a decrease over the period 1990–2008. Thanks to this, the overall occurrence rate in this group has gone down from 2 per million flights to approximately 0.4 per million flights (which is about a factor of 5).

7.4.3   Categories in the Weather Group

This subsection analyses the occurrence rates for the three categories in the group Weather, that is:

1. Windshear or Thunderstorm
2. Turbulence Encounter
3. Icing

Prior to analysing the occurrence rates for these three categories, we first provide an illustrative example accident scenario for (2), that is, *Turbulence Encounter* (Example 10).

Figure 7.12 presents the occurrence rates for the three categories in the group *Weather*. The curves show that for the group Weather a factor two reduction in occurrence rate was realized over the years 1990–2008, and that this reduction is largely due to reductions for turbulence encounter accidents. Thanks to this, the overall accident rate in this group has gone down from about 1.4 per million flights to about 0.6 per million flights.

7.4.4   Categories in the Aircraft Group

Figure 7.13 presents the calculated occurrence rates for the three categories in the group *Aircraft*, that is:

1. Failure powerplant
2. Failure non-powerplant
3. Fire/smoke (non-impact)

Example 10   Turbulence Encounter (NTSB, 2011)

On 6 January, 1995, a McDonnell Douglas MD-88, registered to Wilmington Trust Company, operated by Delta Air Lines Inc., as a scheduled, domestic, passenger flight, encountered turbulence on climbout about 30 miles east of Monroe, Louisiana. The airplane was not damaged. One flight attendant sustained a serious injury but the other occupants were not injured. The flight originated from Monroe, Louisiana, about 4 min before the accident, and diverted to Jackson, Mississippi.

The pilot-in-command stated the airplane encountered light to moderate turbulence on climb-out. The cabin seatbelt signs were illuminated and the flight attendants were briefed to remain seated. The airplane weather avoidance radar was on and no significant weather was present on the radar screen. An overhead compartment had opened up in the cabin area, and a flight attendant got up to secure the overhead compartment. The airplane encountered a couple of seconds of intense moderate turbulence, and the flight attendant collided with the cabin overhead panels.

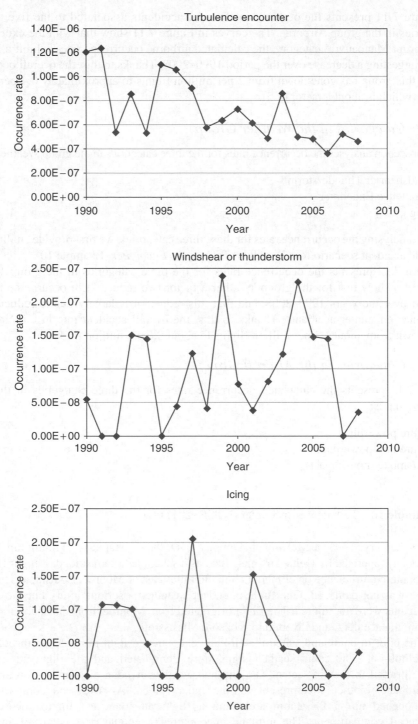

Figure 7.12   Calculated occurrence rates for the categories of the 'Weather' group

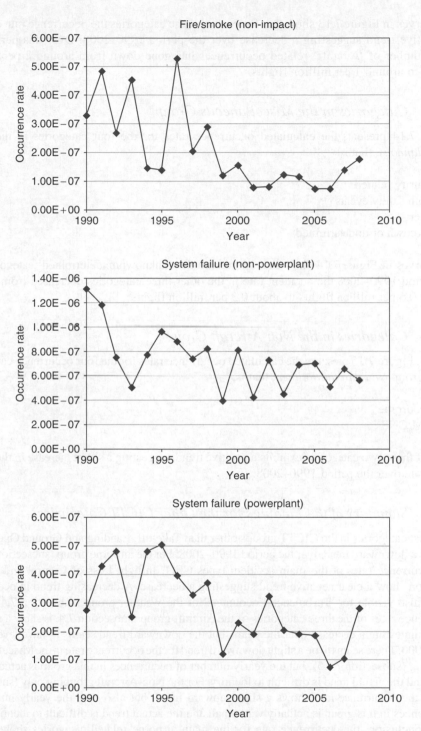

Figure 7.13 Calculated occurrence rates for the categories in the 'Aircraft' group

The curves in Figure 7.13 show that for each of these categories the occurrence rate exhibits a negative trend suggesting a decrease over the period 1990–2008. Over this period, the total number of 'Aircraft' related occurrences has gone down from around 2 per million flights to around 1 per million flights.

7.4.5   Categories in the Miscellaneous Group

Figure 7.14 presents the calculated occurrence rates for the four categories in the group *Miscellaneous*, that is:

1. Security related
2. Cabin safety events
3. Other
4. Unknown or undetermined

The curves in Figure 7.14 show that apart from the 'unknown/undetermined' category, over the period 1990–2008 the accident rate in the other three categories decreased from in total around 0.6 per million flights to about 0.3 per million flights.

7.4.6   Categories in the Non-Aircraft Group

Finally, Figure 7.15 presents the calculated occurrence rates for the four occurrence categories in the group *Non-Aircraft Related*, that is:

1. Aerodrome
2. ATM/CNS

Each of these two categories exhibits a positive trend suggesting a slight increase in the occurrence rate over the period 1990–2008.

7.4.7   Summary of the Findings for the Other CICTT Categories

For most categories in the CICTT groups other than Take-off, Landing and Ground Operations, there is a downward trend over the period 1990–2008. For the airborne group (subsection 7.4.2), the occurrence rates of the main accident types CFIT, Inflight Loss of Control and Mid-air Collision show a clear negative trend suggesting a decrease. A decreasing trend in occurrence rate is also visible for 'Turbulence Encounters' of the Weather group (subsection 7.4.3). The occurrence rates of the three categories of the Aircraft group (subsection 7.4.4) show a negative trend suggesting a decrease over the years, but the downward trend seems to have levelled off since 2000. There seems to be a slight downward trend for the occurrence rate in the Miscellaneous grouping (subsection 7.4.5), but the yearly number of occurrences in this group is actually very small and the actual trend is difficult to identify. For the Non-Aircraft grouping only (subsection 7.4.6), the occurrence rate shows a slight upward trend, but also here the yearly number of occurrences in this group is actually very small and the actual trend is difficult to identify.

In conclusion, the occurrence rate for the main airborne related categories show a clear negative trend suggesting a decrease over the period 1990–2008. Moreover, the improvement

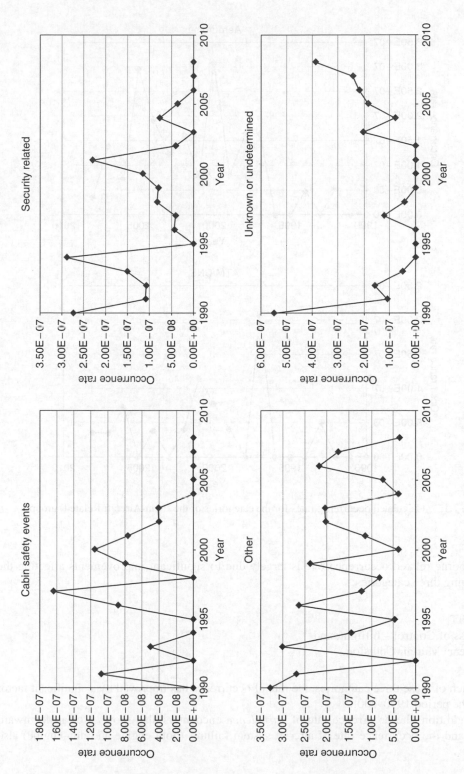

Figure 7.14 Calculated occurrence rates for the categories of the 'Miscellaneous' group

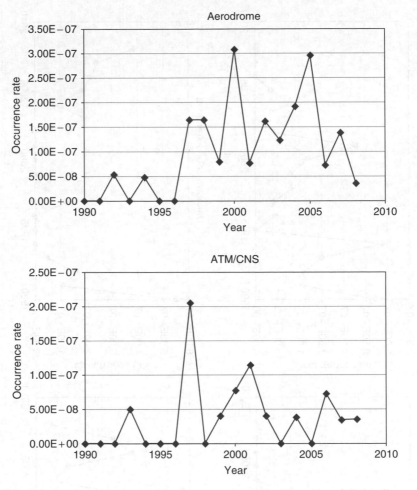

Figure 7.15   Calculated occurrence rates for the categories of the 'Non-Aircraft Related' group

of airborne related occurrence rate is largely due to significant improvements affecting the following three categories:

1. CFIT;
2. Loss of Control – Inflight; and
3. (Near) Mid-air Collision.

For each of these three categories, the rate of occurrence has improved by a significant factor over the period 1990–2008.

In addition, the occurrence rate of *Turbulence* encounters also shows a clear downward trend and the occurrence rate of aircraft system failures (both engine and non-engine) also exhibits a negative trend suggesting a decrease, although there seems to be a levelling off in the improvement trend since 2000.

In contrast with this, there also are five other CICTT categories for which the occurrence rate does not show a positive or negative trend, that is:

1. Abrupt Manoeuvre
2. Windshear or thunderstorm
3. Icing
4. Aerodrome
5. ATM/CNS

Fortunately, the occurrence rates for these categories are significantly lower than those for the five categories mentioned here. Hence, taking all other CICTT categories together, the occurrence rate exhibits a clear negative trend suggesting a decrease over the period 1990–2008.

This finding of significantly reduced occurrence rates in the other CICTT categories is in sharp contrast with the Section 7.3 finding of no improvement in the occurrence rates for the main Take-off, Landing and Ground operations categories. Moreover, the total of take-off, landing and ground operations accidents seems to have taken over the total number of airborne related accidents. Together with a non-decreasing occurrence rates for *Abrupt Manoeuvre*, *Windshear or thunderstorm*, *Icing*, *Aerodrome* and *ATM/CNS*, this means that the total number of aviation accidents might even start to grow with a worldwide growth in commercial aviation.

7.5   Safety Driving Mechanisms

In order to gain a better understanding of the mechanisms available to the aviation sector regarding a further improvement of commercial aviation safety, in this section we review the following safety driving mechanisms:

- Technological developments (subsection 7.5.1)
- Regulation (subsection 7.5.2)
- Competition and reputation (subsection 7.5.3)
- Professionalism and safety culture (subsection 7.5.4).

7.5.1   Technological Developments

Aviation only became possible when a certain level of technological development had been reached. Enabling commercial aviation as a means of mass transportation required further technological development and breakthroughs, such as stressed skin structures, pressure cabins, radio navigation, super critical wings, and so on. Aeronautical research laboratories were established to facilitate the research required for technological advancement (van der Bliek, 1994; Bilstein, 2003). Technological advances that contributed significantly to aviation safety include the following:

- Jet engines, which have a lower failure rate than piston engines due to their inherently simpler design with less accelerating parts.
- Precision navigation systems have improved the flight crew's ability to navigate and have virtually eliminated occurrences where aircraft get lost and run out of fuel before finding a suitable landing place and they help to keep the aircraft on the correct approach path to the landing runway, even in conditions of reduced visibility (Leary, 1992; Abbink, 1996; Enders et al., 1996).
- Fly-by-wire systems and glass cockpits have reduced pilot workload (Daily, 1997; Davis, 1997; Sparaco, 1998).
- Airborne Collision Avoidance Systems (ACAS) has provided a large reduction of mid-air collision risk (NTSB, 1998).
- Terrain Avoidance Warning Systems (TAWS) have dramatically reduced the frequency of controlled flight into terrain accidents (Burin, 2003). The original TAWS concept relied on information from the air data system and radio altimeter. Second generation TAWS use data from satellite systems and databases containing information on a digital elevation model and an aeronautical database, providing greater situational awareness for flight crews. Especially, since the introduction of the second generation TAWS, controlled flight into terrain accidents have been virtually eliminated for those aircraft that are equipped.

Many of these technological developments emerged from continuous efforts to improve aircraft operational efficiency and reduce airline operational costs. ACAS and TAWS are strictly safety related, but they could only be developed because of technological advances such as the development of radar, digital computers and Global Positioning System (GPS).

Aeronautical research and development has been predominantly focussing on the aircraft itself and much less to the supporting infrastructure that is required for aviation. Infrastructure had, with the exception of radar and radio navigation, hardly been a technological problem. Research on Air Traffic Control (ATC) and airport infra-structure has become more important. For instance, air and airport traffic management research simulators have been put in use since the end of the twentieth century, whereas research flight simulators have started to be used half a century earlier.

Examples of safety improvement technologies that have been introduced in ATC are:

- Advanced multi-sensor multi-target surveillance systems have dramatically improved the reliability and decreased the uncertainties in position and velocity information about aircraft under control of ATC (Rekkas and Blom, 1997; Hogendoorn et al., 1999).
- Short Term Conflict Alert (STCA); effectively alerts a controller in case of a predicted infringement of airborne minimum separation criteria within the next 3 min (Eurocontrol, 2009).
- Medium Term Conflict Detection (MTCD) supports the planning controller in a better organization of air traffic plans to be implemented by the tactical controller.
- An early version of Advanced Surface Movement Guidance and Control System (A-SMGCS)[3], which for example provides a runway incursion alert (RIA) to the runway controller in case of a runway incursion event, but not to aircraft crew.

Because the introduction of STCA has been realized under the explicit requirement that it should improve air traffic safety, it has clearly contributed to the reduction of mid-air collision risk. However, the introduction of the early version of A-SMGCS has been done with a two sided objective: to improve safety of accepted operations, and to improve the efficiency of airport operations, for example, by adopting an operation that was considered too unsafe without early A-SMGCS based decision support for the controller. Because of such a double objective, introduction of early A-SMGCS need not have contributed to an overall reduction of runway incursion risks.

For example, with support of early A-SMGCS, crossing of an active runway has been allowed at some major airports in USA and Europe. Often these decisions have been based upon event sequence based safety analysis conducted in the USA (FAA, 2005) and independently also in Europe (Eurocontrol, 2006). Both event sequence based safety analysis studies show that an A-SMGCS based decision support to the controller makes runway crossing safer. However, neither of these two studies addresses the extra risk of a lost aircraft that unintentionally tries to cross an active runway. By using agent based modelling and Monte Carlo simulation, Stroeve et al. (2008; 2011) have shown that these extra risks are significant. The reason is that when a runway crossing happens unintentionally, then a taxiing aircraft will not make a stop prior to starting the crossing, by which the mean taxiing speed is much higher than it is for an intended runway crossing. Hence, the time period between the moment of a runway incursion alert and the latest moment that the controller can effectively instruct a pilot is much shorter (or even negative) in case of an unintended crossing.

7.5.2 Regulation

ICAO is the body that determines the regulatory and operating regimes for international aviation. The standards and recommended practices are spelled out in detail. Neither the birth of ICAO, nor the subsequent standards defined by ICAO, meant that national aviation regulation became superfluous. Many aspects of aviation were (and still are) subject to national requirements, most notable the US Federal Aviation Regulations (FAR) and the European Joint Airworthiness Requirements (JAR) and subsequent European Aviation Safety Agency (EASA) Regulations.

Historically, national regulations on aircraft design, operations and maintenance are specified in much more detail than the corresponding ICAO standards, while for air traffic management, air navigation services and airports the opposite is true. In the early years of commercial aviation there were no air traffic controllers; pilot had to maintain separation with other aircraft by applying see-and-avoid themselves. With the increase of commercial air transport this slowly resulted in mid-air collisions, whereupon in an increasing number of countries, national Civil Aviation Authorities (CAAs) started to extend their Air Navigation Service Provision (ANSP) sections with air traffic controllers helping pilots to maintain safe separation. The role of ICAO was to harmonize ways of working by ANSPs and to set standards for equipment. Because these ANSP sections made part of the CAAs, they also became responsible for the regulation and oversight of ATC related safety issues. Hence, when there was an ATC related safety problem, then the ANSP section both had to identify and resolve the problem, without the presence of an independent national watchdog. Only during the 1990s, developments started to introduce novel working structures for this. One is the introduction of safety management system principles, and the second is to decouple the regulation, the oversight and the service provision from each other. Around 2010 these novel structures are almost working throughout Europe and some countries outside Europe.

Similar problems also apply regarding the relations between airport operators and regulation/oversight. Although some ground handling processes are indirectly regulated by legislation imposed on the aircraft or aerodrome operator, most ground handling processes are not regulated. For example, regulations requiring a service providing organisation to work according to an operating quality system or safety management system do apply to airlines and ANSPs, but not to ground service providers (Balk and Roelen, 2010).

7.5.3   Competition, Reputation and Balancing Objectives

The domains in which national regulations dominate are also those where private companies operate and competition is a prominent factor. In most parts of the world, aircraft manufactures, airlines, repair stations and flight schools are private organizations that are in mutual competition for the favour of customers. This competition is a driver for product improvement on aspects such as punctuality, efficiency, price, noise levels, and so on, and indirectly also safety which can strongly influence a company's reputation. Prevention of reputational damage is a significant driver for safety improvement at airlines and aircraft and system manufacturers. Reputational damage has a direct impact on a company's financial balance sheet and therefore is something that management wants to avoid. Accidents can significantly damage an airline's reputation and the effect of an accident to the airline's stock price and ticket sales is uninsured. The image that certain airlines are 'unsafe' is enhanced by the European Union (EU) 'Blacklist' of airlines (EU, 2005) that are banned from the European Airspace because the EU considers them insufficiently safe. Similarly, accidents can dent the reputation of a certain aircraft type. An example is the Douglas DC-10 that suffered from several highly visible accidents in the first few years of its operational service. The share price of McDonnell-Douglas fell significantly after these accidents (Chalk, 1986) and several airlines cancelled orders, although it cannot be proved that this was directly caused by the accidents. Passengers even avoided the aircraft, although the impact on consumer behaviour recovered very quickly (Barnett and LoFaso, 1983; Barnett et al., 1992).

Reputational damage is much less of a problem for ANSPs and airports. The vast majority of airline passengers are not aware of the ANSPs their flight will be dealing with, and even if they are, they do not have much choice. Passengers also do not select destination or transfer airports because of a safety reputation. A destination airport is selected because of the location, people want or need to be there and that defines the airport they will land on. In some areas there is competition between airports for departure (e.g. between Amsterdam, Brussels and Dusseldorf) but that is purely based on pricing and convenience, safety does not play any role. The EU does not put out a 'blacklist' of ANSPs, countries or destinations, even though in 2011 for several Member States all airlines were banned 'due to the safety deficiencies identified in its system to oversee civil aviation' (EU, 2011). Member States whose airlines are on the blacklist are therefore encouraged to improve oversight of the airlines but they are not directly encouraged to improve oversight of the ANSPs and airports.

There also is another key mechanism that typically plays a role when it comes to the resolution of safety critical problems identified at an airport. In general once such problems are identified, exist a silver bullet does not to resolve the problem right away. This means that a decision has to be made whether the specific operation that carries the high risk is put on halt until a properly improved operation has been developed and validated on safety, or that the specific operation may continue to be used until the improved operation has been implemented on the airport. A well-documented example of such a scenario is described in (Scholte et al., 2008). This example concerns the development of safe taxi routes to and from a novel constructed runway at Amsterdam airport. The initial plan was that taxiing aircraft would cross an active runway. Eventually, new taxi routes have been constructed that avoid active runway crossings, and with a few years delay the use of the new runway could start.

There also may be tension between improving safety and/or environment. For example, under adverse wind conditions, there typically is one runway upon which an aircraft can land most safely, and there is another runway upon which an aircraft can land most environmentally friendly (of course this should fall within the internationally agreed safety margins posed on wind conditions during landing). Of course the flight crew are strongly in favour to go for the safety preferred runway. However, local environmental rules may encourage the ANSP to allocate the environmentally preferred runway.

Recently (Daams, 2011) completed an in-depth policy oriented study regarding the development of solutions for five demanding aviation problems around Amsterdam airport. Each problem was involved finding proper balance between optimizing for safety, environment and efficiency. The study shows that the search for finding a properly balanced solution creates large tensions between various stakeholders involved, such as the ANSP, the airport, airline users of the airport, people living around the airport and Dutch government. These large tensions complicate the development of adequate solutions. In order to avoid the risk that safety benefits are not sacrificed in the tension between multi-stakeholder, (Everdij et al., 2009) has proposed a multi-actor safety validation framework.

7.5.4 Professionalism and Safety Culture

In comparison with other high-risk industries, the rate of catastrophic accidents per exposure in civil aviation ranks amongst the lowest, outperforming, for example, the railways, chemical industry and health care. According to Hudson (2003) and Amalberti et al. (2005) the most important difference between aviation and other industries lies not so much in the tools they use, but in safety culture differences.

Safety culture covers aspects such as communication, organizational learning, senior management commitment to safety, and a working environment that rewards the identification of safety issues (Sorenson, 2002). A 'good' safety culture is facilitated within the aviation industry by training such as crew resource management training and by non-punitive incident reporting that afford protection to the sources of the information. Crew resource management (CRM) with particular emphasis on the merits of participative management was first recommended by the NTSB following an accident in 1978 (NTSB, 1979) and since 1992 there is a requirement for CRM to be included in flight crew training. Non-punitive confidential incident reporting started in earnest in 1976 with the NASA-operated Aviation Safety Reporting System (ASRS) in the USA. Similar initiatives in other parts of the world soon followed. Airlines started to develop their own internal reporting systems for flight crew and in several countries the aviation authorities mandated occurrence reporting. In Europe any operational interruption, defect, fault or any other irregular circumstance that has or may have influenced flight safety and has not resulted in an accident or serious incident has to be reported (EU, 2003).

Recently, both in Europe and in the USA activities have started on improving safety culture (FAA/Eurocontrol, 2008), (CANSO, 2010). This means that although safety culture in aviation often is at a high level, there are areas for improvement. Corresponding safety culture assessments concluded that the safety culture of aviation ground handling companies was generally lower than for other disciplines, suggesting a 'poorer' safety culture (Ek and Akselsson, 2007).

Continuing emphasis of ICAO, FAA and EASA on the introduction of safety management systems (SMS) by aviation service providers also supports a strong safety culture. A positive attitude towards safety culture is a means to improve an effective implementation of an SMS, while clear SMS processes, procedures, documentation and communication are means to support a positive attitude towards safety and thus to improve safety culture.

7.6   Safety Initiatives

The professionalism and dedication to safety of the aviation community is also demonstrated by a number of successful international safety initiatives in which aviation authorities and industry voluntary participate with the sole purpose of improving aviation safety. Some of the most important initiatives are briefly described next.

7.6.1   Initiatives of the Flight Safety Foundation

Flight Safety Foundation (FSF) is an independent, impartial and non-profit international membership organization that was formed in 1947 to pursue the continuous improvement of global aviation safety. In recent years, FSF has been leading numerous flight safety improvement initiatives amongst which the following are considered to be most important:

- CFIT Task Force
- Approach and Landing Accident Reduction (ALAR) Task Force
- Runway Safety Initiative (RSI)

7.6.1.1   CFIT Task Force

CFIT has long been the world's leading cause of commercial aviation fatalities. FSF first helped bring the issue clarity and resources in the early 1990s, when it was the accident type that killed more people than any other in the industry. Essentially, CFIT occurs when an airworthy aircraft under the control of the flight crew is flown unintentionally into terrain, obstacles or water, usually with no prior awareness by the crew.

The FSF-led international CFIT Task Force, created in 1992, set as its five-year goal a 50% reduction in CFIT accidents. The task force included more than 150 representatives from airlines, equipment- and aircraft manufacturers and many other technical, research and professional organizations. The task force believed that education and training are readily available tools to help prevent CFIT accidents. Its main products were training videos and a 'CFIT checklist' which helps pilots and aircraft operators to assess the CFIT risk for specific flights. More than 30000 copies of the checklist have been distributed worldwide.

7.6.1.2   ALAR Task Force

The FSF Approach and Landing Accident Reduction (ALAR) Task Force was created in 1996 as a next phase of CFIT accident reduction launched in the early 1990s. Participants of the

Task Force include representatives from airlines, aircraft manufacturers, system manufacturers, air traffic control organisations, pilots associations and academia, mainly from North America and Europe. The task force presented its final working group reports in November 1998. The report (FSF, 1999) cited data showing that an average of 17 fatal approach and landing accidents had occurred each year from 1980 through 1998 in passenger and cargo operations involving aircraft weighing 5700 kg/12 500 lb or more.

Since the ALAR campaign began, members of the ALAR Task Force have conducted numerous ALAR workshops around the world, and the FSF has distributed more than 40000 copies of the FSF ALAR Tool Kit – a unique set of pilot briefing notes, videos, presentations, risk-awareness checklists and other products designed to prevent approach and landing accidents.

A major update of the FSF ALAR Tool Kit – featuring the findings of analyses of recent accident data, as well as the data-driven findings of the FSF Runway Safety Initiative – was issued in 2010. It is the FSF's intention to periodically update the ALAR Tool Kit to include new information aimed at reducing the risk of approach and landing accidents.

7.6.1.3   FSF Runway Safety Initiative (RSI)

In 2006 the Flight Safety Foundation initiated the RSI to address the challenge of runway safety. This was an international effort with participants from across the aviation community, including the major aircraft manufacturers, FAA, EASA, the International Air Transport Association (IATA), the federation of airline pilots' associations, the Federation of Air Traffic Controllers' Associations and Airports Council International. The RSI group initially reviewed three areas of runway safety: runway incursions, runway confusion, and runway excursions. The RSI group then decided that it would be most effective to focus its efforts on reducing the risk of runway excursions. A runway excursion is defined as an occurrence where an aircraft on the runway surface departs the end or the side of the runway surface and it can occur either on take-off or landing. As part of the RSI a study was conducted of runway excursion accidents from 1995 through March 2008 to investigate the causes of runway excursion accidents and to identify the high risk areas. Based on the analysis the FSF published a list of recommended mitigations. The prevention strategies embrace five areas: flight operations, air traffic management, airport operators, aircraft manufacturers and regulators (Darby, 2009).

FSF received a 1994 Aerospace Laurel from *Aviation Week & Space Technology* magazine and the 1998 Aerospace Industry Award for Training and Safety, from *Flight International* magazine. In addition, for its leadership of the ALAR Task Force, FSF also earned Flight International's 2000 Aerospace Industry Award for Safety.

7.6.2   *Commercial Aviation Safety Team (CAST)*

In February 1997, a USA Commission chaired by Vice President Al Gore (Gore, 1997) has set a national aviation safety goal to reduce fatal accident rates by a factor of five in 10 years. The large transport airplane industry took on the challenging goal of the Gore Commission by forming the Commercial Aviation Safety Team (CAST) which analysed the leading safety issues in the large transport commercial aviation sector. Participants of CAST include the aviation authorities and major aircraft manufacturers from USA and Europe, engine manufacturers, airline associations and pilots and flight attendants organisations.

In CAST, subject matter experts have analysed specific accident categories such as CFIT, Approach and Landing, Loss of Control, Runway Incursion and Turbulence. The analysis process starts with collecting data on accidents and incidents (of the relevant category) that have happened in the past. Each accident or incident is analysed by the team and an event sequence description is generated. After the team has analysed the event sequence for each accident, problem statements and any contributing factors are drafted for the appropriate events. Problem statements are defined as statements that describe what went wrong and why it went wrong, they define an overall deficiency, or describe a potential reason something did or did not occur. Contributing factors are defined as factors both in the crew's environment and personal factors that help explain why a problem occurred. The next step is to define strategies to prevent or mitigate a given problem or contributing factor. These 'interventions' are evaluated for effectiveness and subsequently plans for implementation are developed. In 2007 CAST reported that by implementing the most promising safety enhancements, the fatality rate of commercial air travel in the United States was reduced by 83%.

CAST received the prestigious 2008 Robert J. Collier Trophy for improving safety, as well as a 2006 Laurel Award from *Aviation Week & Space Technology* magazine.

7.6.3   European Action Plan for the Prevention of Runway Incursions

In July 2001 a joint runway safety initiative was launched by the Group of Aerodrome Safety Regulators (GASR), the Joint Aviation Authority (JAA), ICAO and Eurocontrol to investigate specific runway safety issues and to identify preventive actions. Although runway safety includes a number of topics, the Task Force that was subsequently formed specifically addressed the subject of runway incursion prevention. The Task Force's first action was to carry out a survey of incidents at airports to determine causal and contributory factors that led to actual or potential runway incursions. The Task Force then delivered a set of recommendations that was endorsed by all organisations involved and published in 2003 as the European Action Plan for the prevention of runway incursions. Ongoing work continues by a working group of experts whose task it is to raise awareness of the need for immediate action in implementing the recommendations in the action plan (Eurocontrol, 2011).

7.6.4   FAA/Eurocontrol Action Plan 15 on Safety Research and Development

In March 2003, a joint safety initiative was launched by FAA and Eurocontrol within the FAA/Eurocontrol Air Traffic Management (ATM) Action Plan (AP) framework, under the name AP15 on ATM Safety. Overall, AP15's work has involved sharing understanding and discussing common problems, adapting and applying methodologies, and most of all working towards an understanding of the key areas to focus on, in terms both of the critical questions for safety assurance and the main gaps in abilities to answer such questions. Major areas of AP15 work have been as follows:

- Safety Culture
- Integrated Risk Picture
- Human Reliability Assessment

- Safety Data Sharing
- Human Factors and Safety Integration
- Normal Operation Safety Observation Methods
- Advanced Safety Methods
- Resilience
- Safety Metrics

Influential White Papers have been released by AP15 on Safety Assessment Methods (FAA/Eurocontrol, 2005), on ANSP Safety Culture (FAA/Eurocontrol, 2008), on Human Performance and Safety (FAA/Eurocontrol, 2010), and on Observational Safety Survey approaches (Eurocontrol, 2011). These help ANSPs understand the issues in these areas and what needs to be done, or what may be coming in the near to mid-term future. In particular the AP15 group has been instrumental in aligning the Safety Culture approaches of Eurocontrol, FAA and CANSO. Follow-up work of AP15 aims for further strengthening the collaboration between USA and Europe on further ATM safety developments. The joint production of white papers on key ATM safety issues plays an instrumental role in realizing this.

7.6.5   Impact of Safety Initiatives on Safety Improvements

In this subsection we verify the impact of relevant safety initiatives on safety improvements that have been identified for various CICTT categories in Sections 7.3 and 7.4. Because the European Action Plan and the FAA/Eurocontrol AP15 initiatives are rather recent, this subsection considers FSF and CAST initiatives only.

There are four CICTT categories that have been addressed by safety initiatives both by FSF and by CAST. These are:

1. Runway Excursion
2. Undershoot/overshoot
3. Runway Incursion – vehicle/aircraft
4. Controlled Flight into Terrain

The former three fall in the CICTT group Take-off, Landing and Ground Operations grouping. The latter falls in the *Airborne* grouping. The occurrence rate analysis in Sections 7.3 and 7.4 has shown that of these four categories, the latter (Controlled Flight into Terrain) only has seen a systematic improvement (see Figure 7.11). However, this does not apply for any of the former three (see Figures 7.5, 7.6 and 7.9).

There is one CICTT category which is addressed by FSF's ALAR only. This is the category 'Abnormal Runway Contact', which falls in the CICTT group Take-off, Landing and Ground operations. From the occurrence rate analysis in Section 7.3 we know that the rate for this occurrence category has not systematically decreased (see Figure 7.6).

There are three CICTT categories that have been addressed by safety initiatives by CAST only. These are:

1. Loss of Control – Inflight
2. Turbulence Encounter
3. Icing

The former is in the Airborne grouping. The other two fall in the Weather grouping. From the occurrence rate analysis in Section 7.4 we know that for Loss of Control Inflight and Turbulence Encounter a systematic improvement has taken place (see Figure 7.11 and 7.12). However, this does not apply for the *Icing* category (see Figure 7.12).

In conclusion, of the eight CICTT categories for which FSF and/or CAST have/are conducting safety initiatives, only three show systematic improvements in occurrence rate. These are 'Controlled Flight into Terrain', 'Loss of Control – Inflight' and 'Turbulence Encounter'. The first two categories fall within the Airborne grouping, the latter in the Weather grouping.

For the other five no systematic improvements in occurrence rates have been seen. Of these five categories, one (Icing) falls in the Weather grouping, whereas the other four fall in the grouping Take-off, Landing and Ground operations. These four are: Runway excursion, Undershoot/overshoot, Abnormal Runway Contact, and Runway incursion – vehicle/aircraft.

This means that safety initiatives seem to have contributed on the decrease of the rates for two CICTT categories in the Airborne grouping and one CICTT in the Weather grouping. However, there seems to be no impact or contribution found for four occurrence categories in the grouping Take-off, Landing and Ground Operations, and also not for one occurrence category in the Weather grouping.

In order to find a possible explanation for these counterintuitive findings, we take a look at the safety improvement mechanisms identified in subsection 7.5. This shows that for the two categories from the Airborne group, safety improving support systems TAWS and aircraft performance envelope protection systems have been developed and implemented. In contrast with this, for only one of the four categories in the Take-off, Landing and Ground Operations grouping a safety improving system has been developed and fielded, that is, the early A-SMGCS for Runway incursion – vehicle/aircraft. However, Section 7.5 explained that this early A-SMGCS also has been implemented for the improvement of efficiency of airport operations (e.g. crossing an active runway in order to shorten taxi time), and this may very well introduce unintended extra safety risk.

7.7 Conclusion

The analysis conducted in this chapter has shown that occurrence rates of accidents related to Take-Off, Landing and Ground Operations do not show a clear positive or negative trend over the period 1990–2008. This covers the main airport related accident types Runway Excursions, Ground Handling, Abnormal Runway Contact and Ground Collision. For Runway Incursion – vehicle/aircraft/person, there is even a slight increase of the accident rate over this period.

Nevertheless, over this period there still was a clear negative trend suggesting a reduction of the overall accident rate in commercial air transport scheduled flights. This improvement has come from significant reductions in occurrence rates for the Airborne and Aircraft related accident types, in particular Controlled Flight into Terrain, Loss of Control – Inflight, Mid-air Collision, and Aircraft System Failure induced accidents.

If nothing would be done to change the trends for each of these accident types, then the expected consequence of both developments is that the reduction in accident rate tends to fall behind the growth in worldwide air traffic, and would even lead to an increasing number of accidents per year in commercial air transportation.

The recommended way to assure that the future number of accidents in commercial aviation is not going to increase is to act as follows (with airport related categories in italic font):

1. To continue safety initiatives that have shown to be successful in the past. This way it should be possible to continue a decreasing rate of accidents due to *Runway Incursion – vehicle/aircraft/person*, Controlled Flight into Terrain, Loss of control – inflight, (Near) Mid-air Collision, Aircraft system failures (Power Plant and Non Power Plant), and Turbulence Encounter.
2. To strengthen safety initiatives that are ongoing, but that are in need of a boost. This way it should be possible to realize a decreasing rate of accidents due to *Loss of Control – ground, Runway Excursion, Undershoot/overshoot, Abnormal Runway Contact*, Windshear and Thunderstorm, Icing, and Fire/smoke (non-impact).
3. To start safety initiatives that aim to realize a decreasing rate of accidents due to CICTT categories not systematically addressed before, that is; *Ground Collision, Ground Handling, Fire/smoke (post impact), Evacuation, Aerodrome, Runway Incursion – animal*, ATM/CNS, Fuel, and Abrupt Manoeuvre.

This means there is a need for significant increase of safety initiatives addressing airport related safety improvements. The main accident scenarios for Take-off, Landing and Ground operations show that many factors in these accidents are under responsibility of the airport, the ANSP or the ground handling organisations. These factors include issues such as provision of weather information, runway condition, markings and signs of the runway, bird control. Consequently, the safety driving mechanisms for Take-off, Landing and Ground Operations differ a lot from those for Airborne and Aircraft Related aspects.

From the analysis presented in Section 7.6.5, it can be derived that one effective way in realizing safety improvements seems to be developing and implementing adequate technological means. It also has been identified that the introduction of technological means asks for a properly balanced improvement of safety, environment and efficiency. In Section 7.5 it has been identified that multiple stakeholders are typically involved in eventually reaching a balance. Without proper framework to analyse and predict safety consequences of potential changes in future air transport operations (e.g. Everdij et al., 2009; Scholte et al., 2010), environment and efficiency improvements might benefit at the cost of safety improvements. This means that safety initiatives addressing airport related categories (those in italics in the previous list) should also consider ways to assure that their developed improvements are eventually effective in improving safety.

Acknowledgements

The authors would like to thank Professor Konstantinos G. Zografos (Athens University of Economics and Business) and NLR colleague Jelmer Scholte for valuable suggestions in improving this chapter.

Notes

1 V1 means the maximum speed in the take-off at which the pilot must take the first action (e.g. apply brakes, reduce thrust, deploy speed brakes) to stop the airplane within the accelerate-stop distance. V1 also means the minimum speed in the take-off, following a failure of the critical engine, at which the pilot can continue the take-off and achieve the required height above the take-off surface within the take-off distance.
2 For the category Low Altitude operations (LALT), ICAO's ADREP data base contains no accident for any flight that falls within the selection criteria of this study (see subsection 7.2.1).
3 Early A-SMGCS is known as ASDE-X in the USA, which stands for Airport Surface Detection Equipment Model X.

References

AAIB (2006). AAIB Bulletin No: 8/2005 Ref: EW/C2004/11/06, Air Accidents Investigation Branch, UK. [Online] Available from: http://www.aaib.gov.uk [Accessed 20 February, 2013].

AAUI (2009). *Final report on the accident occurred on 25 May 2008 at Brussels Airport on a Boeing B747-209 F registered N704CK*. Ref. AAIU-2008-13, Air Accident Investigation Unit, *Federal Public Service Mobility and Transport*, Brussels, Belgium.

Abbink, F.J. (1996). *Integrated free-flight and 4-D gate-to-gate air traffic management, possibilities, promises and problems*. Technical Report NLR TP 96239U, National Aerospace Laboratory NLR, Amsterdam.

Amalberti, R., Auroy, Y., Berwick, D. and Barach, P. (2005). Five system barriers to achieving ultrasafe health care, *Annals of Internal Medicine*, Volume 142, Number 9, pp. 756–764.

ANSV (2004). *Final report, Accident involved aircraft Boeing MD-87, registration SE-DMA and Cessna 525-A, registration D-IEVX, Milano Linate airport, October 8, 2001*. N. A/1/04, Agenzia Nazioale per la Sicurezza del Volo, Rome, Italy.

Aviation Safety Network (2011). [Online] Available from http://www.aviation-safety.net/ [Accessed 20 February, 2013].

Balk, A.D. and Roelen, A.L.C. (2010). *Risks and regulations in aircraft ground handling*, Technical Report NLR-CR-2009-334, National Aerospace Laboratory NLR, Amsterdam.

Barnett, A. and LoFaso, A.J. (1983). After the crash: the passenger response to the DC-10 disaster, *Management Science*, Vol. 29, issue 11, pp. 1225–1236.

Barnett, A., Menhigetti, J. and Prete, M. (1992). The Market Response to the Sioux City DC-10 Crash, *Risk Analysis*, March 1992, p. 12.

Bilstein, R.E. (2003). *Testing Aircraft, Exploring Space: An Illustrated History of NACA and NASA*. John Hopkins University Press, Baltimore, MA.

Bliek, J.A. van der. (1994). *75 Years of Aerospace Research in the Netherlands*, National Aerospace Laboratory NLR, Amsterdam.

Bloem, E.A. and Blom, H.A.P. (2010). Bayesian analysis of accident rate, trend and uncertainty in commercial aviation, *Proc. ICRAT 2010*, Budapest, June 2–4, 2010.

Boeing (2011). *Statistical Summary of commercial Jet Airplane Accidents, Worldwide Operations 1959–2010*. Boeing Commercial Aircraft, Seattle, WA.

Burin, J. (2003). The CFIT and ALAR challenge: attacking the killers in aviation, *ISASI Proceedings*, Volume 7, No. 1, pp. 97–99.

CANSO (2010). Safety Excellence, The CANSO Standard. *Airspace Magazine*, Quarter 3. [Online] Available from: http://www.canso.org/cms/showpage.aspx?id=330 [Accessed 20 February, 2013].

CAST/ICAO (2011). *Aviation Occurrence Categories, Definition and Usage Notes*, April 2011, Version 4.1.5.

Chalk, A. (1986). Market forces and aircraft safety: The case of the DC-10, *Economic Enquiry*, Vol. 24, issue 1, pp. 43–60.

Daams, J. (2011). *Managing Deadlocks – in the Netherlands Aviation Sector*, PhD Thesis, Delft University of Technology, November 2011.

Daily, J.L. (1997). Boeing safety perspective – addressing safety issues. Paper presented at: *ICAO Regional Accident Investigation and Human Factors Seminar*. April 24, 1997. Panama City, Panama.

Darby, R. (2009). Keeping it on the runway, *AeroSafety World*, Vol. 4, Issue 8, pp. 12–17.

Davis, J. (1997). Automation impact on accident risk. Paper presented at: *ICAO Regional Accident Investigation and Human Factors Seminar*. April 24, 1997. Panama City, Panama.

DGCA (2010). *Report on accident to m/s Kingfisher Airlines ATR-72 aircraft VT-KAC at Mumbai on 10.11.2009*. Air Safety Directorate, Director General of Civil Aviation, New Delhi, India.

Ek, A. and Akselsson, R. (2007). Aviation on the ground: safety culture in a ground handling company, *The International Journal of Aviation Psychology*, Vol. 17, issue 1, pp. 59–76.

Enders, J.H., Dodd, R., Tarrel, R., et al. (1996). Airport safety: A study of accidents and available approach-and-landing aids, *Flight Safety Digest*, Vol. 15, No. 3, Flight Safety Foundation, Alexandria, VA, USA.

EU (2003). *Council Directive 2003/42/EC on occurrence reporting in civil aviation*, Council of the European Union, Brussels, Belgium.

EU (2005). Regulation (Ec) No 2111/2005 of the European Parliament and of the Council of 14 December 2005 on the establishment of a Community list of air carriers subject to an operating ban within the Community and on informing air transport passengers of the identity of the operating air carrier, and repealing Article 9 of Directive 2004/36/EC, *Official Journal of the European Union*, pp. L 344/15 – L 344/22.

EU (2011). Commission implementing regulation (EU) No 390/2011 of 19 April 2011, *Official Journal of the European Union*, pp. L 104/10 – L 104/34.

Eurocontrol (2006). *A-SMGCS Levels 1 and 2 Preliminary Safety Case*, Edition 2.0. [Online] Available from: http://www.skybrary.aero/bookshelf/books/1036.pdf [Accessed 20 February, 2013].

Eurocontrol (2009). *Guidance Material for Short Term Conflict Alert*, Version 2.0. [Online] Available from: http://www.eurocontrol.int/documents/short-term-conflict-alert-guidelines [Accessed 20 February, 2013].

Eurocontrol (2011). *European Action Plan for the prevention of runway incursions*, edition 2.0. [Online] Available from: http://www.eurocontrol.int/ [Accessed 20 February, 2013].

Everdij, M.H.C., Blom, H.A.P., Scholte, J.J., et al. (2009). Developing a framework for safety validation of multi-stakeholder changes in air transport operations, *Safety Science*, Vol. 47, pp. 405–420.

FAA (2005). *Airport Surface Detection Equipment Model X (ASDE-X) Safety Logic, Safety Risk Management Document*, Report version 1.0b, FAA, Washington, DC, September 2005.

FAA/Eurocontrol (2005). *ATM Safety Techniques and Toolbox*, Report of Action Plan-15, Issue 1.1, 10 February, 2005. [Online] Available from: http://www.eurocontrol.int/eec/gallery/content/public/documents/EEC_safety_documents/Safety_Techniques_and_Toolbox_1.0.pdf [Accessed 20 February, 2013].

FAA/EUROCONTROL (2008). *Safety Culture in Air Traffic Management*. Action Plan 15 White Paper. [Online] Available from: http://www.skybrary.aero/bookshelf/books/564.pdf [Accessed 20 February, 2013].

FAA/EUROCONTROL (2010). *Human Performance in Air Traffic Management Safety*, Action Plan 15 White Paper, September 2010.

FAA/EUROCONTROL (2011). *Ensuring Safety Performance in ATC Operations: Observational Safety Survey Approaches*, Action Plan 15 White Paper, July 2011.

FSF (1998). Boeing 757 CFIT accident at Cali, Columbia, becomes focus of lessons learned, *Flight Safety Digest*, Vol. 17, No. 5–6, Flight Safety Foundation, Alexandria, VA, USA.

FSF (1999). Killers in aviation: FSF task force presents facts about approach-and-landing and controlled-flight-into-terrain accidents, *Flight Safety Digest*, Vol. 17, No. 11–12, Vol. 18, No. 1–2. Flight Safety Foundation, Alexandria, VA, USA.

Gore, A.A. (1997). *Final Report to President Clinton*, White House Commission on Aviation Safety and Security, Washington D.C., USA.

Hogendoorn, R.A., Rekkas, C. and Neven, W.H.L. (1999). ARTAS: an IMM based multi-sensor tracker, *Proc. 2nd International Conference on Information Fusion*, July 1999, pp. 1021–1028.

Hudson, P. (2003). Applying the lessons of high risk industries to health care, *Quality and Safety in Health Care*, Vol. 12, pp. 7–12.

ICAO (2001). Aircraft accident and incident investigation, *Annex 13 to the Convention on International Civil Aviation*, 9th edition, International Civil Aviation Organization, Montreal, Canada.

Leary, W.M. (1992). Safety in the air: the impact of instrument flying and radio navigation on U.S. commercial air operations between the wars. In: Leary, W.M. (ed.), *The History of Civil and Commercial Aviation*, Vol. 1, Infrastructure and environment, Smithsonian Institution Press, Washington D.C., USA.

Nativi, A. and Learmount, D. (2001). Taxiway error leads to Milan disaster, *Flight International*, Vol. 160, No. 4802, p. 20.

NTSB (1979). *United Airlines, Inc., McDonnell-Douglas, DC-8-61, N8082U, Portland, Oregon, December 28, 1978*. Aircraft Accident Report NTSB-AAR-79, National Transportation Safety Board, Washington D.C., USA.

NTSB (1998). *We Are All Safer*. SR-98-01, National Transportation Safety Board, Washington, D.C., USA.

NTSB (2003). *Collision with trees on final approach, Federal Express Flight 1478, Boeing 727–232, N497FE, Tallahassee, Florida, July 26, 2002*. Aircraft Accident Report NTSB/AAR-04/02, NTSB, Washington D.C., USA.

NTSB (2010). *Loss of Control on Approach, Colgan Air, Inc., Operating as Continental Connection Flight 3407, Bombardier DHC-8-400, N200WQ, Clarence Center, New York, February 12, 2009.* Aircraft Accident report NTSB/AAR-10/01. National Transportation Safety Board, Washington, D.C., USA.

NTSB (2011). NTSB accident database [Online] Available from http://www.ntsb.gov/ [Accessed 23 August 2011].

Rekkas, C.M. and Blom, H.A.P. (1997). Multi Sensor Data Fusion in ARTAS, *Proc. of IRCTR Colloquium on Surveillance Sensor Tracking*, Delft University of Technology, The Netherlands.

Roelen, A.L.C. and Smeltink, J.W. (2008). *Aircraft damage and occupant fatality profiles for a causal model of air transport safety*, Technical Report NLR-CR-2008-231, National Aerospace Laboratory NLR, Amsterdam.

Scholte J.J., Blom, H.A.P., Bos, J.C. van den, and Jansen, R.B.H.J. (2008). The role of safety validation in ATM concept development: How does it work in practice? *Proc. Eurocontrol Annual Safety R&D Seminar*, Southampton, 22–24 October 2008.

Scholte, J.J., Blom, H.A.P., and Pasquini, A. (2010). Study of SESAR implied safety validation needs, *Proc. ICRAT 2010*, Budapest, June 2–4, 2010, pp. 465–472.

Sorenson, J.N. (2002). Safety Culture: a survey of the state of the art. *Reliability Engineering and System Safety*, Vol. 76, pp. 189–204.

Sparaco, P. (1998). Airbus: Automated transports safer than older aircraft. *Aviation Week and Space Technology*. January 5, pp. 40–41.

Stroeve, S.H., Blom, H.A.P. and Bakker, G.J. (2008). Systemic accident risk assessment in air traffic by Monte Carlo simulation, *Safety Science*, Vol. 47, No. 2, pp. 238–249.

Stroeve, S.H., Blom, H.A.P., and Bakker, G.J. (2011). Contrasting Safety Assessments of a Runway Incursion Scenario by Event Sequence Analysis versus Multi-Agent Dynamic Risk Modelling, *9th USA/Europe Air Traffic Management Research and Development Seminar, ATM2011*, Berlin, 14–17 June 2011.

Transportation Safety Board of Canada (2007). *Runway Overrun and Fire, Air France, Airbus A340-313 F-GLZQ, Toronto/Lester B. Pearson International, Airport, Ontario, 02 August 2005.* Aviation Investigation Report A05H0002, Minister of Public Works and Government Services, Canada.

8

Scheduled Delay as an Indicator for Airport Scheduling Performance

Dennis Klingebiel[A], Daniel Kösters[B] and Johannes Reichmuth[C]

[A]Physics Institute IIIA, RWTH Aachen University, Germany
[B]Frankfurt Airport, FRA Vorfeldkontrolle GmbH, Fraport AG, Germany
[C]RWTH Aachen Institute of Transport Science VIA, and DLR Institute of Air Transport and Airport Research, German Aerospace Center (DLR), RWTH Aachen University, Germany

8.1 Introduction

'Access and equity' is one of 11 global Air Traffic Management (ATM) performance areas (ICAO, 2005). The feasibility of accessing infrastructure elements during operations is evaluated here. At major European airports, access is controlled by allocating airport slots. An airport slot is required for every take-off and every landing and thus can be considered as a temporarily specified operations permit for an airport (Dempsey, 2001). In the European Union (EU), airport slots are allocated within the so called airport coordination following Council Regulation 95/93 (Council Regulation, 1993). Thus, airport coordination is at the center of airport scheduling activities which form the basis of ATM, months before operations (PRC, 2005). It decides on the feasibility of realizing operational infrastructure access as requested.

Airport coordination avoids operational overloads by balancing capacity and demand at a strategic planning level. Here, airport capacity is defined as the declared number of airport slots available for allocation (ACI, 2004). Excess demand may occur if the number of airport slot requests exceeds the number of available slots. Conflict resolution includes both

Modelling and Managing Airport Performance, First Edition. Edited by Konstantinos G. Zografos, Giovanni Andreatta and Amedeo R. Odoni.
© 2013 John Wiley & Sons, Ltd. Published 2013 by John Wiley & Sons, Ltd.

rescheduling (alternative slot times) and rejecting slot requests (Ulrich, 2008). In such cases, airport access cannot be realized as requested.

Coordination results are used to evaluate the feasibility of accessing an airport as requested. Being defined as the difference between requested and allocated slot times, scheduled delays allow the measurement of an airport's scheduling performance (Kösters, 2007). Thus, ATM performance in terms of access and equity is quantifiable. Scheduled delays will occur if an airport does not cope with the slot demand in the strategic planning phase. An airport's capability to accommodate the given demand may be discussed using scheduled delays as performance indicator.

Besides a retrospective analysis of historical data, additional benefits would arise from precise knowledge on the interrelationship between slot demand (measured as the utilization of available slots) and scheduling performance (measured as scheduled delays). Future developments may be forecasted, as well as different input parameters' impact on the scheduling performance analyzed. In addition, actual airport scheduling results can be benchmarked towards a scheduled delay prognosis. Therefore, within this chapter, a model approach is defined which allows the determination and thus, the forecasting of scheduled delays depending on the level of slot utilization. The initial conflict resolution as focal part of the airport coordination is modeled.

This chapter consists of five sections. Following the introduction, Section 8.2 focuses on the background of this research topic. Airport coordination according to Council Regulation 95/93 is introduced. Parameters are defined to discuss an airport's scheduling performance: scheduled delays as performance indicators as well as the interrelationship between scheduled delays and slot demand/slot utilization. In Section 8.3, a deterministic model for calculating scheduled delays is introduced; capacity restrictions and initial slot requests (slot demand) are used as model input parameters. Model validation then is described in Section 8.4. Finally, Section 8.5 focuses on the usage of this model approach. As an example, the impact of varying capacity restrictions and diversified daily demand patterns on the scheduling performance is demonstrated.

With airport coordination data being available for German coordinated airports – DUS (Düsseldorf), FRA (Frankfurt), MUC (Munich), STR (Stuttgart) and TXL (Berlin-Tegel) – for schedule periods summer 2005 and winter 2005/2006, analysis and modeling within this chapter are based on the developments at those airports. Because airport coordination in Germany precisely follows Council Regulation 95/93, it may be assumed that this chapter's research results are transferable to any airport being slot-coordinated according to this regulation.

8.2 Background

8.2.1 Airport Coordination

Airport coordination in the EU and thus in Germany is an administrative process which is based on Council Regulation 95/93. The latter has endorsed the International Air Transport Association (IATA) Worldwide Scheduling Guidelines (IATA, 2007). A national airport coordinator is responsible for allocating airport slots. Although the Regulation specifies relevant procedures precisely, remaining tolerances result in national and local particularities. The following summary corresponds to slot allocation procedures at German coordinated airports.

Article 2 of the Regulation defines the term slot as:

... the permission given by a coordinator in accordance with this Regulation to use the full range of airport infrastructure necessary to operate an air service at a coordinated airport on a specific date and time for the purpose of landing or take-off as allocated by a coordinator in accordance with this Regulation.

Practically, a slot is defined as the right to use the airport infrastructure with an aircraft at a particular time – it is a temporally specified operations permit. The slot time defines the moment of leaving or arriving at the aircraft stand (on block, off block) and is equal to the published flight time usually. At German airports, for coordination a 10 min time span is used as coordination unit. With flights being scheduled in full 5 min steps, two adjacent slot times will be combined and considered within one 10 min coordination interval.

8.2.1.1 Coordination Parameters – Declared Capacity

Airport coordination parameters – predominantly an airport's declared capacity – are relevant input parameters for the slot allocation process. The declared capacity describes a maximum number of slots per unit of time (block period) that can be allocated by the coordinator. Declared capacity values may differentiate between arrival (ARR), departure (DEP) and total movements (TOT). The duration of blocks may vary as well; in addition, several blocks with different duration may be superposed to control the concentration of flights within a certain time period. While 60 min values determine the maximum number of slots, 30 and 10 min values are used optionally to control the concentration of flights. This objective is achieved by applying rolling blocks (10 min steps) in addition. The use of declared capacity values for the whole season means a fixing of the seasonal airport capacity at an early stage. The determination of the declared capacity shall take into account all relevant technical, operational and environmental constraints as well as any changes thereto. It tries to maximize the use of available airport capacity whilst keeping the quality of service during operations at locally acceptable levels.

8.2.1.2 Initial Slot Request, Coordination, and Slot Allocation

About five months before the start of a scheduling season (winter or summer) airlines have to provide details on their planned schedules for the upcoming season to the responsible coordinator. As conflicts may exist at the initial slot request deadline (multiple booking of slots) a balancing of capacity demand and supply is required within the so called initial coordination. Coordinators may retime (reschedule) slot requests to meet capacity restrictions. Then alternative slot times – for the same day usually – are proposed. The second best option to resolve conflicts is a complete rejection of slot requests. This is inevitable if no adequate rescheduling proposal is available.

The initial coordination of slots is carried out using administrative priority criteria as described in Article 8 of the Regulation. The key rule specifies that a series of slots will be assigned to an airline for the upcoming season again if it had been used at least 80% in the previous equivalent season (winter or summer). This rule is known as the use-it-or-lose-it rule, 80/20 rule or grandfather rights. Lower priority is granted to airlines that comply with new entrant criteria at that specific airport. After processing all slot requests the airport coordinator informs the airlines about the results at the initial slot allocation. Now a coordinated schedule is available which meets all relevant capacity restrictions and which represents a feasible reference program for operations.

8.2.1.3 Post Initial Allocation Activities

Airport coordination continues after the initial slot allocation. At first, the worldwide IATA Schedules Conference takes place. Airline representatives as well as airport coordinators may negotiate slots, arrange slot amendments or slot alternatives and exchange slots (between airlines) at this conference. Schedule adjustments are carried out through bilateral discussions between airlines and coordinators. Due to the presence of both airlines and coordinators the conference is the main opportunity for schedules adjustments particularly if there is more than one airport affected.

Council Regulation 95/93 allows for various ways to fine tune schedules. Schedules adjustments may continue not only until the start of the season but until the days of operations. While the aforementioned order of priorities is applied to meet capacity restrictions at the initial allocation, all additional schedule adjustments, slot requests and returns follow a first-come-first-served-strategy, for example, by using waiting lists.

8.2.2 Performance Indicator: Scheduled Delays

A parameter to measure airport performance in terms of 'access and equity' – airport scheduling performance – is to be based on the relevant developments during airport coordination. It needs to allow quantification of feasibility to access the airport as requested, building a basis to discuss the operational airport access with regard to a performance and quality aspect.

Conflict resolution, which is required in periods of excess slot demand, impacts on the feasibility to access airport infrastructure. If the number of slot requests exceeds the number of slots being available, slot requests will be coordinated by either a rescheduling or a complete rejection (see Fig. 8.1 for the proportion slot requests being rejected, allocated but rescheduled and allocated as requested at German coordinated airports in scheduling seasons summer 2005 and winter 2005/2006). For such cases operational infrastructure access cannot be realized as requested: a different access time has to be accepted, or access is denied completely. Thus differences occur between internally planned schedules and allocated slots.

Variations described are to be considered by the indicator. To allow the quantification of such variations, scheduled delays had been defined. Scheduled delays measure the difference between requested and allocated slot times. Being 'hidden' in schedules they remain abstract only. Scheduled delays are calculated as follows:

$$\delta_{s,j} = \left| t_{\text{all},j} - t_{\text{req},j} \right| \quad [\text{min}] \tag{8.1}$$

with
$\delta_{s,j}$ Scheduled delay of allocated slot j
$t_{\text{all},j}$ Allocated slot time of allocated slot j
$t_{\text{req},j}$ Requested slot time of allocated slot j

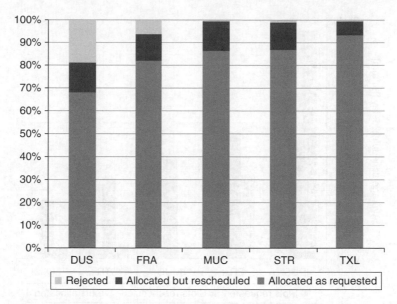

Figure 8.1   Initial allocation: proportion of rejected, allocated but rescheduled, allocated as requested slot requests

Depending on the method of conflict resolution, slot times may be postponed or may be brought forward. Scheduled delays measure the absolute value of differences between requested and allocated slot times only. Within this chapter, scheduled delays are measured at the initial slot allocation only and thus resulting from the initial slot coordination. In general, the performance indicator may be determined at any stage of airport coordination which continues until the days of operations.

Equation (8.1) focuses on the rescheduling of slot requests but disregards slot rejections. Integrating the latter is required to measure the feasibility to access an airport completely. Following the logic of this scheduling performance model approach, a slot rejection would be equal to a slot being delayed infinitely.

Figure 8.2 displays average scheduled delays of requested, rescheduled and allocated slots at German coordinated airports in summer 2005 and winter 2005/2006. In theory, slots that are completely rejected suffer from infinite scheduled delays. To integrate rejected slots in the computation of scheduled delays for requested slots, instead of those infinite scheduled delays a standard value is used per definition: 8 h. This value represents the required temporal shift to displace a slot being requested in the center (14:00 LT – local time) of the daytime period (06:00 to 22:00 LT) to the curfew hours (22:00 to 06:00 LT). This value is used as scheduled delay for all rejected slots.

8.2.3   Slot Utilization and Scheduled Delays

An airport's scheduling performance may be analyzed using scheduled delays as performance indicator. Analogous to common operational delay analyses, such an investigation can be based on various statistical parameters (average delay per slot, proportion of delayed slots,

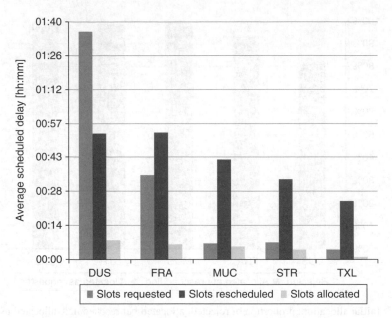

Figure 8.2   Average scheduled delays of requested, rescheduled and allocated slots

delay variability, delay maxima, etc.). Nonetheless, additional information would be useful to address airport capacity and airport design issues by the use of scheduled delays, the interdependency between slot utilization and scheduled delays in particular. Relevant parameters to discuss this relation are introduced in the current section. Additionally, results of an empirical analysis of the airport coordination process at German coordinated airports are demonstrated.

Conflict resolution within the initial coordination is restricted to single days of operations – slot requests will not be rescheduled to precedent or following days usually. Therefore, the interdependency between slot utilization and scheduled delays focuses on complete days of operations (daytime hours 06:00 to 22:00 LT). Slot demand per day (process input) and slots allocated per day (process output) will be put into relation to the number of slots being available for allocation at that particular day. This allows calculating daily slot utilization ratios. Finally, average scheduled delays per day build the counterpart required to demonstrate the interdependency between slot utilization and performance indicator.

Declared capacity values of an airport form the basis of calculating the daily slot supply as the total number of slots being available at one day. The following notation is used for an airport's declared capacity values:

$$C_{i,z,k} \quad [\text{Slots}/\text{time interval}] \qquad (8.2)$$

with
C Declared capacity value
i Type of flight movement (ARR, DEP, TOT)
z Duration of time interval (10, 30, 60 min)
k Temporal position of time interval

Table 8.1 Declared capacities at German coordinated airports; maximum number of available slots (slot supply/slot potential).

	[Local Time]	10 min			30 min			60 min			slot supply/ slot potential 06:00–22:00
		ARR	DEP	TOT	ARR	DEP	TOT	ARR	DEP	TOT	
TXL	06:00–22:00			8						40	640
DUS	06:00–21:00			10						40	635
	21:00–22:00			10						35	
FRA	06:00–14:00	9	9	16	23	25	43	41	43	80	1296
	14:00–21:00	9	9	16	23	25	43	42	45	82	
	21:00–22:00	9	9	16	23	25	43	42	50	82	
MUC	06:00–22:00	12	12	15				58	58	89	1424
STR	06:00 –22:00	6	6	8				30	30	40	640

Source: Airport Coordination Germany (FHKD)

Table 8.1 depicts the number of slots being available for allocation purposes and declared capacities at German coordinated airports in summer 2005 and winter 2005/2006 scheduling season.

Declared capacity values of 60 min for total movements are used to calculate the daily slot supply (daytime hours from 06:00 to 22:00 LT): 10 min and 30 min values control the slot demand concentration, but do not limit the maximum slot supply. Therefore, $\Phi_{\mathrm{TOT},d}$, the number of available slots (total movements) during a specific day d, can be computed as follows:

$$\Phi_{\mathrm{TOT},d} = n_d * C_{\mathrm{TOT},60,k} \quad [\mathrm{Slots/day}] \tag{8.3}$$

with

n_d Number of daytime hours at day d.

Figure 8.3 displays the daily slot supply $\Phi_{\mathrm{TOT},d}$ at German coordinated airports in the period under investigation. In this chapter the terms 'slot supply' and 'slot potential' are used synonymously.

Besides airport declared capacity values, the number of slot requests (for ARR, DEP and TOT) is the second relevant input parameter for airport coordination:

$$D_{\mathrm{req},i,z,k} \quad [\mathrm{Slots/time\ interval}]. \tag{8.4}$$

Daily slot supply $\Phi_{\mathrm{TOT},d}$ and the number of slot requests per day $D_{\mathrm{req},\mathrm{TOT},d}$ allow determination of the slot utilization factor $\rho_{\mathrm{req},\mathrm{TOT},d}$ at the initial slot request deadline:

$$\rho_{\mathrm{req},\mathrm{TOT},d} = \frac{D_{\mathrm{req},\mathrm{TOT},d}}{\Phi_{\mathrm{TOT},d}} \quad [-] \tag{8.5}$$

with

$\rho_{\mathrm{req},\mathrm{TOT},d}$ Utilization of available slots at initial request

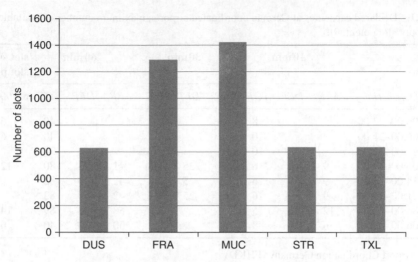

Figure 8.3 Daily slot supply/slot potential $\Phi_{TOT,d}$

$\rho_{req,TOT,d}$ measures the system load and indicates demand excess on a daily basis. For $\rho_{req,TOT,d} > 1$ some slots at least have to be rejected as the total number of slot requests exceeds the total number of available slots (slot supply).

As a result of the initial coordination, slots will be allocated. The following notation represents the total number of slots being allocated at the initial allocation:

$$D_{all,i,z,k} \left[\text{Slots / time interval} \right]. \tag{8.6}$$

Analogous to Equation (8.5), a slot utilization factor $\rho_{all,TOT,d}$ may be determined being based on the number of allocated slots:

$$\rho_{all,TOT,d} = \frac{D_{all,TOT,d}}{\Phi_{TOT,d}} \quad [-] \tag{8.7}$$

with

$\rho_{all,TOT,d}$ Utilization of available slots at initial allocation.

$\rho_{all,TOT,d}$ indicates the process output. If no rejections of slot requests are required during the initial coordination it will be $\rho_{all,TOT,d} = \rho_{req,TOT,d}$.

After defining slot utilization factors $\rho_{all,TOT,d}$ and $\rho_{req,TOT,d}$, an adequate scheduling performance indicator is required to demonstrate the interdependency between those parameters. As conflict resolution focuses on full days of operations, slot utilization factors $\rho_{all,TOT,d}$ and $\rho_{req,TOT,d}$ use a full day time basis accordingly. For a reasonable matching, the performance indicator would need to feature this same attribute. Therefore, average scheduled delays per allocated slot are calculated as follows:

$$\overline{\delta}_{s,d} = \frac{\sum_{1 \le j \le D_{all,TOT,d}} \delta_{s,j}}{D_{all,TOT,d}} \quad [\text{min/slot}] \tag{8.8}$$

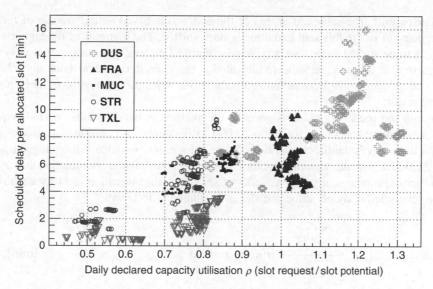

Figure 8.4   Average scheduled delays versus slot utilization at initial request

with

$\bar{\delta}_{s,d}$ Average scheduled delay per allocated slot at day d.

In Figure 8.4, average scheduled delays (Eq. 8.8) are related to slot utilization factors at the initial request (Eq. 8.5) at German coordinated airports for every day in the scheduling seasons summer 2005 and winter 2005/2006. The dimension of required demand coordination in the form of rescheduled slot requests depends on the slot utilization and thus on the level of capacity scarcity and slot shortage. The empirical results confirm the intuitive expectation that when slot utilization increases then also average scheduled delays per allocated slot will rise.

8.3   Definition of a Model to Predict Scheduled Delays

In the following section, we introduce a model to determine scheduled delays. Although featuring a minor stochastic element, the overall approach is considered to be deterministic. Therefore, it is referred to as 'deterministic model' in this chapter.

The model approach concentrates on modeling the airport coordination process with the following input parameters only: Declared capacity values and initial slot requests. Every single 10 min coordination unit will be considered separately and unidirectionally one after the other. In the case of excess demand in one unit, surplus slots are shifted to the following one. If a displaced slot is allocated at any time on that particular day at least, its temporal displacement will be measured as a scheduled delay. If it is not allocated but rejected at the end of the day, its temporal displacement will not be measured.

To balance demand and capacity in the case of excess demand, the airport coordinator uses priorities such as grandfather rights or the new entrant privilege. Due to a lack of information on this background of conflict resolution, this chapter's model does not differentiate between priority classes. Here, slots are *a priori* indistinguishable. In the case of excess demand in a certain coordination unit, at least one requested slot needs to be shifted into the next coordination unit. The selection of this or these slot(s) is based on the amount of time that every requested slot in this unit had already been shifted. The higher it is, the higher a slot's priority is to be shifted again. For a peak-wise demand pattern, this ensures that slots which are requested in off-peak periods are allocated at the requested time without any scheduled delay. The selection will be randomized only if this selection strategy does not provide a clear-cut result.

Although the model approach is generic, the analysis at hand focuses on slots requested within the daytime hours only (06:00 to 22:00 LT). To minimize possible inaccuracies resulting from the model's unidirectional coordination, it will be run twice: Chronologically (from 06:00–22:00 LT with the temporal location of the day's last 10 min coordination unit at 21:50 LT) and anti-chronologically (from 22:00–06:00 LT). Using Equation (8.8) for each run the average scheduled delay per allocated slot is calculated as follows:

$$\overline{\delta}_{s,06:00->21:50,model} = \frac{\sum_{j=1}^{n} \delta_{s,j}}{D_{all,TOT,d}} \quad [min] \quad and \quad \overline{\delta}_{s,21:50->06:00,model} = \frac{\sum_{j=1}^{n} \delta_{s,j}}{D_{all,TOT,d}} \quad [min].$$

The arithmetic mean of the two runs is considered to be the final result of the modeled coordination:

$$\overline{\delta}_{s,d,model} = \frac{\overline{\delta}_{s,06:00->21:50,model} + \overline{\delta}_{s,21:50->06:00,model}}{2} \quad [min].$$

German airports use a 10 min time span as smallest coordination unit. If $k = 06:00, 06:10, \ldots, 21:50$ denotes the temporal location of every coordination unit,

$$\mathbf{D}_{req,i,d} = \left(D_{req,i,k=06:00,d}, D_{req,i,06:10,d}, \ldots, D_{req,i,21:50,d} \right) \quad [Slots/time\,interval] \quad (8.9)$$

is the number of requested slots at day d with $i \in \{ARR, DEP\}$. The kth rolling block is defined as the sum of all slot requests within the next z minutes

$$f_{z,i}(k) = \sum_{j=0}^{z/10-1} D_{req,i,k+j*00:10} \quad [Slots/time\,interval]$$

with $f_{z,TOT}(k) = f_{z,DEP}(k) + f_{z,ARR}(k)$.

Some airports differentiate declared capacity values C [Slots/time interval] by arrivals and departures as well as by time interval durations z. Furthermore, declared capacity values may change in the daytime. Allocated slots have to comply with coordination parameters ($i \in \{ARR, DEP, TOT\}$) at any time; with

$$C = (C_{i,z,k}) = \begin{pmatrix} C_{ARR,10,k} & C_{ARR,30,k} & C_{ARR,60,k} \\ C_{DEP,10,k} & C_{DEP,30,k} & C_{DEP,60,k} \\ C_{TOT,10,k} & C_{TOT,30,k} & C_{TOT,60,k} \end{pmatrix} \quad [Slots/time\,interval] \quad (8.10)$$

the slot availability in the kth coordination unit is then

$$\Delta = (\Delta_{i,z,k}) = \begin{pmatrix} \Delta_{\text{ARR},10,k} & \Delta_{\text{ARR},30,k} & \Delta_{\text{ARR},60,k} \\ \Delta_{\text{DEP},\ 10,k} & \Delta_{\text{DEP},30,k} & \Delta_{\text{DEP},60,k} \\ \Delta_{\text{TOT},10,k} & \Delta_{\text{TOT},30,k} & \Delta_{\text{TOT},60,k} \end{pmatrix} \ [\text{Slots/time interval}] \qquad (8.11)$$

where $\Delta_{i,z,k} = f_{z,i}(k) - C_{i,z,k}$.
In every single coordination unit k;

$\Delta_{\text{ARR},k} := \max(\max(\Delta_{\text{ARR},10;k};\ \Delta_{\text{ARR},30,k};\ \Delta_{\text{ARR},60,k}),0)$ arrivals,

$\Delta_{\text{DEP},k} := \max(\max(\Delta_{\text{DEP},\ 10,k};\ \Delta_{\text{DEP},30,k};\ \Delta_{\text{DEP},60,k}),\ 0)$ departures, and

$\Delta_{\text{TOT},k} := \max(\max(\Delta_{\text{TOT},10,k};\ \Delta_{\text{TOT},30,k};\ \Delta_{\text{TOT},60,k}) - \Delta_{\text{ARR},k} - \Delta_{\text{DEP},k},\ 0)$ arrivals or departures, need to be shifted into the next interval.

$$\Delta_i > 0 \text{ corresponds to excess slot demand,} \qquad (8.12)$$

$$\Delta_i < 0 \text{ corresponds to excess slot supply.} \qquad (8.13)$$

Slots are requested as on block and off block times (gate times), but the coordinator uses runway times to check compliance with capacity restrictions. Therefore, arrival slot times (scheduled on block) are adjusted by a standard taxi time of -5 min, departure slot times (scheduled off block) are adjusted by a standard taxi time of $+5$ min at German airports (except FRA, MUC: $+10$ min) (Kösters, 2007).

Figure 8.5 represents the functional principle of modeling the initial coordination. The slot demand ($D_{\text{req,ARR},d}$ and $D_{\text{req,DEP},d}$) is allocated according to the available capacity (slot supply), within all coordination units with their temporal location $k \in \{06\text{:}00, 06\text{:}10, \ldots, 21\text{:}50\}$ and vice versa. A slot's time displacement will be measured as scheduled delay $\delta_{s,j}$ (Eq. 8.1) if the slot is allocated at any time on that day at least.

8.4   Validation of the Model Approach

The model approach can be validated by available airport coordination data (summer 2005 and winter 2005/2006 scheduling seasons at DUS, FRA, MUC, STR and TXL).

Figure 8.6 shows the arithmetic mean of 100 iterations of the scheduled delay per allocated slot per day $\bar{\delta}_{s,d,\text{model}}$ (Eq. 8.8): For identical input parameters (declared capacity, slot requests) 100 runs of the model are performed. The standard deviation is taken into account as a statistical error to evaluate the significance of the model's stochastic element (see Fig. 8.5: Selection in random order). However, the statistical errors are quite small. This result forms the basis of referring to the approach as 'deterministic model' apart from its – minor – stochastic component.

The difference between modeled and actual (real) average scheduled delays $\bar{\delta}_{s,d,\text{model}} - \bar{\delta}_{s,d,\text{real}}$ is illustrated in Figure 8.7; $\bar{\delta}_{s,d,\text{real}}$ is defined according to (Eq. 8.8) (see also Fig. 8.4).

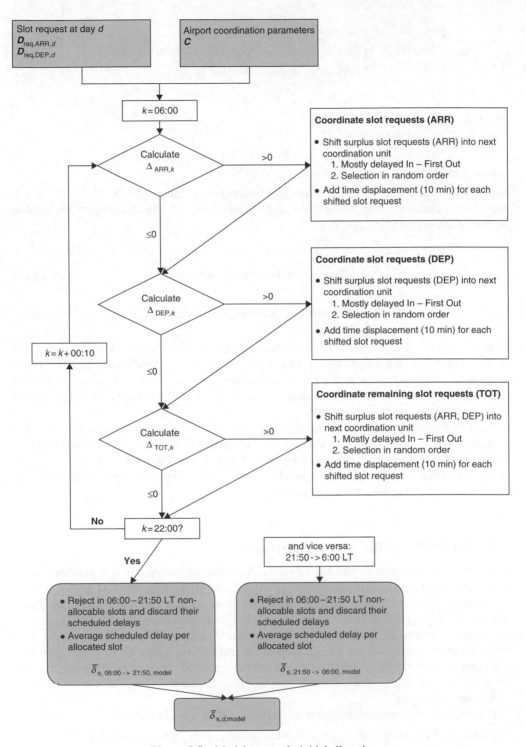

Figure 8.5 Model approach: initial allocation

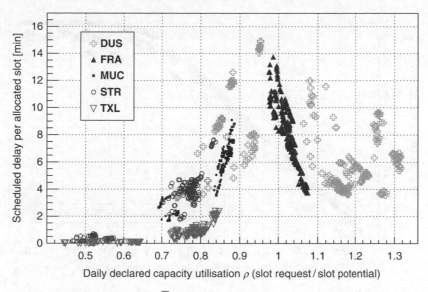

Figure 8.6 Modeled scheduled delays $\overline{\delta}_{s,d,model}$ based on declared capacity utilization at the initial request

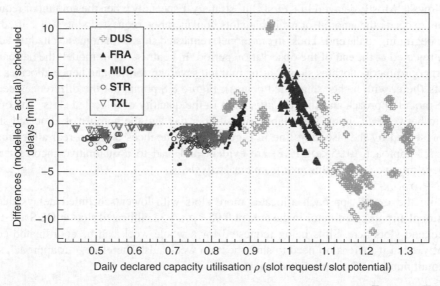

Figure 8.7 Difference between modeled and actual average scheduled delays: $\overline{\delta}_{s,d,model} - \overline{\delta}_{s,d,real}$

We specify the following three areas of interest for the utilization ρ of available slots at initial request (Eq. 8.5):

1. $0.95 < \rho \leq 1.32$

   It is obvious that there is a significant difference between modeled and actual scheduled delays at daily declared capacity utilization levels exceeding 0.95. In contrast to the expectation, modeled scheduled delays decrease due to the fact that our model approach

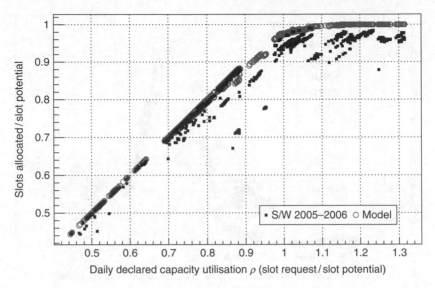

Figure 8.8 Number of allocated slots: Model approach versus initial allocation

follows a 'Mostly delayed In – First Out' strategy. Especially when the amount of requested slots exceeds the amount of available slots considerably, requested slots are shifted from the beginning to the end. Thus, in our model, requested slots with highest scheduled delays are rejected at the end of the observation period. In contrast to our model, the national airport coordinator allocates slots bearing in mind several priorities and hence allocates fewer slots (here, with higher scheduled delays). Figure 8.8 points out the differences between the model approach and the real allocation in the quantity of allocated slots. The optimal ratio between an airport's available capacity per day and the number of allocated slots is represented by the bisecting line in Figure 8.8. Differences between real allocation and model approach ($0.95 < \rho \leq 1.32$) are evident and lead to a qualitative incorrect result. Therefore, we exclude this area from our analysis.

2. $0.84 < \rho \leq 0.95$

   While the model approach allocates more slots with lower scheduled delays than the national airport coordinator in the area of $0.95 < \rho \leq 1.32$, allocated slots at DUS are mainly weekend slots here. Slots being requested for a whole week can be coordinated consistently, even if there is no need to shift them on weekends. Therefore, our approach shows a small number of quantitative discrepancies.

3. $\rho \leq 0.84$

   The results of the model approach are consistent with actual data except for an approximately constant systematic uncertainty of $\sigma_{\overline{\delta}_{s,d,model}} = -1{:}15 \pm 0{:}02$ [min].

   This value is derived by a linear regression on the resulting difference between modeled and actual average scheduled delays ($\overline{\delta}_{s,d,model} - \overline{\delta}_{s,d,real}$) with the method of least squares for $\rho \leq 0.84$ (sum of squares/degrees of freedom $= 550.6/847$).

In conclusion, it may be assumed that model accuracy is limited as a result of excluding relevant parameters which form the basis of initial conflict resolution decisions. Being the focal parameter of the initial coordination and conflict resolution decisions this is related to historical priorities of slot requests (grandfather rights) in particular. Thus, available airport coordination data does not allow modeling the decision process completely.

Nevertheless, limited data restrains this model approach's scope of application to a daily declared capacity utilization of $\rho < 0.95$. Acceptable results can be achieved for $\rho < 0.95$. A utilization of $\rho < 0.95$ is basically the field of interest if estimated scheduled delays should be used for airport planning and design issues.

8.5 Application of the Model Approach

Scheduled delays allow quantifying and measuring an airport's scheduling performance. Indicating its incapability and insufficiency to cope with the slot demand, this criterion can be used to deal with airport capacity and design issues.

Besides a retrospective scheduling performance analysis, existence of a model to predict scheduled delays extends the field of possible usage of the indicator. Future developments may be forecasted to determine maximum capacity utilization ratios still complying with predefined scheduling performance standards. This allows quantifying the required slot supply and thus minimum declared capacity values. Actual coordination results can be benchmarked towards an ideal model case to detect reasons for variations. Finally, by using the model the impact of different initial coordination input parameters on the scheduling performance may be analyzed. The latter will be under consideration in the following.

So far the model approach is limited to the initial slot demand and declared capacity values as input parameters. Thus, in this section, we discuss the effects of different slot demand profiles (subsection 8.5.1) and varying capacity restrictions (subsection 8.5.2) on the scheduling performance. Characterizing the initial slot demand by a typical demand profile, we focus on the relation between scheduled delays and capacity utilization ρ for a given airport.

8.5.1 Analyzing the Impact of Different Demand Profiles on the Scheduling Performance

Here, 'demand profile' is defined as the course of demand volumes over the day. At different airports demand profiles may vary depending on airport function (e.g., hub vs. origin/destination airport), on capacity utilization resulting in constant demand volumes, on market segments (intercontinental, international, domestic flights) and on varying passenger needs (business, leisure, tourist). The alternation of peak and off-peak periods within the demand profile is characteristic at every airport. Although slight differences occur from day to day and between weekdays and weekends in particular, 'typical demand profiles' exist over one season at least.

As an example, Figure 8.9 illustrates the bandwidth of total slot requests in rolling hours at MUC. Only slots requests from 06:00–22:00 LT are considered here.

The following Section 8.5.1.1 describes how to deduce a 'typical demand profile' with actual data for each German slot coordinated airport. This allows demonstrating the relation between scheduled delays and capacity utilization ρ for each typical demand profile (Section 8.5.1.2).

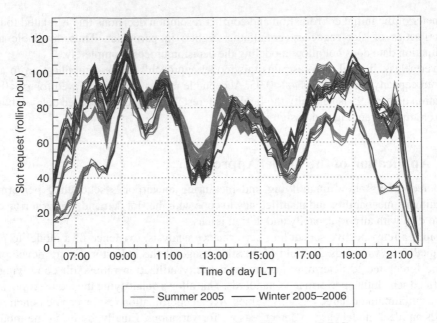

Figure 8.9 Total slot request in rolling hours (MUC)

8.5.1.1 Typical Demand Profiles

Since slot requests differ between summer and winter season significantly, we concentrate on one season only (here, the summer season). This is not a limitation, since the coordination of an actual summer season depends on the equivalent previous season. The demand profile is optimized for airport planning purposes and coordination activities by taking only weekdays into account.

Assuming that an airport's slot demand has a local structure, the magnitude of slot requests for a single day depends on individual factors such as the day of the week, peak seasons (e.g., holidays) and also inaccessible airline decisions. Scaling every single demand curve $D_{req, ARR, d}$ and $D_{req, DEP, d}$ with $D_{req, i, d} = (D_{req, i, 06:00}, D_{req, i, 06:10}, \ldots, D_{req, i, 21:50})$ (Eq. 8.9) for each considered day d by its total slot request $D_{req, TOT, d}$ (Eq. 8.4) and taking the arithmetic mean in every coordination unit k (06:00, 06:10, … , 21:50) results in a standardized:

$$\text{'typical demand profile' } \mathbf{TDP} = \begin{pmatrix} \mathbf{TDP}_{ARR} \\ \mathbf{TDP}_{DEP} \end{pmatrix} \quad [\text{Slots/time interval}]. \quad (8.14)$$

As slot requests are summed up in ten minutes intervals according to a coordination unit of 10 min (06:00 … 21:50), $\mathbf{TDP}_i$ ($i \in \{ARR, DEP\}$) is a vector with $\dim(\mathbf{TDP}_i) = \dim(D_{req, i, d}) = 96$. Arrival and departure daily slot requests are evaluated separately regarding to arrival and departure coordination parameters.

If $\mathbf{TDP}_i$ describes the typical demand profile vector, the standardized typical slot request $(TDP_i)_k$ in the kth coordination unit ($k = 06:00, 06:10, \ldots, 21:50$) is

$$(TDP_i)_k = \frac{1}{n} \sum_{d=1}^{n} \frac{D_{req, i, k, d}}{D_{req, TOT, d}}, \quad i \in \{ARR, DEP\} \quad (8.15)$$

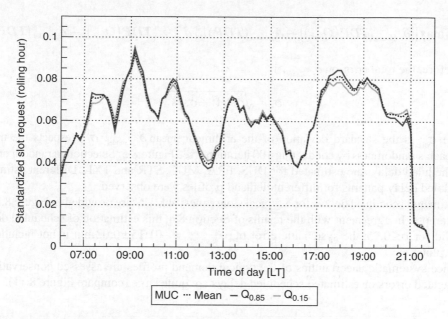

Figure 8.10 Typical demand profile (rolling hours) at MUC

where the sum runs over all days d of a scheduling season, that is, n is the number of days in this season, and depends on the total slot request at a specific day d

$$D_{req,TOT,d} = \sum_k D_{req,ARR,k,d} + D_{req,DEP,k,d}. \tag{8.16}$$

The use of the arithmetic mean requires the estimation of errors on $\boldsymbol{TDP}_i$. The typical demand profiles $\boldsymbol{TDP}(Q_{0.15})$ and $\boldsymbol{TDP}(Q_{0.85})$ concerning 0.15- and 0.85-quantiles of given demand $D_{req,i,d}$ are considered as systematic errors.

A typical demand profile at MUC is shown in Figure 8.10. For a better overview, the typical demand profile is represented with standardized total movements in rolling hours and therefore is not standardized itself.

8.5.1.2 Prediction of Scheduled Delays

The expectation is that different demand profiles lead to specific estimated scheduled delays. The model approach is now applicable to typical demand profiles $\boldsymbol{TDP}$ which are multiplied by the slot request $D_{req,TOT,d}$. The daily declared capacity utilization ρ is defined as in (Eq. 8.5)

$$\rho_{req,TOT,d} = \frac{D_{req,TOT,d}}{\Phi_{TOT,d}} \quad [-] \tag{8.17}$$

with the number of available slots per day $\Phi_{TOT,d}$ according to (Eq. 8.3).

Taking systematic errors on $\boldsymbol{TDP}_i$ into consideration, the propagated systematic uncertainty σ_{sys} on estimated scheduled delays $\delta_{s,model}$ is:

$$\sigma_{sys} = \max\left(\left| \delta_{s,model}\left(\mathbf{TDP}(Q_{0.15})\right) - \delta_{s,model}(\mathbf{TDP}) \right|, \left| \delta_{s,model}\left(\mathbf{TDP}(Q_{0.85})\right) - \delta_{s,model}(\mathbf{TDP}) \right| \right).$$

Therefore, the total error on $\overline{\delta}_{s,d,model}$ is:

$$\sigma_{\pm} = \sqrt{\sigma stat^2 + \sigma sys^2}$$

where σ_{stat} is the standard deviation of the arithmetic mean $\overline{\delta}_{s,d,model}$. σ_{stat} respects the model approach's randomness by conducting 100 iterations (see Figure 8.5: Selection in random order).

Scheduled delays are estimated for DUS, FRA, MUC, STR and TXL. Different estimated scheduled delay patterns for different demand profiles were observed.

Estimated scheduled delays for a typical slot request at MUC are provided in Figure 8.11 as an example. In agreement with the results of Section 8.4, this estimation of scheduled delays is valid for $\rho \leq 0.95$; the systematic error of $\sigma_{\overline{\delta}_{s,d,model}} = -01:15 \pm 0:02\,\text{min}$ is not included in these plots.

Since systematic uncertainties of the typical demand profiles are assessed conservatively, propagated errors on estimated scheduled delays are quite large (compare Figure 8.11).

8.5.2 Analyzing the Impact of Declared Capacity Values on the Scheduling Performance

Subsection 8.5.1 describes the influence of varying daily demand profiles (with given declared capacity restrictions) on scheduled delays; the following subsection evaluates the effect of different declared capacity values C (Eq. 8.10) (with given slot demand patterns) on scheduled delays.

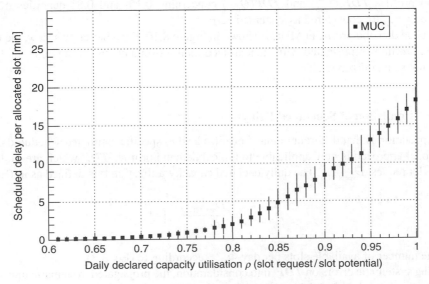

Figure 8.11   Predicted scheduled delays at MUC

Various combinations of declared capacity values C are used at German coordinated airports (Table 8.1). In the following, it is assumed that declared capacity values are constant through a specific day d, that is, the dependencies on k and d are not written explicitly. As $C_{TOT,60}$ differs for investigated German airports, we consider the:

$$\text{'Restrictiveness' } R_{i,z} := \frac{60\,\text{min} \cdot C_{i,z}}{z[\text{min}] \cdot C_{TOT,60}} \tag{8.18}$$

for each declared capacity value $C_{i,z}$ with $i \in \{ARR, DEP, TOT\}$ and $z \in \{10, 30, 60\}$ min (see Eq. 8.10).

Different declared capacity values $C_{i,z}$ for different airports are comparable and can be evaluated as 'more restrictive' or 'less restrictive':

For example, at MUC the ratio between 10 min declared capacity value and 60 min declared capacity value is $R_{TOT,10,MUC} = \dfrac{6 \cdot C_{TOT,10,MUC}}{C_{TOT,60,MUC}} = 1.01$, while at DUS this ratio is calculated as follows: $R_{TOT,10,DUS} = \dfrac{6 \cdot C_{TOT,10,DUS}}{C_{TOT,60,DUS}} = 1.5$.

Thus, MUC is significantly more restrictive with regard to its 10 min declared capacity value ($R = 1.01$) compared to DUS ($R = 1.5$).

Assessing maximal effects of coordination parameters on scheduled delays, we concentrate only on restrictive values of coordination parameters. Practically, the lowest analyzed ratio R (Eq. 8.18) of a coordination parameter at German airports (Table 8.1) is used in the following.

Let us define: $D^*_{req,TOT} = 1600$ [Slots/day] (Eq. 8.16), $C^*_{TOT,60} = 100$ [Slots/hour] (Eq. 8.10). Using $C^*_{TOT,60} = 100$, the following capacity values are calculated, being based on capacity restrictions at German airports (Table 8.1):

	deduced from:
$C^*_{TOT,10} = 17$ ('restrictive')	MUC
$C^*_{TOT,10} = 25$ ('nonrestrictive')	DUS
$C^*_{ARR,10} = C^*_{DEP,10} = 10$	FRA
$C^*_{TOT,30} = 54$	FRA
$C^*_{ARR,60} = C^*_{DEP,60} = 53$	FRA

with, for example, $C^*_{TOT,10} = R_{TOT,10,MUC} \cdot \dfrac{C^*_{TOT,60} \cdot 10\,\text{min}}{60\,\text{min}} = 17$.

Since the estimated scheduled delay is measured as a relative quantity (estimated scheduled delay *per slot*), the chosen magnitude of total requested slots does not have any effect on our results and is for rounding purposes only.

After having defined the declared capacity values C^* as one of the two input parameters for our model approach, we scale the (normalized) typical demand profiles (Eq. 8.14) to the slot request of $D^*_{req,TOT} = 1600$ movements a day (Eq. 8.16). Therefore, the diversified slot demand patterns for German coordinated airports are respected in our model. The effect of varying the declared capacity values on estimated scheduled delays for a given demand pattern can now be studied using our model approach.

With respect to the combinations of declared capacity values used at German coordinated airports, we define the following scenarios:

1. $C^*_{TOT, 60}$
2. $C^*_{TOT, 60}$, $C^{* \text{ (nonrestrictive)}}_{TOT, 10}$
3. $C^*_{TOT, 60}$, $C^*_{TOT, 10}$ (restrictive)
4. $C^*_{TOT, 60}$, $C^*_{TOT, 10}$, $C^*_{ARR, 10}$, $C^*_{DEP, 10}$
5. $C^*_{TOT, 60}$, $C^*_{TOT, 10}$, $C^*_{ARR, 10}$, $C^*_{DEP, 10}$, $C^*_{TOT, 30}$
6. $C^*_{TOT, 60}$, $C^*_{TOT, 10}$, $C^*_{ARR, 10}$, $C^*_{DEP, 10}$, $C^*_{TOT, 30}$, $C^*_{ARR, 60}$, $C^*_{DEP, 60}$
7. $C^*_{TOT, 60}$, $C^*_{TOT, 10}$, $C^*_{ARR, 10}$, $C^*_{DEP, 10}$, $C^*_{ARR, 60}$, $C^*_{DEP, 60}$

with all other capacity values being set to infinity.

Estimated scheduled delays per slot for a given scenario of declared capacities and derived demand patterns (subsection 8.5.1.1) for declared capacity utilizations $0.6 \leq \rho \leq 0.95$ (Eq. 8.5) can now be estimated using our model approach.

In order to compare different demand profiles, scheduled delays per slot are summed up for $0.6 \leq \rho \leq 0.95$. As those absolute cumulative values are not relevant here, scenario 1 is normalized to the sum of estimated scheduled delays per slot in $0.6 \leq \rho \leq 0.95$.

Figure 8.12 describes the (relative) change $\left(\dfrac{scenario\, x}{scenario(x-1)} \right)$ of the sum ($0.6 \leq \rho \leq 0.95$) of the estimated scheduled delays per slot relative to the preceding scenario $(x-1)$, where x is the number of the scenario. As an example, using MUC's restrictive 10 minutes total declared capacity value (this is scenario 3) in combination with FRA's demand profile would lead to an on average ~4 times higher scheduled delay per allocated slot compared to the application of DUS's 10 min declared capacity value (this is scenario 2).

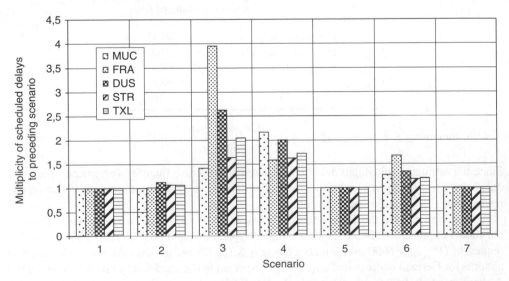

Figure 8.12 Influences of additional capacity restrictions: Multiplicity of scheduled delay per slot in scenario x with respect to the preceding scenario $x-1$

It is obvious that a declared capacity value $C_{TOT,10}$ increases scheduled delays if it is too restrictive (scenario 3). A nonrestrictive 10 min coordination parameter as used in DUS has nearly no effect on scheduled delays (scenario 2). Additional 10 min arrival and departure coordination parameters can double the scheduled delay (scenario 4) and additional 60 min arrival and departure coordination parameters at least rise scheduled delays (scenario 6). In contrast, 30 min declared capacity value $C_{TOT,30}$ has no effect on scheduled delays (scenarios 5 and 7).

8.6   Conclusion

Scheduled delays allow indicating an airport's incapability and insufficiency to cope with the slot demand. By a retrospective scheduling performance analysis scheduled delays may contribute to addressing airport capacity and therefore also design issues.

In this chapter, a model was introduced to predict scheduled delays depending on the level of declared capacity utilization. This extends the field of possible usage of the indicator. Future developments may be forecasted to determine minimum declared capacity values still complying with desired scheduling performance standards. Also actual coordination results could be benchmarked towards an ideal model case. Here, the impact of relevant initial coordination input parameters, the demand profile, and declared capacity restrictions on the scheduling performance have been analyzed.

Due to limited data availability the introduced model approach excludes some crucial parameters forming the base for the initial coordination: Priority categories of all slot requests in particular. Therefore, model accuracy is sufficient for a limited capacity utilization range only ($\rho \leq 0.95$). Future research needs to aim for full coverage of the decision process for conflict resolution during the initial coordination. Additional slot allocation data including priority categories would be required. To calibrate and to finally validate a model approach being enhanced accordingly, data is needed to cover several scheduling seasons at different coordinated airports.

References

Airport Council International – ACI. (2004) *Study on the Use of Airport Capacity.* An ACI Europe study, pp. 1–2, Brussels.

Council Regulation (EEC) No 95/93 of 18 January 1993 on common rules for the allocation of slots at Community airports. (1993) *Office Journal L* 14, 22.1.1993, p. 1, Brussels, as amended in 2002, 2003, 2004.

Dempsey, P. S. (2001) Airport Landing Slots: Barriers to Entry and Impediments to Competition. In: *Air and Space Law*, Volume 26, Number 1, pp. 20–48.

International Air Transport Association – IATA. (2007) *Worldwide Scheduling Guidelines.* 14th Edition, Montreal.

International Civil Aviation Organization – ICAO. (2005) *Global Air Traffic Management Operational Concept.* Doc. 9854, 1stE, pp. D1–D2. Montreal.

Kösters, D. (2007) Airport scheduling performance – an approach to evaluate the airport scheduling process by using scheduled delays as quality criterion. *Proceedings of ATRS Annual World Conference*, Berkeley.

Performance Review Commission – PRC. (2005) *Report on punctuality drivers at major European airports.* Report commissioned by the Performance Review Commission, pp. 5–17, Brussels.

Ulrich, C. (2008) How the present IATA slot allocation works. In: Czerny, A., Gillen, D., Forsyth, P. and Niemeier, H.-M. (eds) *Airport Slots – International Experiences and Options for Reform*, pp. 9–21, Ashgate, Hampshire.

9

Implementation of Airport Demand Management Strategies: A European Perspective

Michael A. Madas[A] and Konstantinos G. Zografos[B]

[A]Transportation Systems and Logistics Laboratory (TRANSLOG), Department of Management Science and Technology, Athens University of Economics and Business, Greece
[B]Department of Management Science, Lancaster University Management School, Lancaster University, UK

9.1 Introduction

Over the past 30 years, exceptionally high air transport growth has been evident worldwide. The overwhelming increases in demand in conjunction with severe political, physical, and institutional constraints in providing sufficient capacity have resulted in a serious gap between demand and capacity. Characteristically, the planned capacity at the 138 Eurocontrol Statistical Reference Area (ESRA) airports was expected to increase by 41% in total by 2030, while the corresponding demand was foreseen to exceed airport capacity by as many as 2.3 million flights (or 11%) in the most-likely growth forecast scenario for 2030 (Eurocontrol, 2008). Under such a scenario, 14–39 European airports will need to operate at full capacity 8 h per day to serve the demand, similarly to what most severely congested airports of the world did during the pre-economic crisis era (Eurocontrol, 2008).

Although traffic growth has been significantly slowed due to the recent global economic downturn, most recent forecasts provide evidence of recovery albeit with around five years of growth lost as compared to the pre-2008 trend forecasts (Eurocontrol, 2008; Steer Davies Gleave, 2011). Irrespective of the actual timing for rebound, the recovery of the air transport market will bring again into the forefront severe capacity shortages and give rise to sharp

Modelling and Managing Airport Performance, First Edition. Edited by Konstantinos G. Zografos, Giovanni Andreatta and Amedeo R. Odoni.
© 2013 John Wiley & Sons, Ltd. Published 2013 by John Wiley & Sons, Ltd.

congestion and delay phenomena. Even amidst the crisis, demand exceeded capacity for most or even throughout the entire day at the busiest European airports, while the percentage of flights delayed reached almost 39% at an average delay per delayed flight exceeding 28 min (Steer Davies Gleave, 2011; Eurocontrol, 2012).

The increasing imbalance between airport capacity and traffic resulted in congestion and delay figures that have drawn the attention of aviation policy makers and triggered policy discussions that bring into the forefront the challenging dilemma: Demand/Congestion Management or Capacity Enhancement? The solution to the congestion and delay problem can be viewed through reducing the ratio of demand (traffic) to capacity (supply), but there is much controversy over which part of the ratio should receive higher priority. The rapidly increasing congestion and delay phenomena in conjunction with problems or complications in expanding existing capacity render a pure supply-side solution rather impossible. Because of the severity of the problem, demand management solutions aiming to control congestion through the allocation of scarce airport capacity (expressed in slots[1]) have lately received a great deal of consideration from policy makers, airport users, and operators.

Demand management has been long recognized as a principal instrument to deal with congestion and capacity shortfalls or internalize transport externalities (e.g., delays, noise, emissions, safety/risks) imposed by various modes on the transportation system and the society at large. In an airport context, demand management exhibits several variations and peculiarities. First, it covers a large variety of decisions related to the allocation and use of interdependent airport components/elements such as runways, apron stands and gates, concessions, and so on. Second, it deploys a wide spectrum of measures ranging from pure administrative rules such as the diversion of general aviation traffic in "reliever" airports and air traffic perimeter rules (U.S. Congress, 1984; Fisher, 1989; TRB, 2001) to pure economic, market-based or hybrid instruments such as the implementation of airport congestion-based pricing schemes (Brueckner, 2002; Fan, 2003; Morrison and Winston, 2007), as well as slot trading and auctions (DotEcon, 2001; TUB, 2001; NERA, 2004; Mott MacDonald, 2006; Ball et al., 2007). Third, it deals with various policy objectives with regards to airport externalities either through the rationing of capacity and congestion management (e.g., slot allocation), or the assessment, mitigation and pricing of noise (e.g., noise charges or surcharges, noise limits/ traffic curfews) (Forsyth, 1997) or emission-related externalities (e.g., emission charges, quantity controls, inclusion of the air transport sector in the EU Emission Trading Scheme) (Mendes and Santos, 2008). Fourth, it deals with both congestion and scarcity of airport infrastructure. In principle, strategies focusing on the efficient and fair allocation of capacity are meant to treat scarcity of airport infrastructure (mainly runway slots) in various ways (e.g., grandfathering, trading, auctions, congestion-based pricing). Once capacity is allocated, congestion may develop for several reasons. In this case, administrative procedures/rules (e.g., hourly limits, traffic diversion) or pure pricing schemes (e.g., peak hour pricing) can be applied in order to deal with congestion by somehow handling the temporal profile of demand (e.g., traffic shifts throughout the day). The scope of this chapter covers airport demand management approaches and strategies aiming to deal with both scarcity and congestion by strategically allocating scarce runway capacity in the form of slot allocation[2].

Despite a comprehensive amount of policy and research proposals in this field (discussed in Sections 9.2 and 9.3), attempts to bring forward demand management initiatives were not widely adopted and have not flourished into policy practice mainly due to industry inertia forces, practical complexities, and political hesitance. Furthermore, public opposition from established industry actors and voices doubting the actual necessity or effectiveness of demand management initiatives has also raised a barrier to potential implementation. On the other hand, policy makers exhibit some reluctance to introduce a new policy regime that will signify a substantial departure from the status quo with many and – to some extent – unpredictable implications or side effects on the operation of the global air transport system. Finally, a possible reason behind this gap between theory and practice might be the limited guidance as to how various demand management instruments and measures can be realistically integrated and operationalized within an overall strategic policy framework for the allocation of scarce airport capacity.

This chapter presents the various policy and research proposals on the allocation of scarce airport capacity with particular emphasis on European Union's airports. Its ultimate objective is twofold: (1) to contribute in the discussion/justification of the real need and motivation behind the adoption of a new policy regime and (2) to formulate an integrated policy roadmap that will guide the implementation process for such a new regime at different types of EU airports. In addition, alternative approaches that can potentially act as improvement complements (rather than substitutes) to the existing policy practice are also discussed. The remainder of this chapter consists of six thematic sections. Section 9.2 provides an overview of the current state of affairs in terms of the historical evolution of relevant policy practices in European airports. Section 9.3 discusses the main research efforts, proposals, and market-based allocation mechanisms examined in the literature for the allocation of scarce airport capacity. Section 9.4 provides some arguments in favor of the adoption of a new policy regime, while Section 9.5 proposes a strategic policy framework guiding the adaptation and implementation process of the new regime at both local airport level and the broader airport network. Section 9.6 identifies and briefly discusses some alternative complementary approaches to the existing policy practice, and finally Section 9.7 presents the concluding remarks.

9.2 Current Practice

Access to airports is typically "controlled" – if at all – through a landing fee that is proportional to the maximum take-off weight (MTOW) of the aircraft irrespectively of the actual infrastructure utilization and congestion levels. Currently, at most airports, aircraft are charged a small (relative to total operating costs of the airlines) ante fee that is uniform throughout the day. The historical justification for this fee was the ability-to-pay principle (i.e., higher charge for large aircraft carrying more passengers as compared to smaller aircraft and general aviation), as well as the notion of cost-relatedness. The cost-relatedness principle implies an accounting rather than economic concept for the definition of costs, which specifies that "airport charges should be based on the cost of facilities and services provided by the airport, allowing for a reasonable return on capital, the proper depreciation of assets, as well as the efficient management of capacity" (European Commission, 1998).

Typically, charges are designed and set in order for the airport authority to break even. Given the cost-relatedness principle and the fact that large airport operators have been highly successful in boosting non-aeronautical revenues over the years, there is a paradoxical situation in which aeronautical charges (mainly landing fees) have hardly increased (in many cases have even declined in real terms) in the world's largest and busiest airports since 1986 (Fan, 2003). As a result, landing fees are not at a level that automatically clears the market at least in congested airports experiencing excess demand for some part of the day (Fan, 2003). Since the typical charging mechanism cannot clear the market, as would be the case in other "normal" private goods, the industry has developed alternative means and procedures (dating back to a system created by IATA in 1947 and gradually updated over the years) to handle the conflicts of interest involved in the allocation of scarce airport capacity or, at least, to establish a clear and concrete slot allocation mechanism. Traditionally, capacity at most congested airports of the world is expressed in slots and allocated within the framework of voluntary guidelines developed and evolved over the years under the auspices of IATA with local interpretations and adaptations (IATA, 2010). Since 1993, the existing IATA-based slot allocation regime has been adapted, complemented, and further updated by regulation (European Commission, 1993; 2004; 2009) for the European Union (EU) airports. According to the IATA-based system and its complementary version of the EU regulation, these guidelines set out administrative procedures, appoint schedule/slot coordinators in the airport or national level, organize biannual scheduling conferences, and include a set of rules and criteria to be applied for the allocation of airport capacity.

The traditional IATA-based system of slot allocation, complemented by EU regulation, is built upon a principal and overriding rule, which essentially acknowledges an incumbent airline's "grandfather's right" (i.e., historic slot holding) to a particular slot time at an airport where the slot was used in the previous equivalent season. These rights continue to be active until an air carrier ceases to utilize it or surrenders it back to the airport slot pool. The pool of slots that are not "grandfathered", in addition to newly available slots, are allocated according to priorities and rules dictated by IATA (e.g., 50% of the slot pool goes to new entrants, slot usage patterns, flight regularity, carrier characteristics). The allocated slots, in turn, become "grandfathered" for the next season, that is, the specific users obtain the first "right" on them provided that they have used them over 80% of the time during the previous year (i.e., "use-it-or-lose-it" rule). Slots that are allocated to an airline – whether "grandfathered" or newly available – may be exchanged for slots held by another airline, on a strict one-for-one, non-monetary basis. In 2008, the European Commission published an interpretative Communication (European Commission, 2008) which, among other issues, recognizes slot exchanges with monetary considerations (commonly referred to as secondary trading) as a legitimate option to reflect differences in values between exchanged slots. The latter essentially has left much room for – more or less formal – monetary slot exchanges.

Demand management in the sense of IATA rules-driven slot allocation has been the dominant access control mechanism practiced by the majority of busiest (or schedule coordinated) airports outside the United States[3]. Most of these airports have put into use more or less sophisticated instantiations of the IATA rules and allocation criteria to handle traffic and control its growth in one way or another. During the last decade or so, the European Commission pursues a radical reform of the existing IATA-based slot allocation regime that will mostly fit the definition of market and/or hybrid demand management instruments (e.g., slot trading, congestion pricing, auctions) towards easier new entrants' access to the market, better use of airport capacities, and the establishment of "market values" for airport capacity. To that end, an industry-wide dialogue and consultation with stakeholders has been launched in Europe, while simultaneously several studies and research efforts (discussed in the next section) have come to describe, assess, and ultimately bring forward market-based approaches.

The outcome of this lengthy consultation process has been recently reflected on a proposal announced by the European Commission in December 2011 (European Commission, 2011). This proposal ("Airport Package") contains a dedicated legislative proposal on slots that introduces changes such as formally allowing secondary trading, reforming market access rules for new entrants, tightening the minimum slot usage rules, and reinforcing the independence of coordinators. The new legislative proposal seems to pave the way to the introduction of market-based mechanisms (initially through secondary trading) in the future. On the other hand, it only introduces incremental, targeted improvements that do not admittedly justify for a long-awaited, radical reform of the EU slot regulation.

9.3   Review of Existing Policy Proposals

The role of the traditional IATA system along with the enhanced version introduced by EU Regulation 95/93 is crucial and effective in airports operating under capacity or experiencing only temporary capacity shortages. However, this is not the case for airports already operating at maximum capacity and experiencing severe congestion problems throughout the day. The rapidly increasing congestion and delay figures in most of the busiest European airports in conjunction with the inefficiencies of the current slot allocation regime have stimulated substantial research interest and various proposals towards market-based capacity allocation.

On the European side, the European Commission has initially commissioned a number of studies with the aim to assess the implementation of the Regulation 95/93 (and its subsequent amendments) on the allocation of slots at European Community airports. The Coopers and Lybrand study (Coopers and Lybrand, 1995) assessed the extent to which the EU Regulation has been implemented across the Community. Furthermore, it investigated the effectiveness of the Regulation and identified problems and possible modifications required for the improvement of its efficiency. In 2000, another study was commissioned by the European Commission (PwC, 2000) in order to assess the state of implementation of the various aspects envisaged in the Regulation 95/93. Both studies have been used in order to identify problems, necessary modifications, and areas of improvements of the first version of the Regulation (1993) with the ultimate objective to prepare and release an improved version of the Regulation amending Regulation 95/93 (eventually appearing in early 2004).

In a subsequent study, the Technical University of Berlin (TUB, 2001) analyzed the current practice of slot allocation and originally introduced market-based instruments and strategies in the policy landscape of Europe. In particular, the study explored the following strategies: (1) the "Big Bang" involving the withdrawal of grandfather rights at a certain reference date and the auctioning of all slots at once, (2) the "Gradual Approach" standing for a gradual withdrawal of grandfather rights accompanied by improved slot recycling mechanisms and a centralized auction for the reallocation of slots returned to the pool, (3) a "More Complete Pricing Policy" envisaging the adaptation of the airport take-off and landing fees to the respective scarcity situation in accordance with the economic principle of cost-relatedness, and (4) the "Slot Trading" introducing a "buy-and-sell" environment for exchanging slots under monetary terms.

Almost immediately after the amended EU Regulation (793/2004) has been put into force, the European Commission launched an industry-wide consultation on the potential implementation of market mechanisms, while simultaneously commissioning a study to assess the effects of market-based slot allocation schemes (NERA, 2004). This consultation document proposed a drastic revision of the current status and practice by introducing market-driven slot allocation instruments in various contexts: (1) secondary trading accompanying either the existing administrative procedures of primary allocation or even a market-based primary allocation mechanism, (2) higher posted prices accompanied by secondary trading, (3) auctioning of pool slots complemented by secondary trading, and (4) auction of 10% of slots complemented by secondary trading.

In parallel to research proposals, some European Union Member States (e.g., UK, Italy, France, The Netherlands, Germany) have considered technical amendments and proposals. The UK Government has instituted a consultation study (UK Department of the Environment, Transport and the Regions, 2000; DotEcon, 2001) stating its intention to support and lead a proposal for a fundamental change on the current system of slot allocation. The proposed mechanism was based on a concession system with a time limitation of slot rights subject to regular auctions combined with secondary trading. In a similar direction, the UK Civil Aviation Authority has prepared a technical consultation paper setting out guidelines and proposals for the implementation of a formalized system of secondary trading in slots (CAA, 2001). This widely discussed alternative (DotEcon, 2001; TUB, 2001; NERA, 2004) practically involves the establishment of a "buy-and-sell" environment, where slots will be traded in monetary terms under specific legal provisions, thus enabling carriers to better adapt their slots' portfolio in the global airport network.

A more recent study (Mott MacDonald, 2006) investigated the potential acceptability and evaluated the implications from the introduction of secondary trading mechanisms for runway slots at congested European Community airports. This study presented the relevant experience of secondary trading in US and the "grey" London airports' market as well as other industries (e.g., capacity rights for natural gas and electricity, emissions trading), developed comprehensive forecasts of slot demand to 2025 (both with and without secondary trading), and performed a series of economic impact assessments based on slot forecasts. Furthermore, it gave consideration to a range of possible amendments to the current regulation with emphasis on the impact assessment of an increased (from 80–90%) "use-it-or-lose-it" rule, the auctioning of newly created slots, and a recycling program for increasing the mobility of historic slot usage rights.

Along the same lines, another study (Steer Davies Gleave, 2011) provided an updated evaluation of the operation of the EU regulation on airport slots and assessed various options for revision to the Regulation that are mostly related to the following: (1) improvement of the independence of the coordinator, (2) strengthening slot monitoring, usage, and enforcement rules, (3) revision of the "new entrant" definition to further facilitate market access and promote competition, and (4) introduction of market mechanisms both for primary allocation of slots and/or secondary trading. As far as the latter point is concerned, the study examines the following options: (1) withdrawal of grandfather rights and auctioning of the slots only for a small number of the most congested EU airports, (2) auctioning of all newly created/available slots, and (3) establishment of a formal secondary trading environment for all slots. Some of these options have been eventually included in the recent legislative proposal by the European Commission (European Commission, 2011).

Demand management measures at their more aggressive market-based variations have lately received a great deal of consideration by the air transport research community. Auctions have recently gained increasing acceptance as a hybrid demand management measure for congested airports. Many researchers believe that auctions can be applied in the airport context similarly to their application in other industries (e.g., radio spectrum, emissions, railway track capacity) (Nilsson, 2003) and went even further by designing and assessing auction-based allocation models at specific congested airports (Le et al., 2004; Ball et al., 2007). In most cases, however, the operation of an aftermarket (e.g., secondary slot trading) has been considered to be a necessary supplement of the auction process in order to mitigate possible slot complementarity problems. In a pure pricing form, congestion-based pricing schemes have been extensively examined as an access control mechanism. In one way or another, congestion-based pricing schemes envisage the complete removal of slots and grandfather rights. Each carrier will be able to operate at any time by paying the corresponding scarcity surcharges that will vary throughout the day based on congestion and will be set at a level being able to internalize marginal social costs or at least this portion of delay costs imposed on others (TRB, 2001; Brueckner, 2002; Fan, 2003; Morrison and Winston, 2007). Other research efforts (Madas and Zografos, 2006, 2008) have capitalized on the individual mechanisms/instruments discussed above in order to define and assess comprehensive slot allocation strategies introducing a varying application of market-driven allocation mechanisms (e.g., decentralized auctions, centralized trading, full secondary trading) or pure pricing schemes (e.g., congestion pricing).

Recognizing the crucial importance of airport congestion management towards sustainable air transport in the future, many researchers have recently elaborated on possible causes and alternative remedies to congestion in conjunction with the associated performance outcomes of such congestion management options. These basically address congestion pricing and slot trading in various formats and assess their potential effectiveness and applicability under different market power, demand, and cost conditions at both single and airport network level (Basso and Zhang, 2010; Czerny, 2010; Verhoef, 2010). Finally, side impacts and possible implications of such options on both airlines and airports (e.g., anti-competitive behavior, aircraft size, flight frequency, capacity investments) were further elaborated in recent literature (Flores-Fillol, 2010; Fukui, 2010; Zhang and Zhang, 2010).

The approaches discussed in the literature range from minor adjustments or enhancements of the status quo to more radical or aggressive revisions of the current slot allocation regime contemplating primary and/or secondary slot trading, auctions of (part of or the entire) slot pool, and congestion pricing. In their majority, research proposals introduce a drastically new, market-driven capacity allocation mechanism with clear and implicit barriers, though, involved in the practical implementation process. On the other hand, there is industry consensus that the existing system cannot (and will not definitely in the future) cope with the current and, most importantly, the forecasted traffic volumes. The existing slot allocation regime needs to be drastically improved, while small, periodical adjustments or local adaptations of the status quo may not suffice. A number of policy attempts or initiatives motivated by relevant research work have emerged with the purpose of introducing pure market-driven schemes but they have not materialized until now. One major reason seems to be the long standing industry debate concerning the real need and potential impacts of such a "policy shift". This is further discussed in the following section.

9.4   Is a New Regime Really Necessary?

This section aims to provide some arguments and quantitative evidence on the real need and motivation behind the adoption of a new demand management regime at EU airports. This has been pursued through the following: (1) identification of main problems, limitations, and inefficiencies of the existing IATA-based slot allocation system, (2) assessment of potential quantitative impacts (delay and cost savings) of a potential congestion management regime at different types of European airports, and (3) assessment of the pricing effectiveness of the existing system in controlling access to busy airports. These are further discussed in the subsequent sub-sections.

9.4.1   Mismatch and Misuse

The primary area of inefficiency of the existing slot allocation system in Europe is its reduced capability to balance growing traffic with scarce capacity (mismatch) especially at the busiest and most congested European airports. Demand currently exceeds by far capacity for most or even throughout the entire day at the busiest European airports (Steer Davies Gleave, 2011). At the same time, misuse (early and late slot returns, overbidding, "off slot", "no shows") of existing capacity emerges as a new "thorny" issue that has also its share on the inefficient allocation and use of an already insufficient resource (ACI Europe, 2004). Even at airports where demand for slots exceeds capacity, over 10% of the allocated slots are not eventually utilized (Steer Davies Gleave, 2011). Some quantitative evidence of the economic magnitude of this problem has been provided by ACI Europe who estimated that unused slots due to late slot returns not allowing for substitution or redistribution of slots account for approximately €20 million fewer revenues per season only for large, congested European airports (ACI Europe, 2009). In addition, several slot misuse patterns have been also observed such as the use of high proportion of small aircraft – thus limiting the number of passengers that can be accommodated – and the loss of direct short-haul connections of small communities/regions to hub airports in favor of longer-haul and more profitable services (Steer Davies Gleave, 2011).

Unfortunately, although slot misuse reduces the effective capacity and sharpens the capacity shortage problem, mitigation measures against misuse (e.g., regulatory directions, sanction schemes) have been quite ineffective if applied at all. This is basically reflected on: (1) the limited compliance of Member States with the EU regulatory obligation to introduce sanctions for slot misuse into national law, (2) the inadequate/insufficient application of the regulatory requirements related to slot usage monitoring, and (3) the limited application of criteria (by the Coordinator) for slot withdrawal in case of underutilization (by airlines) (Steer Davies Gleave, 2011). In any case, unused or misused slots, regardless of causality, provide an indication of the poor efficiency of the allocation process itself, and they admittedly have a high economic and disruptive impact on airports, airlines, passengers, and the society at large.

9.4.2   Poor Allocation Efficiency

There is clear evidence that the observed mismatch and severe slot misuse are further magnified by inefficiencies of the slot allocation mechanism *per se* (Zografos et al., 2012). The allocation process undertaken by slot coordinators produces really poor allocation outcomes since it fails to properly match slots requested with slots eventually allocated to airlines. This can be also attributed to the fact that slot coordinators are asked to deal with and manage empirically a rather complicated allocation process with multiple criteria, rules, and priorities. As a result, the problem size and complexity give rise to serious allocation inefficiencies that are initially reflected on substantial deviations between requested and allocated slots. Quantitative evidence from three Greek regional airports has shown that the efficiency of the initial allocation outcome produced by the existing slot coordination process can be improved in a range between 14–95% (Zografos et al., 2012). This can be achieved by the use of optimization aiming to minimize the difference between the allocated and the requested slot time, thus better accommodating airlines' preferences. At the outset, more preferable slots imply more satisfied demand, and most importantly, allocated slots, being closer to initial requests, will be more efficiently used or, at least, will be eventually operated.

9.4.3 Declared Capacity Considerations

One of the fundamental concepts and principal input parameter for the slot allocation mechanism is declared capacity, which specifies the number of slots available at an airport per unit of time. The declared capacity notion represents an artificial, administrative measure of capacity which is determined rather than computed. Despite its utmost importance and substantial influence on the efficiency of the allocation process, the declared capacity determination process has not been sufficiently examined in the literature (Railsback and Sherry, 2006; Koesters, 2007). Current practice on the determination of declared capacity at EU airports has received much criticism with the main arguments being the following (PwC, 2000; Odoni and Morisset, 2010): (1) there is ample room for local interpretations and adaptations (e.g., coordination parameters, technical, operational, and environmental constraints, capacity constraining factors), (2) the setting of strikingly low capacity levels that are basically determined as a percentage of the capacity under Instrument Flight Rules (IFR), and (3) the empirical or *ad hoc* process[4] (rather than technical, based on "commonly recognized methods" as suggested by the Regulation) often applied by coordinators in determining the declared capacity.

As a result, the following two major categories of problems arise: (1) lack of a harmonized way of interpreting, determining/computing, and managing declared capacity, and (2) declared capacity is set at too high or too low levels. It is important to note here that the level of declared capacity has serious impacts on the allocation efficiency (Zografos et al., 2012), while simultaneously affecting the classical trade-off between acceptable level of service and utilization levels of scarce airport resources. At the outset, the determination of declared capacity signifies a very important and highly impacting decision to leave it based solely on an administrative – and to some extent empirical or even arbitrary – determination process.

9.4.4 Barriers to New Entrants

Despite several revisions of the EU slot regulation towards reforming market access to new entrants – mainly through the (re)definition of the "new entrant" term – new entrant rules have been rather ineffective. First, the slot pool provides, by definition, very limited opportunities for new entrants to grow their operations since newly available slots due to capacity expansion are extremely scarce, while at the same time the system of historical preference ("grandfather rights") offers very little incentives for incumbent airlines to withdraw slots even if they cannot use them effectively (slot hoarding, "babysitting") (ACI Europe, 2004). The latter further deteriorates the previously discussed slot mismatch and severe misuse problems. Second, various policy efforts to broaden the definition of "new entrant" have eventually led to an over-fragmentation of the schedule at congested airports due to the allocation of a small numbers of slots to a large number of carriers. This, in turn, undermines economies of scale in airline operations and the schedule viability in its entirety. Overall, the existing, grandfather-based slot allocation system suffers from a very low mobility (turnover) of slots. As a result, it is quite difficult for new entrants to establish a competitive and sustainable foothold at busy airports and challenge the dominant position of incumbent airlines (Steer Davies Gleave, 2011).

9.4.5   Potential Impacts

Recent research (Madas and Zografos, 2010) provided some quantitative insights into the remarkable impacts from the potential implementation of a new congestion management regime. The impact assessment exercise was implemented with the use of DELAYS model (Malone and Odoni, 2001) for a number of European airports exhibiting significantly different traffic, congestion, and delay patterns (i.e., Amsterdam-Schiphol, London-Heathrow, Brussels, Düsseldorf, Copenhagen, Athens, Venice). The DELAYS model was used to assess the potential operational and economic impacts of congestion management (or the lack of it) by quantifying the delays caused by additional runway movements or delay savings due to reduced runway movements (i.e., marginal delays). The idea behind the use of marginal delays is that the delay one aircraft imposes on another can be substantial, and therefore landing fees should be charged to account for that delay (Morrison and Winston, 1989, 2007) (see also Section 9.4.6). The marginal delay figures were then converted to monetary terms (i.e., marginal delay costs) by considering direct operating costs and passenger costs (through the time spent waiting) (Oxford Economic Forecasting, 2006).

According to the results of the impact assessment exercise, some of the representative European airports experience severe delays whose a snapshot is reflected on the marginal delay figures presented in Figure 9.1. For example, it is interesting to see that one additional movement at Amsterdam-Schiphol Airport (AMS) during the 17 busiest hours of its typical day of operations would cause an additional delay to practically each of the subsequent movements during the rest of the day that would result on average in roughly 1 aircraft-hour of additional total delay. The same figure corresponding to London-Heathrow Airport (LHR) would be roughly 3.9 aircraft-hours of additional total delay with lower values being evident at large and medium-sized airports (e.g., Brussels-BRU,

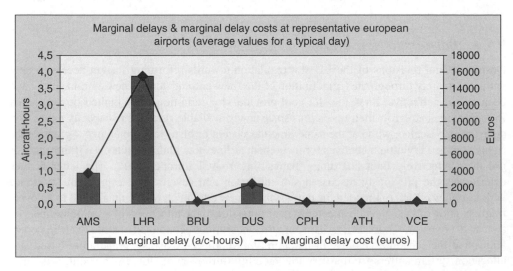

Figure 9.1   Marginal delays and costs at representative European airports

Düsseldorf-DUS) or even negligible values appearing at smaller airports (e.g., Copenhagen-CPH, Athens-ATH, Venice-VCE).

Delays were then translated into costs assuming that each minute of delay would entail direct operating costs for airlines and waiting costs for passengers (Oxford Economic Forecasting, 2006). The cost quantification of marginal delays demonstrates that the aggregate impact of such an incremental movement is remarkable and significant – at a varying extent though – in economic terms as well[5]. In other words, an additional movement during the 17 busiest hours would impose on airlines and passengers a total cost ranging €100–300 (at non-congested European airports) and €3000–15 000 (at heavily congested European airports) (Figure 9.1). Furthermore, it has been also shown that the extent and severity of marginal delays and their derived costs vary substantially with the airport, while simultaneously being in direct and close association with the airport congestion levels (demand-to-capacity ratios). Airports experiencing high demand-to-capacity ratios (above 80%) like LHR, AMS, and DUS, are those suffering from the most severe delays and their associated costs (ranging from €2500 to €15000 of total marginal delay costs). In general, a demand-to-capacity ratio of approximately 60% seems to constitute the maximum threshold above which total marginal delays and their costs start to boom (Madas and Zografos, 2010).

Extrapolating the estimated marginal costs (computed on the basis of a typical day) to an annual basis would give some outstanding figures on the order of one to a few million Euros in most of the busiest European airports. Certainly, marginal delays and their associated costs can be also seen from the viewpoint of less delays and cost savings as a result of one less flight. Consequently, marginal delays and their cost quantification can be reasonably interpreted as the potential impact of an attempt to control demand through any kind of congestion management scheme.

9.4.6 Pricing Effectiveness of Existing System

The previously discussed congestion and delay externalities could be theoretically eliminated through an appropriate pricing system that would price the utilization of a scarce airport infrastructure – in the form of one landing or take-off operation – by one user at a level equal to the total marginal cost imposed on all subsequent operations (Morrison and Winston, 1989, 2007; Fan and Odoni, 2002). Externalities like congestion and delays and their associated costs are the result of an inappropriate pricing system under which the cost internalized by each user (in the form of airport landing fees) lags behind the total marginal cost imposed on the system as a whole (i.e., users, passengers). The comparison between marginal delay costs and landing fees (Figure 9.2) reveals that the actual pricing system of scarce airport infrastructure is quite far from perfect (Madas and Zografos, 2010). Ideally, a perfect balance (i.e., 50: 50) between marginal delay costs and landing fees would constitute the definition of the perfect pricing system since it would imply full internalization of marginal delay costs through the landing fee. As it can be inferred from Figure 9.2, substantial imbalances exist between marginal delay costs and landing fees, thus rendering some (mainly the largest and busiest) airports extremely underpriced. The latter provides additional justification of the need for a "policy shift" to

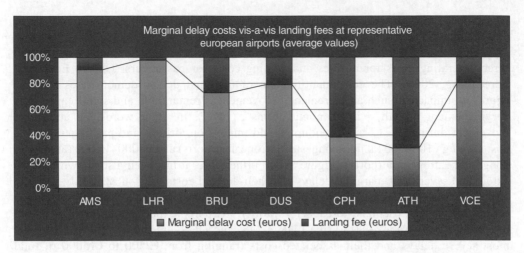

Figure 9.2   Marginal delay costs vis-à-vis landing fees at European airports

a new congestion management regime with a specific pricing scheme capable of dealing with congestion and delay costs.

9.5   From Theory into Policy Practice

Research proposals aiming to bring forward new congestion management measures, instruments, or schemes were not widely adopted and have not flourished into policy prac- tice. One possible reason behind the observed gap between theory and policy practice is the lack of guidance as to how these various instruments and measures can be integrated and operationalized within an overall strategic policy framework for the allocation of scarce airport capacity. This section aims to formulate a policy roadmap that will guide the implementation and adaptation process for a new congestion management strategy/regime at different types of EU airports. The proposed strategic policy framework borrows some theoretical principles of congestion pricing and introduces a simple, open, and adaptive framework for allocating capacity through congestion-based surcharges complemented by reservation fees and the traditional weight-based landing fees. Furthermore, it exhibits some clear advantages over the existing rule-based procedures or even market-driven mechanisms in certain cases since:

- It confronts directly with the severe congestion problem of airports by means of varying congestion fees aiming to internalize congestion costs,
- it assigns some quasi-market valuations upon scarce airport capacity,
- it is simple, inexpensive, and easily implementable since it does not require substantial organizational and institutional arrangements,
- it does not involve drastic regulatory amendments and it is fully compatible with the existing IATA worldwide scheduling procedures, and
- it reduces the burden of administering slots and substantially enhances carriers' scheduling flexibility.

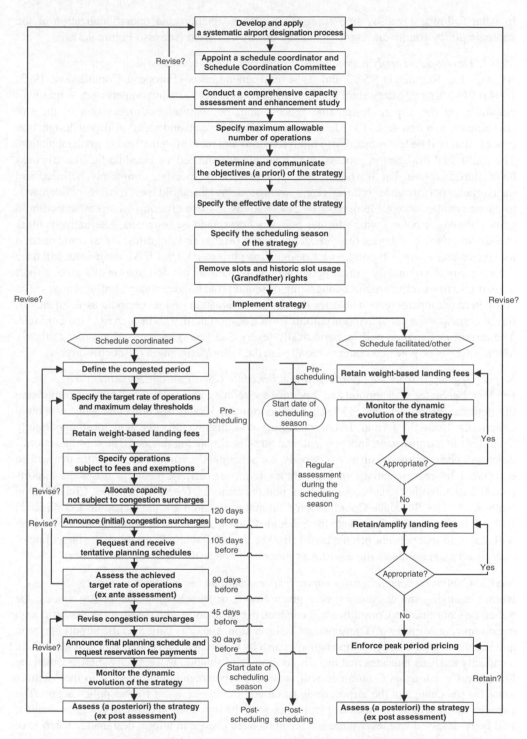

Figure 9.3   Strategic policy implementation roadmap

In what follows, a step-by-step approach for the adaptation and operationalization of the strategic policy framework is presented and briefly discussed (see also Figure 9.3):

Step 1: Develop and apply a systematic airport designation process
As with EU Regulation 95/93 and its several amendments (European Commission, 1993; 2004), EU Member States shall continue to play an active – mainly supervisory – role with emphasis on the airport designation process and the appointment/supervision of the slot coordinator. At a first stage, EU Member States should develop and apply an airport designation process that will be less subjective to interpretations and more systematized to implementation. The traditional designation process (e.g., schedule coordinated vs schedule facilitated) could be a starting point, but it should be supplemented by specific, commonly defined, and measurable performance criteria (e.g., average delay threshold per movement, demand-to-capacity ratio, acceptable number of cancellations[6]). These criteria will represent threshold values above (or below) which further policy action should be triggered. Alternatively, other classification methodologies (e.g., cluster analysis) (Madas and Zografos, 2008) could be used to develop some airport typologies for designation purposes. Each EU Member State will have the authority to voluntarily trigger policy action by changing the designation of a given airport in response to extreme/irregular congestion or sudden (but not temporary) traffic changes. The competent authorities responsible for the airport designation and its periodic assessment will be the corresponding civil aviation authorities or transport ministries in each EU Member State. The airport designation will be periodically (every 2–3 years) re-assessed in order to timely identify trends or potential changes required in the status/designation of specific airports.

Step 2: Appoint a schedule coordinator and Schedule Coordination Committee
EU Member States shall appoint and supervise a schedule coordinator for all airports comprising the national airport network. The schedule coordinator should be competent with substantial experience from the former IATA scheduling coordination process, and most importantly should act in a transparent, independent, and non-discriminating manner. The schedule coordinator will be supported by an advisory body, the Schedule Coordination Committee that will be composed by representatives of user groups (e.g., air carriers regularly using the airports concerned) and other stakeholders (e.g., airport operators, ATC service provider, Civil Aviation Authority). The Schedule Coordination Committee (chaired by the Schedule Coordinator) will still have a consultative role to the Schedule Coordinator. Most importantly, the Committee will act as an independent pricing board (Ball et al., 2007) that will establish the target rate of operations and coordinate the adjustment process of congestion fees (discussed later).

Step 3: Conduct a comprehensive capacity assessment and enhancement study
Before changing the designation of a given airport or launching further policy actions, the Schedule Coordination Committee will conduct, in consultation with the Member State (through the competent Ministry of Transport or Civil Aviation Authority), the specific airport operator, and regular airport users, a comprehensive capacity assessment and enhancement study. Where a capacity analysis indicates that there is no short-term solution to a capacity shortage, then the Schedule Coordination Committee will provide a recommendation/proposal to the Member State for the change of the airport designation and the triggering of further policy action. The capacity study will be conducted at least once with the introduction of the new regime, while it will be re-assessed whenever there is a recommended change in airport designation (Step 1) or considerable changes in the airport's infrastructure with possible implications on capacity.

Step 4: Specify maximum allowable number of operations
Based on the aforementioned capacity assessment study and previous operating experience of
the airport, the Civil Aviation Authority will have the ultimate control to specify the maximum
allowable number of runway operations for safety purposes at each airport within its (basically
national) jurisdiction (FAA, 2001).

Step 5: Determine and communicate the objectives (a priori) of the strategy
The objectives of the strategy subject to implementation should be explicitly specified and
communicated in advance to all airport stakeholders and affected parties. Furthermore,
performance targets or impacts should be early identified as input for the ex-post assessment of
the strategy. Among them, one could see indicators like average delay, number of cancellations,
passenger costs, airline costs, passenger enplanements, aircraft movements, average load factor,
service to small communities or thin routes, short versus long-haul services, unscheduled operations, new airline entrants, low-cost airlines, degree of competition and so on.

Step 6: Specify the effective date of the strategy
The date that the new strategy will take effect should be specified and communicated to all
affected parties and stakeholders. This will be specified at a centralized level (e.g., European
Union) and be part of the necessary regulatory amendment for the introduction of the new
strategy. For the specification of the effective date of the strategy, the following issues should
be also taken into consideration: (1) transition from and synchronization with the existing
IATA scheduling procedures, the current HDR regime, or other demand management schemes
applied in airports worldwide, (2) synchronization with the strategies implemented in other
EU airports governed by the same overall framework, but probably different instantiations of
the same strategy. The expiration date of the strategy will be signified by either a possible
change in the airport designation or the expiration date that might be explicitly specified for
this particular airport (or groups of airports) for the transition to a new regime.

Step 7: Specify the scheduling season of the strategy
The scheduling season of the strategy will be also specified at a centralized level (e.g., European
Union) and will be similar to the existing regulatory framework, that is, six months split into
winter and summer scheduling seasons. All parameters and decisions about capacities, fees,
surcharges, and schedules will be in effect within the framework of the scheduling season.

Step 8: Remove slots and historic slot usage (grandfather) rights
Slots and grandfather rights will be totally and immediately removed at the effective date of
the strategy. Each carrier will be able to operate at any time by paying the corresponding
access fees (weight-based landing fee) and congestion surcharge. Under this strategy, access
will be open to all users and price (in the form of variable or non-variable congestion
surcharges) will be solely used as the airport access control mechanism. The removal of all
slots will be centrally enacted through EU legislation, while slots will be generally replaced
by their US-equivalent operating authorizations (FAA, 2001).

Step 9: Implement strategy
This step will signify the implementation and adaptation of the new congestion management
regime at the local airport context. From this step forward, the overall framework will break
down to a series of sub-steps pertaining to the specific strategy operationalized at each
designated airport group or other airport typology adopted.

Based on the existing designation of EU airports (schedule coordinated vs schedule facilitated), the following sub-steps are proposed for schedule coordinated airports (Step 9) (Figure 9.3):

Step 9.1.1: Define the congested period
The Congested Period will consist of all hours of the day during which the demand for access at the particular airport is close to or even above the maximum allowable number of operations. The Congested Period will be purposely "expanded" by at least 1 h before or after the identified congested period. This inflation aims to avoid an insufficient rescheduling of flights in the close proximity of the Congested Period that would simply shift delays during the day. The Congested Period will be defined by the Schedule Coordination Committee in cooperation with the local airport operator.

Step 9.1.2: Specify the target rate of operations and maximum delay thresholds
The identification of the target rate of operations is of paramount importance for the successful application of this strategy since it is the most influential control in determining the level of delays experienced at the airport. A target rate of operations in terms of runway movements per unit of time will be established by the Schedule Coordination Committee during the Congested Period of each airport. The target rate of operations will be invariable (or slightly variable) within the day. A slight variability of the target rates might be introduced to account for local airport peculiarities or policy objectives. Most importantly, some variability of target rates is strongly recommended to airports congested during only some parts of the day. This would purposely specify some lower target rates immediately before and after the currently experienced traffic peaks as a time buffer to allow for a system recovery or preparation before or among various daily peaks. Moreover, lower target rates before and after peaks would keep airlines from squeezing flights just before or after the Congested Period in order to circumvent the congestion pricing process. The target rates will be specified in direct association with the maximum allowable number of operations, as well as the resulted delay figures. Finally, this step will also specify a maximum delay threshold above which further policy action will be triggered towards revising the strategy either in its entirety or its integral elements (e.g., level of congestion surcharges, target rate of operations, congested period).

Step 9.1.3: Retain weight-based landing fees
Existing weight-based landing fees will continue to exist at each airport as the time-tested way to recover operating and capital costs in accordance with the cost-relatedness principle foreseen in the existing regulatory framework.

Step 9.1.4: Specify operations subject to fees and exemptions
All landings at the particular airport during any time of the day will be subject to the existing weight-based landing fees specified by the airport operator. In addition, all aircraft movements operated during the Congested Period would be charged (on top of landing fees) the corresponding congestion surcharge. The operations exempted from the congestion surcharge will be specified by the competent supervisory state authority (basically Civil Aviation Authority or Ministry of Transport). Such exemptions will pertain to flights operated outside the Congested Period, flights subject to international bilateral agreements, or other policy-driven exemptions. Unscheduled operations will be also subject to congestion surcharges, while a multiplier factor (e.g., 1.2–1.5) can be considered as a disincentive on top of congestion surcharges due to the additional scheduling uncertainty and stochastic delay that unscheduled operations introduce in the price-capacity adjustment process.

Step 9.1.5: Allocate capacity not subject to congestion surcharges
The state authority (Civil Aviation Authority or Ministry of Transport) responsible for the determination of fee exemptions will also determine the number of operating authorizations that should become available per hour of the Congested Period. Interested carriers will be asked to submit their interest and scheduling details for operating authorizations falling into this category 120 days before the start date of the next scheduling season. The Schedule Coordination Committee will collect all requests for fee-exempted operating authorizations and assess them in terms of eligibility, operational feasibility, and possible violation of the fixed number of available operating authorizations. In case that the requested number of operating authorizations does not exceed availability, all applicant carriers will be notified about the approval of the requested operating authorizations. Otherwise, the Schedule Coordination Committee will allocate the available number of operating authorizations based on a lottery among users that are eligible for this category.

Step 9.1.6: Announce (initial) congestion surcharges
Traditional landing fees will be supplemented by a scheme of congestion surcharges aiming to discourage the actual operation of flights beyond the targeted level of operations. Congestion surcharges will apply only during the Congested Period and will be announced by the Schedule Coordination Committee 120 days before the start date of the next scheduling season. The surcharges will vary throughout the day based on congestion and will be set at (or close to) a level being able to internalize congestion costs. At a simplified version, a flat surcharge (or even multiplier to weight-based landing fees) can be alternatively applied. Although this would appear to be a "second-best" fee (coarse tolling) (Fan, 2003), it is not expected to have a serious impact in the end of the process due to the fact that this is only aimed to be a first round of fees subject to many subsequent rounds of adjustments (FAA, 2001). An interesting observation is that even in the absence of detailed knowledge about the demand curve, coarse tolling results in congestion fees that are quite near to optimal levels (Fan, 2003). Furthermore, it is strongly recommended that congestion surcharges will be initially set at a relatively high level (with downward prospects) in order to trigger more drastic or, at least, sufficient price responses from carriers since the beginning of the process. The authority responsible for the establishment of congestion surcharges will be the Schedule Coordination Committee in the form of an Independent Pricing Board (Ball et al., 2007) with the ultimate mission to set congestion surcharges that will clear the market for airport access by adjusting fees up or down in order to meet the mandated level of operations.

Step 9.1.7: Request and receive tentative planning schedules
All airport users will be requested by the Schedule Coordination Committee to express their interest for operating authorizations during the Congested Period based on the initial congestion surcharges. Practically, interested carriers will submit a tentative planning schedule with all scheduling details by considering the congestion surcharges applied during the particular hour of the Congested Period. The planning schedule will be submitted to the Schedule Coordination Committee 105 days before the start date of the next scheduling season.

Step 9.1.8: Assess the achieved target rate of operations (ex ante assessment)
The Schedule Coordination Committee will collect all tentative planning schedules submitted by applicant carriers and will consolidate them in an overall planning schedule. This will summarize the requests for operating authorizations for each hour (or smaller time window)

of the Congested Period and will be announced to all interested parties 90 days before the start date of the next scheduling season. The ultimate objective of this exercise will be to assess before the start of the scheduling season (ex ante assessment) whether (and to what extent) the targeted rates of operations can be achieved. Given that an absolute matching will be difficult to achieve in practice, some reasonable deviations of the consolidated planning schedule from the targeted rates of operations will be deemed acceptable as long as safety standards are not compromised and deviations are not persisting for more than few (2–3) hours in a row. In this case, the Schedule Coordination Committee will notify all carriers about the approval of the requested operating authorizations, otherwise, a next round of adjustments will be triggered.

Step 9.1.9: Revise congestion surcharges
In case that the requested number of operating authorizations is substantially above or below targeted rates of operations, the Schedule Coordination Committee will propose a revised scheme or levels of congestion surcharges aiming to bridge the gap between targeted operations and actually requested operating authorizations. The new pricing adjustments/revisions will practically signify the starting of a feedback loop that will be exited only after some balance has been achieved even with a reasonable deviation from the targeted rates of operations. The length of these pricing adjustment sessions will be of about two weeks with an ultimate deadline of the process to be not later than 45 days before the start date of the next scheduling season.

Step 9.1.10: Announce final planning schedule and request reservation fee payments
After an acceptable planning schedule has been achieved, the final requests for operating authorizations (from the latest pricing adjustment session) will be "frozen" and announced to their prospective users at least 45 days before the start date of the next scheduling season. Before the final approval of operating authorizations, each carrier will be asked to deposit a reservation fee for each awarded operating authorization during the Congested Period. The reservation fee will be equivalent to some percentage (10–30%) of the finally posted congestion surcharge. In addition, it will not be recovered in case of non-use ("no-show").

Step 9.1.11: Monitor the dynamic evolution of the strategy
The actual implementation of the strategy during the scheduling season will be closely monitored by the airport operator in cooperation with the Schedule Coordination Committee. Their basic responsibility will be to monitor the actual utilization of operating authorizations and the regular assessment of experienced congestion levels. The Schedule Coordination Committee will retain the right to revise the congestion surcharges at any level and any time during the Congested Period in case that congestion and delays start to build up beyond certain acceptable operational thresholds. In case of additional requests for authorizations, additional flights (beyond the ones included in the awarded planning schedule) will be treated as unscheduled operations that will be subject to congestion surcharges in conjunction with a multiplier factor (e.g., 1.2–1.5) applied on top of congestion surcharges.

Step 9.1.12: Assess the strategy (ex post assessment)
The close monitoring of the actual implementation of the strategy will be complemented by an ex-post assessment of the strategy that will basically aim to evaluate the degree of accomplishment of certain agreed performance targets. The ex-post assessment of the strategy will provide useful experience, evidences, and lessons learnt for the implementation and potential adjustments of the strategy for the next scheduling season. Finally, it will provide input or even trigger the process for a possible revision of an airport's designation.

As far as the schedule facilitated or non-designated airports are concerned, the following sub-steps are proposed (Figure 9.3):

Step 9.2.1: Retain weight-based landing fees

Airport access will be open to all users and each carrier will be able to operate at any time by only paying the corresponding landing fees. Traditional weight-based landing fees will continue to be in effect at small regional airports with negligible or no congestion at all. In such airports, weight-based landing fees will be maintained at their existing form, that is, based on the Maximum Take-Off Weight (MTOW) of the aircraft.

Step 9.2.2: Monitor the dynamic evolution of the strategy

The actual implementation of the strategy will be closely monitored by the airport operator in cooperation with the Schedule Coordination Committee. The Schedule Coordination Committee will retain the right to intervene into the determination of landing fees in case that these are proven to be insufficient or inefficient. Finally, the Schedule Coordination Committee will be empowered to put forward the escalation of congestion management measures in case that congestion and delays start to build up beyond certain acceptable thresholds.

Step 9.2.3: Amplify landing fees

In case that a particular airport violates the established delay or other performance thresholds, the Schedule Coordination Committee will be entitled to replace weight-based landing fees by flat, nonweight-dependent fees. In this case, all operations will be charged with an identical landing fee irrespectively of the aircraft weight[7]. The justification of such a fee is that traditional weight-based fees may lead to a paradoxical situation where carriers will be encouraged to fly lighter aircraft, a fact that has undoubtedly played a decisive role on deteriorating airside congestion and delays. Consequently, the application of flat landing fees will practically provide the appropriate incentive to carriers to use heavy aircraft and upgauge their average load factors. These amplified landing fees will be specified by the Schedule Coordination Committee in consultation with the airport operator and will be applicable either throughout the entire day or some part thereof. Finally, in case delays still persist, the Schedule Coordination Committee will be authorized to further strengthen the pricing impact.

Step 9.2.4: Enforce peak period pricing

In the event that landing fees, in their traditional or amplified version, did not succeed to reduce delays to acceptable levels, the Schedule Coordination Committee will be authorized to migrate to a peak period pricing scheme that is closer to or an intermediate step before shifting to a congestion-based pricing strategy described for schedule coordinated airports. Peak period pricing will be applied at airports experiencing delay problems due to some time-limited traffic peaking during the day with sufficient capacity during most of the day. In this case, peak period pricing will involve higher landing fees during the specified peak hours in the form of a multiplier factor (1.5–3) applied on top of existing landing fees. Existing landing fees will remain unchanged during off-peak hours.

Step 9.2.5: Assess the strategy (ex post assessment)

By the end of the scheduling season, an ex-post assessment of the strategy will be performed. This assessment will be performed by the airport operator in cooperation with the Schedule Coordination Committee and will basically aim to evaluate the degree of accomplishment of established performance targets. The ex-post assessment of the strategy will provide useful input for the implementation and potential adjustments of the strategy for the next scheduling season. Finally, it will provide input or even trigger the process for a possible revision of an airport's designation when delays reach levels that may justify a more drastic strategy.

The implementation of the proposed congestion-based policy framework introduces a low risk, first step for the pricing and the rationalization of use of scarce airport resources. It exhibits some clear advantages over administrative procedures or other proposed market-based strategies in that: (1) it assigns some market valuations on scarce airport capacity, (2) it is simple and inexpensive to implement, (3) it does not involve drastic regulatory amendments, and (4) it substantially enhances carriers' scheduling flexibility. Most importantly, a congestion pricing regime confronts directly with the severe congestion problems of airports by means of varying congestion fees and enjoys high acceptability albeit without compromising implementability. Furthermore, it introduces an open and adaptive framework, which fits directly with the worldwide scheduling approach but can be also easily customized with local airport needs and characteristics through the appropriate congestion fee schemes and levels.

On the other hand, there are technical difficulties (e.g., fee scheme, use of revenues), political concerns (e.g., social exclusion of small communities, competition effects, operations subject/exempted from fees, acceptability of congestion and reservation fees, role of slot coordinators and supervising authorities), and legal/institutional complications (e.g., required regulatory amendments, possible discrimination/equity issues, revenue neutrality) that have to be overcome at the policy and/or technical level. Most importantly, there should be a commonly agreed and technically sound methodology for the definition of core technical elements of the strategy with emphasis on the structure, components, and levels of fees, the definition of congested periods, the targeted rate of operations, as well as the maximum (acceptable) delay thresholds or other performance indicators. For the latter, it is necessary to establish both ex ante and ex post assessment procedures for monitoring both anticipated and actualized strategy results and undertaking corrective actions in the event of deviations from the stated strategy objectives.

9.6   Improvement Complements to Existing Policy Practice: Directions for Future Research

The common denominator of various policy proposals discussed in the literature is that they introduce – though at varying degrees– some form of a new capacity allocation mechanism or procedures necessitating more or less drastic regulatory reforms. Such regulatory reforms are often blocked due to technical practicalities, transition complexities, as well as strong industry opposition and inertia forces. In response to these policy implementation barriers, there have been recent research efforts aiming to act as complements – rather than substitutes – to the existing slot allocation practice. These basically introduce improvements in the internal allocation efficiency of the IATA-based mechanism without seeking any departure from the existing slot allocation regime. As a result, they are in full alignment with the existing IATA worldwide scheduling procedures being therefore simple, immediately implementable and not requiring any organizational or regulatory amendments.

This interesting stream of research focuses on the allocation of scarce capacity from a slot scheduling point of view. Traditionally, such approaches deal with the development of analytical models aiming to control the distribution of traffic by (re)scheduling movements at either network (i.e., en route slots) or airport level (i.e., airport slots) (Andreatta et al., 1995). At the strategic[8], single-airport scheduling level, which is of relevance to this chapter, Koesters (2007)[9] developed a heuristic procedure to calculate scheduled delays – the difference between requested and allocated slot times – for various levels of declared capacity, demand and slot utilization. He mainly focused on the analysis of the interrelationship between slot demand and scheduled delays with the ultimate objective of predicting scheduled delays depending on various levels of slot utilization. Zografos et al. (2012) developed and tested a strategic, single-airport optimization model implementing the existing EU/IATA scheduling rules, coordination procedures, and operational constraints with the ultimate objective to allocate scarce airport capacity (series of slots) more efficiently at schedule coordinated airports. This objective is achieved by better accommodating airlines' preferences, that is, minimizing the difference between the requested slot time and the slot time eventually allocated to airlines. Furthermore, the results of the model were assessed and compared vis-à-vis the allocation outcome produced according to current slot coordination practice in three regional Greek airports. The proposed model brought out a large room for improvement of the efficiency of the existing allocation process in a range between 14–95% (Zografos et al., 2012).

Starting from the observation that deviations between requested and allocated slots result in undesired costs or revenue losses for airlines, Pellegrini et al. (2012) developed a secondary trading mechanism that complements the primary IATA-based allocation, but introduces a single-market, secondary alternative to the bilateral trade recognized as a legitimate slot exchange option in EU since 2008 (European Commission, 2008). The idea behind this secondary trading mechanism is that airlines suffer from "shift costs" when they have to shift their flight schedules forward or backward to deal with the allocation of slots differing from the originally requested (ideal) departure and arrival bundles of slots (Pellegrini et al., 2012). Therefore, they proposed a secondary slot trading mechanism as a combinatorial exchange and modeled it as an integer linear programming model. The objective of the model is to allocate coherently and simultaneously slots at both origin and destination airports by minimizing airlines' shift costs due to sub-optimal slot allocation. The proposed combinatorial exchange is shown to offer significant cost reduction to airlines as compared to current practice, while at the same time introducing two interesting features, namely the simultaneous allocation at both origin and destination airports and the en route capacity considerations that often degrade the allocation outcome.

In addition to the strategic allocation of scarce airport capacity, the efficient specification of capacities system-wide (airports and air sectors) constitutes a strategic challenge for future research. Barnhart et al. (2012) stress the need for a comprehensive evaluation of the impact of different levels of capacity specification that will, *inter alia*, provide insights into the optimal trade-off between schedule and queuing delay. As a matter of fact, one of the fundamental concepts and principal parameters for slot allocation in EU airports is declared capacity, which specifies the number of scheduled movements available at an airport per unit of time. The declared capacity notion represents an artificial, administrative measure of capacity that is determined rather than computed in most cases. Despite its utmost importance and substantial influence on the efficiency of the allocation process (Zografos et al., 2012), the declared capacity determination process has not been sufficiently examined in the literature (Railsback and Sherry, 2005).

Current practice on the determination of declared capacity in EU airports has received much criticism (PwC, 2000; Odoni and Morisset, 2010): (1) there is ample room for local interpretations and adaptations (e.g., coordination parameters, technical, operational, and environmental constraints, capacity constraining factors, "established" capacity), (2) the setting of strikingly low capacity levels that are basically determined as a percentage of the capacity under Instrument Flight Rules (IFR), (3) the empirical or *ad hoc* process (rather than technical, based on "commonly recognized methods" as suggested by EU Regulation) often applied by coordinators in determining the declared capacity, and (4) limited consultation with stakeholders (e.g., airport operator, airlines, ATC representatives). As a result, the following two major categories of problems arise: (1) lack of a harmonized way of interpreting, determining/computing, and managing declared capacity, and (2) declared capacity is set at too high or too low levels.

The level of declared capacity has serious impacts on the allocation efficiency, while simultaneously affecting the classical trade-off between acceptable level of service and utilization levels of scarce airport resources. As far as the allocation efficiency is concerned, the level of declared capacity represents essentially the degree to which the allocation problem is constrained and, consequently, the potential ability of the allocation mechanism to satisfy airlines' requests for slots by properly matching slots requested with slots eventually allocated to airlines ("distance" measure). In that respect, it has been characteristically demonstrated that an increase (with the opposite being also true) in capacity improves disproportionately the allocation outcome (Zografos et al., 2012). As a result, the level of declared capacity basically determines not only the number (i.e., "quantity") of slots available for allocation to airlines, but also the "quality of slots" in terms of the "distance" from original slot requests. On the other hand, there is a clear trade-off between declared capacity levels and operational efficiency. For example (with the opposite being also true), higher declared capacity leads to increased frequency, competition, access, and ultimately utilization of scarce infrastructure, often at the expense of delays ("delay" measure) and reliability of the airport system. At the outset, the determination of declared capacity signifies a very important and highly impacting decision to leave it based solely on an administrative – and to some extent empirical or even arbitrary – determination process.

Ongoing research work of the authors (Madas and Zografos, 2012) recognizes the substantial influence of declared capacity on both allocation (e.g., schedule delays) and operational efficiency (e.g., operational delays) and draws explicit consideration on the declared capacity determination process *per se* by: (1) contributing in the harmonization of the declared capacity determination process and (2) eliminating the artificial declared capacity metric in favor of an optimization approach that will capture the strong interdependencies between declared capacity levels, allocation efficiency, and operational efficiency. The latter produces essential benefits for policy makers and interest groups in that: (1) it enables competent authorities and regulators to make well-informed decisions on the specification of declared capacity at appropriate/realistic levels, and (2) it develops "optimal" schedules in terms of both allocation ("distance" measure) and operational efficiency ("delay" measure) not only for individual airlines but also for the airport system as a whole.

A common element of the proposals discussed previously is their explicit intention to introduce improvements and complement rather than substitute the existing slot regime. Despite some reasonable advantages of alternative market-based allocation mechanisms, the proposed options are in full alignment with the existing IATA worldwide scheduling procedures and are, therefore, simple, immediately implementable and do not require any organizational or regulatory amendments. As a result, they have sensibly gained the increasing attention of the research community and pave the way to future research directions/solutions that will effectively deal with the trade-off between efficiency and feasibility of implementation. On top of operational efficiency and feasibility of implementation, fairness and equity considerations needs to be also carefully addressed by any scheme or strategy aiming to deal with congestion and allocate demand among several classes of customers/users of congested airport systems (Andreatta and Lulli, 2009). Future research is expected to build on recent research efforts by focusing on the use of optimization techniques and models with a dual objective: (1) the specification of airport capacity at appropriate levels with simultaneous consideration of multiple criteria (e.g., allocation efficiency, operational efficiency, equity, environmental impacts) and (2) the efficient allocation and utilization of this capacity from a strategic, airport scheduling point of view.

9.7 Conclusions

During the last decade, there is an industry consensus that the existing IATA-based system for allocating scarce airport capacity cannot cope with existing – and even more – forecasted traffic, hence urging for a radical reform of the slot regulation. A substantial amount of research work has been documented in the literature with the aim to review and critically assess alternative capacity allocation approaches ranging from minor adjustments or enhancements of the status quo to more radical or aggressive revisions of the current slot allocation regime. Despite a comprehensive amount of policy and research proposals in this field, attempts to bring forward more radical revisions towards market-driven allocation mechanisms were not widely adopted and have not flourished into policy practice mainly due to industry inertia forces, technical complexities, as well as opposition from established industry actors doubting the necessity or effectiveness of such initiatives. Another contributing factor to the gap between theory and policy practice seems to be the lack of guidance as to how these instruments and measures can be integrated and operationalized within an overall strategic policy framework.

In response to the identified gap, this chapter provided arguments in favor of the adoption of a new congestion management regime. Furthermore, it formulated a policy roadmap aiming to guide the implementation process and propose specific instantiations/adaptations of a new congestion-based pricing strategy applicable to different types of EU airports. The implementation of a congestion-based pricing strategy constitutes a first step towards the efficient pricing and the rationalization of use of scarce airport resources. It exhibits some clear advantages over competing approaches in that it confronts directly with the congestion problem, it is simple and inexpensive to implement, it does not involve drastic regulatory amendments, and it can be easily adapted to both worldwide scheduling procedures and local airport needs. On the other hand, there are practical difficulties (e.g., fee scheme, use of revenues), political concerns (e.g., social exclusion of small communities, competition effects, fairness, acceptability), and legal/institutional complications (e.g., equity issues, revenue neutrality) that have to be overcome. At the same time, other important practicalities need to be also addressed at both policy and technical fronts such as the structure, components, and levels of fees, the definition of congested periods, the targeted rate of operations, as well as the maximum delay thresholds.

The chapter concludes with the discussion of recent efforts and future research directions towards improving or complementing – rather than substituting – the existing slot allocation practice from the strategic slot scheduling point of view. Such approaches basically introduce improvements in the internal allocation efficiency of the IATA-based mechanism by means of optimization models aiming to strategically control the distribution of traffic at the airport level. These are quite attractive from both the policy and research perspective since they bring promises for substantial improvements of the efficiency of the existing allocation process. At the same time, they are in full alignment with the existing worldwide scheduling procedures being therefore simple and immediately implementable. Most importantly, they effectively deal with the trade-off between efficiency and feasibility of implementation since they build upon existing custom and practice in order to produce some notable improvements in the efficiency of the capacity specification and allocation process.

Acknowledgements

Part of the research work presented in this chapter has been supported by the Hellenic General Secretariat of Research and Technology (GSRT), under contract SPADE-2/NC (NC-1375-01).

Notes

1   According to the European Commission Regulation 95/93 (European Commission, 1993), a slot signifies "the permission given to a carrier to use the full range of airport infrastructure necessary to operate an air service at a slot-controlled airport on a specific date and time for the purpose of landing or take-off".

2   Demand management schemes aiming to mitigate noise and emission-related externalities are not covered here but might be viewed as either stand-alone measures or add-on mechanisms to the congestion management scheme.

3   The corresponding US practice is based on a largely non-interventionist approach, where demand is not administered and IATA system does not apply at all, while airlines schedule their flights by considering only the expected delays and secondarily the negligible, weight-based landing fees. Exceptions to this regime are the airports (New York region airports, Washington/Reagan, and Chicago/O'Hare) governed by the High Density Rule (HDR), for which some scheduling limits are only applied. For a more elaborated discussion of the relevant US experience, the reader is referred to the subsequent chapter (Chapter 10) of this book by Ball et al.

4   That is often vulnerable to governmental/political intervention in certain cases (e.g., Greece, Spain) (PwC, 2000).

5   Please note that the estimated costs do not account for externalities other than congestion and delays (e.g., noise, emissions).

6   The specific time horizon (e.g., daily, hourly, annual) and reference base (e.g., typical, peak day) needs to be defined as well.

7   An alternative, milder approach would be to apply a fee in the form of (alpha*old-fee)+(1-alpha)*flat-fee, where old-fee is the conventional, weight-based landing fee and flat-fee the new, non weight-dependent fee. The parameter alpha could compromise between the two extremes [0,1].

8   Dealing with 2–12 months before operation.

9   The reader is also referred to Chapter 8 of this book by Klingebiel et al.

References

Airport Council International (ACI) Europe (2004) *Study on the Use of Airport Capacity*. Brussels, Belgium, pp. 1–15.

Airport Council International (ACI) Europe (2009) *ACI Europe position on the proposed revision of the Council Regulation (EEC) No 95/93 on common rules for the allocation of slots at Community airports*. Presentation on the TRAN Meeting at the European Parliament, March 25, Strasbourg, France.

Andreatta, G., Odoni, A.R., Richetta, O. (1995) Models for the Ground Holding Problem. In: *Large Scale Computation and Information Processing in Air Traffic Control*, (eds) by Odoni, A.R. and Bianco, L., Springer-Verlag, Berlin, Germany, pp. 125–168.

Andreatta, G. and Lulli, G. (2009) Equitable Demand Management Strategies for Different Classes of Customers. *International Journal of Pure and Applied Mathematics*, 57(1), 1–22.

Ball, M.O., Ausubel, L.M., Berardino, F., et al. (2007) Market-Based Alternatives for Managing Congestion at New York's LaGuardia Airport. In: *Proceedings of the AIRNETH/GARS Research Workshop*, The Hague, April 13.

Barnhart, C., Fearing, D., Odoni, A., and Vaze, V. (2012) Demand and capacity management in air transportation. *EURO Journal of Transportation and Logistics*, 1(1–2), 135–155.

Basso, L.J. and Zhang, A. (2010) Pricing vs. slot policies when airport profits matter. *Transportation Research Part B – Methodological*, 44(3), 381–391.

Brueckner, J.K. (2002) Internalisation of airport congestion. *Journal of Air Transport Management*, 8(3), 141–147.

Civil Aviation Authority (CAA) House (2001) *The Implementation of Secondary Slot Trading*. Consultation paper prepared for the UK Government, CAA House, London, UK.

Coopers and Lybrand (1995) *The Application and Possible Modification of Council Regulation 95/93 on Common Rules for the Allocation of Slots at Community Airports*. Study commissioned by the European Commission.

Czerny, A.I. (2010) Airport congestion management under uncertainty. *Transportation Research Part B – Methodological*, 44(3), 371–380.

DotEcon Ltd (2001) *Auctioning Airport Slots*. Report for HM Treasury and the Department of the Environment, Transport, and the Regions, London, UK.

Eurocontrol (2008) *Long-Term Forecast: IFR Flight Movements 2008–2030*. Forecast prepared as part of the Challenges of Growth 2008 project, Brussels, Belgium.

Eurocontrol (2012) *CODA Digest: Delays to Air Transport in Europe (December 2011)*. Report prepared by Eurocontrol's Central Office for Delay Analysis (CODA), Brussels, Belgium.

European Commission (1993) European Council Regulation (EEC) No 95/93 of January 1993 on common rules for the allocation of slots at Community airports. *Official Journal of the European Union*, L 014, 0001-0006, Brussels, Belgium.

European Commission (1998) *Fair Payment for Infrastructure Use: A Phased Approach to a common transport infrastructure charging framework in the EU*. 446 final, White Paper presented by the European Commission, Brussels, Belgium.

European Commission (2004) *European Council Regulation (EC) No 793/2004 of April 2004 amending Council Regulation (EEC) No 95/93 on common rules for the allocation of slots at Community airports*. Official Journal of the European Union, L138, 50-60, Brussels, Belgium.

European Commission (2008) *Communication from the Commission to the European Parliament, the Council, the European Economic and Social Committee and the Committee of the Regions on the Application of Regulation (EEC) No 95/93 on common rules for the allocation of slots at Community airports*. COM(2008) 227 final, April 30, 2008, Brussels, Belgium.

European Commission (2009) European Council Regulation (EC) No 545/2009 of June 2009 amending Council Regulation (EEC) No 95/93 on common rules for the allocation of slots at Community airports. *Official Journal of the European Union*, L167, 24–25, Brussels, Belgium.

European Commission. (2011) *(Airport Package) Proposal for a Regulation of the European Parliament and of the Council on common rules for the allocation of slots at Community airports*. COM(2011) 827 final, December 1, 2011, Brussels, Belgium.

Fan, T.P. and Odoni, A.R. (2002) A Practical Perspective on Airport Demand Management. *Air Traffic Control Quarterly*, 10(3), 285–306.

Fan, T.P. (2003) *Market-based Airport Demand Management – Theory, Model and Applications*. Doctoral Dissertation, Center for Transportation and Logistics, Engineering System Division, Massachusetts Institute of Technology, Cambridge, MA.

Federal Aviation Administration (FAA) (2001) *Notice of alternative policy options for managing capacity at LaGuardia airport and proposed extension of the lottery allocation*. Federal Register, 66(113), June, 31731–31748; Docket FAA-2001-9852.

Fisher, J.B. (1989) Managing Demand to Reduce Airport Congestion and Delays. Transportation Research Record, *Official Journal of the Transportation Research Board*, 1218, 1–10.

Flores-Fillol, R. (2010) Congested hubs. *Transportation Research Part B – Methodological*, 44(3), 358–370.

Forsyth, P. (1997) Price regulation of airports: principles with Australian applications. *Transportation Research Part E – Logistics and Transportation Review*, 33(4), 297–309.

Fukui, H. (2010) An empirical analysis of airport slot trading in the United States. *Transportation Research Part B – Methodological*, 44(3), 330–357.

International Air Transport Association (IATA). (2010) *Worldwide Scheduling Guidelines*. 19th Edition, Montreal, Canada.

Koesters, D. (2007) *Study on the Usage of Declared Capacity at Major German Airports*. Study in cooperation with the Eurocontrol Performance Review Unit (PRU), Aachen, Germany.

Le, L., Donohue, G., and Chen, C.H. (2004) Auction-Based Slot Allocation for Traffic Demand Management at Hartsfield Atlanta International Airport: A Case Study. *Transportation Research Record,* Official Journal of the Transportation Research Board, 1888, 50–58.

Madas, M.A. and Zografos, K.G. (2006) Airport Slot Allocation: From Instruments to Strategies. *Journal of Air Transport Management*, 12(2), 53–62.

Madas, M.A. and Zografos, K.G. (2008) Airport Capacity vs. Demand: Mismatch or Mismanagement? *Transportation Research Part A – Policy and Practice*, 42(1), 203–226.

Madas, M.A. and Zografos, K.G. (2010) Airport Slot Allocation: A Time for Change? *Transport Policy*, 17(4), pp. 274–285.

Madas, M.A. and Zografos, K.G. (2012) Towards a New Approach for the Declared Airport Capacity Determination. *Unpublished manuscript.*

Malone, K. and Odoni, A.R. (2001) *The Approximate Network Delays Model*. Operations Research Center, Massachusetts Institute of Technology, Cambridge, MA.

Mendes, L.M.Z. and Santos, G. (2008) Using economic instruments to address emissions from air transport in the European Union. *Environment and Planning A*, 40(1), 189–209.

Morrison, S.A. and Winston, C. (1989) Enhancing the performance of the deregulated air transportation system. In: *Brookings Papers on Economic Activity*, Microeconomics, 61–123.

Morrison, S.A. and Winston, C. (2007) Another Look at Airport Congestion Pricing. *American Economic Review*, 97(5), 1970–1977.

Mott MacDonald (2006) *Study on the Impact of the Introduction of Secondary Trading at Community Airports*. Volume I, Technical Report prepared for the European Commission (DG TREN) by Mott MacDonald's Aviation Group, UK.

National Economic Research Associates (NERA) (2004) *Study to Assess the Effects of Different Slot Allocation Schemes*. Technical Report prepared for the European Commission (DG TREN), London, UK.

Nilsson, J. (2003) *Marginal cost pricing of airport use: The case for using market mechanisms for slot pricing*. VTI not at 2A-2003, Swedish National Road and Transport Research Institute (VTI), Technical report prepared for the EU-funded Research Project MC-ICAM.

Odoni, A.R. and Morisset, T. (2010) Performance Comparisons between U.S. and European Airports. *Proceedings of the 12th World Conference on Transport Research (WCTR)*, July 11–15, 2010, Lisbon, Portugal.

Oxford Economic Forecasting (2006) *Economic Contribution of the Aviation Industry in the UK*. London, U.K.

Pellegrini, P., Castelli, L., and Pesenti, R. (2012) Secondary trading of airport slots as a combinatorial exchange. *Transportation Research Part E – Logistics and Transportation Review*, 48(5), pp. 1009–1022.

PricewaterhouseCoopers (PwC) (2000) *Study of certain aspects of Council Regulation 95/93 on common rules for the allocation of slots at Community airports*. Technical Report prepared for the European Commission.

Railsback, P. and Sherry, L. (2006) A Survey of Rationales and Methods for Determining Declared Airport Capacity. *Proceedings of the Transportation Research Board (TRB) 85th Annual Meeting*, January 22–26, 2006, Washington, D.C.

Steer Davies Gleave (2011) *Impact Assessment of Revisions to Regulation 95/93*. Study prepared for the European Commission (DG MOVE), London, UK.

Technical University of Berlin (TUB) (2001) *Possibilities for the Better Use of Airport Slots in Germany and the EU*. Technical Report prepared by the Technology University of Berlin, Department of Infrastructure Economics, Workgroup for Infrastructure Policy, Berlin, Germany.

Transportation Research Board (TRB) (2001) *Aviation Gridlock: Understanding the Options and Seeking Solutions*. Transportation Research E-Circular, E-C029, National Research Council, pp. 41–54, Washington, D.C.

UK Department of the Environment, Transport and the Regions (2000) *The Future of Aviation*. The Government's Consultation Document on Air Transport Policy, London, UK.

US Congress, Office of Technology Assessment (1984) *Airport System Development*. OTA-STI-231, Washington, D.C., 109–120.

Verhoef, E.T. (2010) Congestion pricing, slot sales and slot trading in aviation. *Transportation Research Part B – Methodological*, 44(3), 320–329.

Zhang, A. and Zhang, Y. (2010) Airport capacity and congestion pricing with both aeronautical and commercial operations. *Transportation Research Part B – Methodological*, 44(3), 404–413.

Zografos, K.G., Salouras, Y., and Madas, M.A. (2012) Dealing with the Efficient Allocation of Scarce Resources at Congested Airports. *Transportation Research Part C – Emerging Technologies*, 21(1), pp. 244–256.

10

Design and Justification for Market-Based Approaches to Airport Congestion Management: The US Experience

Michael O. Ball[A], Mark Hansen[B], Prem Swaroop[A] and Bo Zou[C]

[A]*Robert H. Smith School of Business and Institute for Systems Research, University of Maryland, USA*
[B]*National Center of Excellence for Aviation Operations Research, Department of Civil and Environmental Engineering, University of California, Berkeley, USA*
[C]*Civil and Materials Engineering, University of Illinois at Chicago, USA*

10.1 Introduction

Air transportation delays in the US and around the world represent a well-known burden to society and are the subject matter of both technical and public policy debates. A recent study (Ball et al., 2010) estimated the total economic impact of air transportation delays on the US economy in the year 2007 to be $28.9 billion. The most obvious and often called-for actions are investments in the expansion of system capacity either in the form of infrastructure, for example, new runways and airports, or new capacity enhancing technology and systems. On the other hand, the fact that delays have fluctuated, sometimes substantially, with yearly demand variations suggests that controlling, in some way, the demand placed on the system could also yield (possibly substantial) reductions in delay. The US National Airspace System (NAS), as well as the worldwide air transportation system are queuing systems. To be sure, these are very complex queuing systems. However, they exhibit classic queuing system features in that they have capacities and as demand approaches capacity delays increase at a greater than linear rate. Thus,

Modelling and Managing Airport Performance, First Edition. Edited by Konstantinos G. Zografos, Giovanni Andreatta and Amedeo R. Odoni.
© 2013 John Wiley & Sons, Ltd. Published 2013 by John Wiley & Sons, Ltd.

analyzing the "delay problem" at a very basic level, one can consider two possible solutions: increase capacity or reduce demand. While increasing capacity can be a very expensive, politically controversial, and/or technically challenging proposition, controlling demand can involve controlling the behavior of individuals or companies operating in a competitive, free market environment.

Thus, while the latter approach may superficially seem much cheaper, it can involve much higher social and political hurdles. Not surprisingly the national approach to controlling demand can vary depending on underlying national and social norms. For example, slot controls, which restrict the level of scheduled demand at an airport, exist at virtually all large European airports but are relatively rare in the US.

This chapter broadly addresses certain issues related to both the justification and design of market mechanisms for airport access control. Section 10.2 both discusses some of the basic issues and reviews recent attempts in the US to implement slot controls with provisions for some allocation taking place via auctions. Ultimately these proposals were not implemented. While many political forces were at work in this outcome, it can also be argued that the political discussions had not been informed by scientific research related to the economic tradeoffs underlying the implementation of slot controls. Recent research by ourselves and others have attempted to rectify this situation. In Sections 10.3 and 10.4, we review this work. In Section 10.5, we consider important design questions that arose during the deliberations in the US and we provide at least partial answers to these questions.

10.2   Background

10.2.1   Airport Operations and Slot Controls

The number of operations (arrivals and departures) per hour that can be supported by an airport is limited by its system of runways. Thus, as the number of scheduled operations (demand) approaches the airport's runway capacity, classic queuing system phenomena are exhibited. That is, when demand reaches to point when it persistently and severely exceeds capacity, delays become excessive. On the other hand, airline competitive behavior and other factors can lead to a socially sub-optimal level of operations. Specifically, classic economic phenomena, for example, "the tragedy of the commons", can lead to misuse of the available capacity. In this context, the airlines and other flight operators form a community of users of a common resource, airport capacity. The users gradually start to overuse that resource. In this case, the detrimental effect of overuse is high delays. However, no single user can unilaterally solve the problem: A reduction in use by one user will have minimal impact on overall delay and, in any case, capacity freed up will quickly be used by the remaining users. Thus, while collective actions by all users would improve overall social welfare, there is no incentive for individuals to take actions that would lead to a better overall solution. This type of phenomenon is observed in many settings and, when it exists, it generally indicates the need for a coordinating policy or mechanism.

The need for, and possible forms of, policy innovation to reduce flight volumes at congested airports has been studied for several decades. Several authors have computed marginal delay costs for flights in busy periods, and found these costs to be higher than the actual charges paid for these operations (Carlin and Park 1970, Morrison 1983, Hansen 2002, Ashley and Savage 2010). This body of research provides strong economic arguments for some external controls on the use of airport capacity. At virtually all major European airports, airport slot controls have been implemented using the International Air Transport Association (IATA) administrative rules – which are adapted and complemented by EU Regulation 95/93 and its several amendments, most importantly, 793/2004. The interested reader is referred to Chapter 9 for a detailed description of the EU slot allocation practice. On the other hand, in the US, slot controls have historically been used very sparingly (currently slot controls exist at four US airports). They were first imposed in 1969, in response to growing congestion at major US airports. Under the so-called high density rule, FAA imposed hourly quotas on IFR operations at five airports, including the three commercial airports serving the New York region, plus Chicago O'Hare and Washington National. Slots were allocated by a scheduling committee composed of representatives from the airlines serving the slot controlled airports. The slot limits were seen as a temporary fix when congestion was severe and capacity expansion could not provide timely relief. The limits were continually extended, however, while allocations methods were adapted to industry developments such as airline deregulation and the emergence of new entrant carriers. They have come to be accepted as an acceptable (though not desirable) means of managing congestion at the five original high density airports, but nowhere else.

The reluctance to use slot controls in the US is also evidence in the hourly limits on the number of slots. While the level of operations at European airports is set at a level consistent with poor weather conditions (IMC – instrument meteorological conditions), for those airports with slot controls in the US, it is set close to the good weather capacity (VMC – visual meteorological conditions). This liberal policy allows more maximum exploitation of available capacity on good days, but at the cost of severe disruptions on days with adverse weather. Thus, comparing US and Europe leads immediately to two questions:

1. Should more US airports have slot controls or alternatively are slot controls used too extensively in Europe?
2. How should the level of operations be set at slot controlled airports?

A recent paper, Odoni et al. (2009), sheds light on these questions by providing a detailed comparison of Newark Liberty International Airport (EWR) and Frankfurt/Main International Airport (FRA). These two airports have strong physical similarities so the comparison seeks to highlight differences that are generic to differences in the US and European approaches. Evidence certainly indicates that traffic levels at EWR have been allowed to grow too large (in fact, the study considered a time period prior to the most recent imposition of slot controls). The general approach in the US of having fewer slot controls at higher levels is termed "laissez-faire". It is argued that this approach had led to excessive delays at times. On the other hand, the more conservative European approach could at times have led to under-utilization of available capacity. When considering the alternatives of setting slot levels close to VMC capacity versus IMC capacity, one must also take into account differences in climate, that is, the percentage of time such conditions exist.

Once a decision is made to institute slot controls, one is faced with the questions of exactly how to institute such controls. In both the US and Europe so-called administrative measures have been used. These have granted a form of ownership of the slots to air carriers based on their historical use of the airport. This ownership remains in place as long as those carriers use their slots with a certain minimum frequency (use-it or lose-it rules). Such rules tend to preserve airline schedules and market shares at an airport, create entry barriers, and over time may make slot use suboptimal. In fact, one argument in favor of the relatively sparing use of slots in the US is that administrative slot controls reduce competition. At the same time, it may be possible to use market mechanisms in conjunction with slot controls, for example, a combination of slot controls and auctions, to gain the benefits of slot controls while maintaining a competitive airport environment. This approach to demand management was taken in the recently proposed US rule makings for the New York area airports as discussed next.

Congestion pricing has also been proposed as means of curtailing demand at busy times. While the theory of congestion pricing seems as applicable to airports as it is to other transport facilities, in practice there are special challenges to implementing it in this setting. Most proposals have involved setting an additional per flight surcharge for operations at certain times of day. There are legal obstacles to such surcharges because federal law requires that aeronautical charges by airport operators be set on a cost-recovery basis. Conceptually, one should set congestion charges to bring the level of scheduled operations in line with capacity. Doing so requires that one estimate ex ante the relationships between the prices and the level of operations: in general this is a difficult problem. Furthermore, as charges are increased in one time window demand will move to another making the price setting problem multi-dimensional.

Congestion surcharges have been analyzed starting with Levine (1969), who argues this approach would be better than administrative slot allocation. Morrison (1983) proposes a methodology for simultaneously optimizing landing charges and investment levels. Doganis (1991) examines the airline schedule impacts from peak pricing at London Heathrow Airport, and finds that they accorded with expectations. A more recent evaluation by Schank (2005), however, finds that, in practice, peak runway pricing faces significant institutional barriers, in particular small aircraft operators who claim that such policies "discriminate" against them.

Several researchers have compared the effectiveness of pricing and slot controls. Brueckner (2009) finds that atomistic pricing, which charges each flight its marginal congestion cost even though some of that cost is borne by flights of the same airline, is less efficient than slot controls so long as the number of slots is optimally chosen. Czerny (2010) argues that demand and congestion cost uncertainty, which may lead to a suboptimal choice for the number of slots or the congestion price, favors congestion pricing, that is, the pricing errors that result from imperfect information are less harmful than errors in setting the number of slots. Ball et al. (2007) report on gaming simulations of congestion pricing and slot auction policies for New York LaGuardia. While the simulation indicated that both schemes are feasible, it confirmed the challenge of setting congestion prices. At the same time, congestion pricing was seen to have certain advantages, including increased carrier scheduling flexibility, and reduced incentive for airlines to hoard slots.

Traditional administrative slot controls, set at the proper level, can certainly lead to a socially optimal level of operations and congestion. However, administrative controls have the potential disadvantage of stifling competition by preserving a pre-existing market structure through grandfather rights. This disadvantage can be eliminated in concept through an effective secondary market. A secondary slot market has existed in the US for many years. Analysis of this marketplace, however, has shown it to be at best, only partially effective (Fukui 2010), and, at worst, a near-failure (Berardino 2007, 2009). In particular, it is shown that nearly all major transactions have been carried out under conditions of bankruptcy or seller distress. Using an analytic model (Verhoef 2010), Verhoef considers both congestion charges and a secondary slot market. He concludes that since airlines, particularly dominant carriers, internalize some congestion costs, congestion charges should not be set at the true marginal cost level. Further, it is shown that secondary markets must be designed with care as it is possible under broad conditions that monopoly or dominant carriers will be able to increase their dominance. Recently a secondary market for airport slots has been allowed in Europe. Various analyses have been carried out to gain an understanding of the potential impact on this new marketplace (see for example, De Wit and Burghouwt 2007).

The impact of any strategy for reducing peak period airport traffic will be mediated by individual airline scheduling decisions. The decisions are very complicated, since they involve management of expensive and highly constrained resources, including aircraft and flight crew. Nonetheless, the published schedule is widely recognized to be "the single most important product of an airline" and is usually viewed as the starting point in a sequential process that also includes fleet assignment, maintenance routing, and crew scheduling (Barnhart and Cohn 2004). While offering a schedule that is convenient to passengers is of primary concern, competitive factors can distort this process to some degree, in a manner analogous to the Hotelling (1929) ice cream vendor problem, wherein competing vendors maximize their market shares by both locating at the center of the beach. Borenstein and Netz (1999) and Salvanes et al. (2005) find that schedules in more competitive airline markets tend to be more clustered as competing airlines seek to match one another's schedules.

10.2.2   Recent Public Policy Initiatives in the US

While slot controls at certain US airports have existed since the institution of the High Density Rule (HDR) in 1969, the passage of the Wendall H. Ford Aviation Investment and Reform Act (AIR-21) in 2000 marks the starting point for recent policy making. AIR-21 called for the elimination of slot controls at New York's John F Kennedy International Airport (JFK) and LaGuardia Airport (LGA) by January 1, 2007 and at Chicago O'Hare Airport (ORD) beginning July 1, 2002. Although AIR-21 called for the elimination of slot controls at all New York airports, the Federal Aviation Administration (FAA) anticipated the potential need to replace the HDR rules and caps with an alternative. Ball et al. (2007) describe the results of a broad research program funded by the US Department of Transportation and the FAA that investigated several issues related to airport congestion management. The research, which included two multi-day strategic simulations that brought together major stakeholders from industry and government, defined and analyzed various mechanisms for limiting airport demand including the use of slot auctions.

In a proposed 2006 rulemaking (FAA 2006), the FAA sought to require airlines serving LGA to maintain a certain average gauge (seat capacity). Airlines failing to attain the average gauge standard would lose slots for their smaller-gauge flights until the standard was attained. The proposal was based on the idea that larger aircraft would allow more passengers to benefit from the limited available slots. Several airlines, as well as the operator of the New York airports (the Port Authority of New York and New Jersey), were strongly critical of this approach, however, arguing that it was overly disruptive and prescriptive, and did not take into account airport-specific constraints.

The FAA next proposed a slot allocation policy for LGA, and soon after JFK and Newark Liberty International Airport (EWR), based primarily on grandfather rights, but with auctioning of a limited number of slots (FAA 2008d,b). In its final rule for LGA, each carrier currently holding slots would have lost 15% of its slots in excess of 20 (FAA 2008c). The slots would be relinquished over a five-year period, with two thirds of them auctioned and the remaining one third retired, decreasing the hourly cap from 75 to 71. Similar rules, albeit with relinquishment of 10% of slots in excess of 20 and no retirements, were set forth for JFK and EWR (FAA 2008a). These rules were challenged in court by the Air Transport Association and the Port Authority, who argued that FAA lacked legal authority to conduct slot auctions. The DC Court of Appeals issued a stay delaying the plan, and this, in combination with Congressional action caused FAA to rescind the rule in 2009. An observation from the debate that took place during this time is that the potential benefits of slot controls had not been well-quantified and had not been well-understood by the traveling public. This observation is, in part, the motivation for recent research (Swaroop et al. 2011, Le 2006, Odoni et al. 2009, Vaze and Barnhart 2012), which we review here.

While it was unable to implement the policies just described, the FAA did feel compelled to implement simple caps on the number of operation at each of the three major airports in the New York Region (FAA 2008e,f). These caps remain in effect as of 2011. Neither the setting of caps nor the allocation of slots in these episodes was based on economic analysis. The caps, for example, were not set by comparing the costs and benefits of various slot levels, and slot allocation procedures were not designed to find the highest and best uses of the slots.

10.3   The Fundamental Question: Economic Justification for Slot Controls

While in Europe slot controls are broadly accepted, in the US there is substantial public debate about the widespread necessity of slot controls and reluctance to set slot levels under the airport's VMC capacity. Consequently, it is important to generate compelling economic arguments for the usefulness of slot controls and also to understand the economic tradeoffs for varying the slot levels.

A basic starting point for modeling these tradeoffs is the simple observation that the principal motivation for slot control is reduction of congestion, which is generally manifested by delayed flight departures and arrivals. These are the most evident delays experienced by passengers. We refer to them as *queuing delays* since they result from classic queuing system behavior, with large delay increases resulting as NAS (or airport) demand approaches NAS (or airport) capacity. Thus the economic benefit of slot control is a reduction in the costs associated with queuing delays. On the cost side there are potentially multiple components involving: (1) passenger disutility; (2) operational adjustments within the airlines, for example, aircraft and crew re-scheduling; and (3) an airline's actions in the market to combat likely competitive reactions, for example, re-evaluating load factor and fares on re-scheduled flight.

Probably the most direct cost incurred by slot controls results from the forced reduction in the number of scheduled operations. That is, if the number of available slots in a time window is reduced, then the impacted airlines will either be forced to reduce the number of scheduled operations or move some scheduled operations to less preferred time windows. In the former case, a reduction in the number of scheduled flights implies airlines must reduce the frequency of service they offer in certain markets. Such reductions result in increasing *schedule delay*.

Schedule delay is a well-known phenomenon in transportation systems – first recognized and modeled as a service quality measure in the airline industry by Douglas and Miller (1974). It measures the degree to which passengers must adjust their planned departure time to accommodate the schedule offered by a transportation service. For example, if a passenger

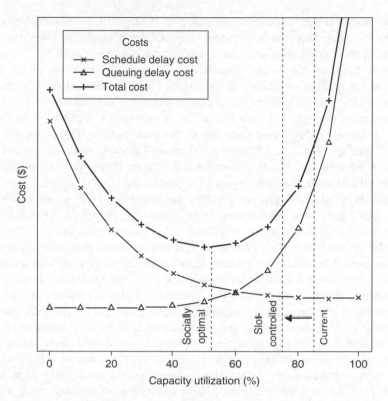

Figure 10.1 Schematic representation of cost curves for queuing delay and schedule delay versus capacity utilization

wished to depart at 9 am but there were only flights offered at 8 am and 10 am, then that passenger might choose the 8 am flight and we would say the passenger suffered one hour of schedule delay. It should be clear that as frequency of service offered in a market decreases, schedule delay increases. It is also intuitive that, in the latter case, when an airline moves a flight to a less desirable time window, most typically the flight would be moved from a higher demand period to a lower demand period, which also increases overall schedule delay.

Figure 10.1 illustrates the basic tradeoff between queuing delay cost and schedule delay cost. The x-axis is given in units of the fraction of available capacity at which the operations level is set, for example, by the hourly slot levels. The queuing delay cost increases rapidly as the airport capacity is approached, while the ex-ante schedule delay cost decreases as the higher service offerings increase the airport capacity utilization. In Figure 10.1, the optimal level of operations – from the passenger welfare perspective – is identified by the minimum point on the total cost curve which denotes the sum of the two cost components. If it is the case that many US airports currently are "over-scheduled", then they would operate at a point far to the right of this optimal point. Verifying this would seem to give strong support for instituting slot control at more airports and for setting the existing slot control levels lower than they are now set. We note that in general it is difficult to estimate schedule delay (and its associated cost), ex post or ex ante, as schedule delay depends on the distribution of preferred passenger departure times. In our related work, Swaroop et al. (2011), we have made progress in this direction. Specifically, we estimate the slopes of the two curves illustrated in Figure 10.1 for several airports using a combination of models. Our results do in fact support the hypothesis that many US airports are over-scheduled. We now summarize these results and the underlying models.

In line with the discussion above Swaroop et al. (2011) describe two alternative models that represent the reaction of airlines to reductions in the available slots in a time window: *FlightMove*, which brings scheduled operations in line with a slot limitation by moving flights to a less-congested time window, and *FlightTrim*, which brings scheduled operations in line with a slot limitation by eliminating some flights. We perform both *FlightMove* and *FlightTrim* based on an "average" day schedule aggregated across Tuesdays, Wednesdays, and Thursdays in August 2007 for each of the 35 "Operational Evolution Partnership" (henceforth OEP35) airports. The OEP35 airports are the largest 35 airports in the US, these are listed in Table 10.1.

Given that the *FlightMove* process involves the interactive behavior of multiple airlines and potentially many cost and revenue considerations, an exact prediction of which specific flights airlines would move is difficult. As a consequence, in the study we adopt a more agnostic approach by randomly generating a series of small perturbations to the existing schedule that eventually yields a new schedule that conforms to the slot limits. As this process is stochastic, we simulate it multiple times and look at the average results. In each simulation run, we randomly select a time period where scheduled flights exceeds the slot level, and use multinomial draws to determine the number of flights to be moved for each market in that period. Each selected flight is then randomly moved either to the neighboring time periods, or allowed to remain in the original time-period. This is repeated until the new schedule meets the slot control restrictions. In this procedure, we find that the schedule delay cost associated with each move is non-linear in the length of a move and the same flight segment could be moved multiple times over the iterations. Simply adding the cost of each move would not be accurate to represent the total *FlightMove* cost. We instead employ a linear programming transportation model that minimizes the total cost of all moves. The results can be viewed as a lower bound of the total cost given the original and new schedules, and the most likely way in which the new schedule would be reached. The whole process is repeated multiple times. Further details of the *FlightMove* can be found in Swaroop et al. (2011).

In the case of *FlightTrim*, we adopt a successive trimming procedure. We divide one operation day (16 hours) into four 4-h time windows. For every city-pair market and 4-h window, we drop one "representative" flight at each step of the trimming procedure if excess demand exists. This flight is a "representative" one since it consists of portions from all markets. The portion for a given market is equal to the proportion of flights in that market over the total number of flights in the same window. Under the assumption of uniform distribution of passengers' preferred departure (arrival) time in each 4-h period and that flights are divisible, the associated passenger schedule delay cost increase for each market and each time period can be calculated. We repeat this until the schedule satisfies the slot restrictions. The passenger schedule delay costs are then summed over different trimming steps, markets, and time windows, to produce the total passenger schedule delay cost increment. For the purpose of preserving service to small communities, markets with less than three flights in a given time window are exempted from the *FlightTrim* process.

The passenger schedule delay cost increment resulting from either *FlightMove* or *FlightTrim* must then be compared with the passenger queuing delay savings that result from the schedule adjustments. To quantify these savings an econometric airport delay model is developed. To do this, we first use the FAA Aviation System Performance Metrics (ASPM)

Table 10.1 List of the US Operational Evolution Partnership (OEP) 35 airports

Airport code	Airport	City	State
ATL	Atlanta Hartsfield Intl	Atlanta	GA
BOS	Boston Logan Intl	Boston	MA
BWI	Baltimore-Washington Intl	Baltimore	MD
CLE	Cleveland Hopkins Intl	Cleveland	OH
CLT	Charlotte Douglas Intl	Charlotte	NC
CVG	Cincinnati-Northern Kentucky Intl	Covington-Cincinnati, OH	KY
DCA	Washington Reagan Natl	Washington	DC
DEN	Denver Intl	Denver	CO
DFW	Dallas-Ft Worth Intl	Dallas-Ft Worth	TX
DTW	Detroit Metropolitan Wayne County	Detroit	MI
EWR	Newark Intl	Newark	NJ
FLL	Ft Lauderdale-Hollywood Intl	Ft Lauderdale	FL
HNL	Honolulu Intl	Honolulu	HI
IAD	Washington Dulles Intl	Washington	DC
IAH	George Bush Intercontinental	Houston	TX
JFK	John F Kennedy Intl	New York	NY
LAS	Las Vegas McCarran Intl	Las Vegas	NV
LAX	Los Angeles Intl	Los Angeles	CA
LGA	La Guardia	New York	NY
MCO	Orlando Intl	Orlando	FL
MDW	Chicago Midway	Chicago	IL
MEM	Memphis Intl	Memphis	TN
MIA	Miami Intl	Miami	FL
MSP	Minneapolis-St Paul Intl	Minneapolis	MN
ORD	Chicago O'Hare Intl	Chicago	IL
PDX	Portland Intl	Portland	OR
PHL	Philadelphia Intl	Philadelphia	PA
PHX	Phoenix Sky Harbor Intl	Phoenix	AZ
PIT	Pittsburgh Intl	Pittsburgh	PA
SAN	San Diego Intl-Lindburgh Field	San Diego	CA
SEA	Seattle-Tacoma Intl	Seattle	WA
SFO	San Francisco Intl	San Francisco	CA
SLC	Salt Lake City Intl	Salt Lake City	UT
STL	Lambert-St Louis Intl	St Louis	MO
TPA	Tampa Intl	Tampa	FL

database to calculate deterministic queuing delay for each of the OEP35 airports on each day in 2007. Deterministic queuing delay is sensitive to flight schedule changes associated with different slot control policies. We then estimate an econometric model of average flight arrival delay with the deterministic queuing delay and its higher order terms as part of the explanatory variables. Other explanatory variables include the portion of time under Instrument Flight Rules (IFR) conditions and its quadratic term, wind speed, temperature, airport acceptance rate (AAR), the number of airports connected to the airport of interest, and a series of airport and monthly time dummies. The model is estimated using the Prais–Winsten procedure with panel corrected standard errors. The estimated results are consistent with the conventional wisdom, and produces good prediction of the observed queuing delay. With this estimated model new flight arrival delays can be predicted for each new schedule generated from *FlightMove* or *FlightTrim*. Specifically, we use the new schedule to calculate the new deterministic queuing delay, and then use the econometric model to predict the corresponding new average flight delay. Comparing the new average flight delay with the predicted original average flight delay gives the average flight delay saving, which is multiplied by the number of passengers per flight and a standard value of travel time, and then summed over all flights, to yield the total passenger queuing delay cost savings. Interested readers can refer to Swaroop et al. (2011) for more details about the model and delay cost computation.

The passenger schedule delay cost increment and queuing delay cost savings are computed and compared for each OEP35 airport under three scenarios, which set slot levels at 80, 90, and 100% of the peak airport capacity averaged over Tuesdays, Wednesdays, Thursdays in August 2007 (this is generally close to the VMC capacity). The peak hours are defined as between 6 am and 10 pm. The peak airport capacity is measured as the maximum declared Airport Arrival Rate (AAR) for the specified days, as reported in ASPM database of the FAA. We assume that as long as *FlightMove* is feasible, airlines would always prefer *FlightMove* to *FlightTrim*, because *FlightMove* preserves baseline demand without requiring changes in fleet mix. Four of the 35 airports justify use of *FlightTrim* model, as explained later. We find that the results from multiple simulation runs in *FlightMove* are very stable with small cost deviations of both schedule delay costs and queuing delay cost savings, suggesting the robustness of our proposed *FlightMove* schedule generating procedure.

We identify good candidates for slot control as those airports with positive net benefits from schedule adjustments (i.e., queuing delay cost saving – schedule delay cost increase > 0) and with daily passenger queuing delay cost saving above $10 000. This leads to 16 airports (ATL, CLE, CLT, DCA, DTW, EWR, IAD, JFK, LAX, LGA, MSP, ORD, PHL, PHX, SAN, SEA) for which slot control is recommended. Thus, we find that slot control should be far more widespread than it is presently.

Our results further reveal that the number of slots should be limited to a number that is often less than the peak airport capacity. Based on the marginal change in schedule delay cost and queuing delay cost, which in a rough sense is equivalent to the slopes of the corresponding curves in Figure 10.1, we find that the maximum net passenger benefits are achieved when slot limits are set at 100% of the airport capacity level for PHL and SAN; 90% for ATL, DCA, EWR, LGA, LAX, ORD, PHX, and SEA; and 80% for CLE, CLT, DTW, IAD, JFK, and MSP. For all airports except EWR, JFK, LGA, and ORD, the limits can be attained from *FlightMove*. When slot controls are imposed at 90% level, at least one of the 4-h time windows at EWR, JFK, LGA, and ORD will encounter insufficient capacity to service the scheduled demand. As a result, *FlightTrim* is applied at 80% and 90% levels at these airports. For the four airports that currently have slot controls (DCA, JFK, LGA, ORD), our analysis suggests the current slot levels are set too high: a slot limit set a 90% of capacity would be most beneficial for DCA, EWR, and LGA; while at JFK the limit should be 80% of capacity.

It is perhaps surprising to have airports, such as CLE, MSP, SAN, and SEA, which are normally not considered highly congested, on the list of logical candidates for slot control. This occurs because of pronounced peaks at these airports which could be reduced by spreading flights to less congested periods when slot control is imposed. This is illustrated in Figure 10.2, which shows schedules for some highly, mildly, and least congested airports.

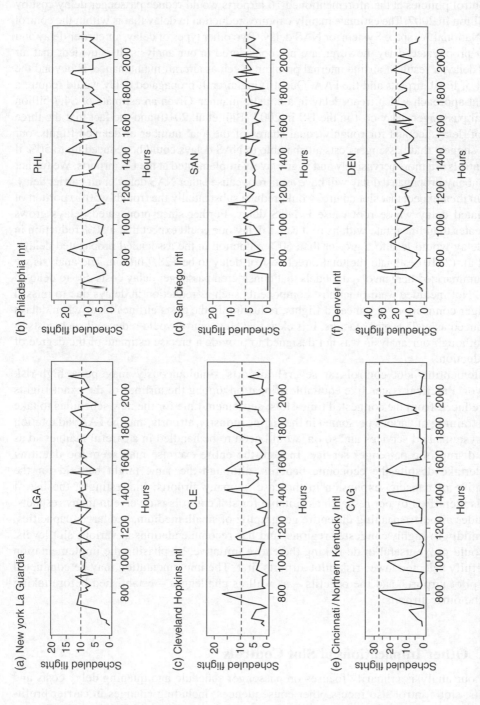

Figure 10.2 August 2007 Aggregated Arrival Schedules for several airports. Congestion levels decrease from top to bottom. The dotted line shows the peak arrival capacity (*AAR*).

As a first-order estimate of annual benefits, we conclude that implementing the best slot control policies at the aforementioned 16 airports would reduce passenger delay cost by $0.8 billion in 2007. The estimate mainly captures reduction in delay that is within the control of the National Airspace System, or NAS delay. Two other types of delay, air carrier delay and aircraft propagated delay, however, are not considered in our analysis. It is evident that air carrier delay, a result of airline internal problems such as aircraft maintenance, is beyond the control of local airports and the FAA. Quantifying aircraft propagated delay would require a different approach which treats delay in a holistic manner. Given an estimate of $4.7 billion for total passenger delay cost in the US in 2007 (Ball et al. 2010) and the fact that the three types of delay account for roughly equal shares of the total number of delayed flights, our results suggest that passenger costs attributable to NAS delays could be reduced up to 50% if slot controls are more pervasively and aggressively implemented at the US airports. We further note that any propagated delay will have as its root cause either NAS delay or air carrier delay. One can then expect that slot control would reduce substantially the (roughly 50%) portion of propagated delay whose root cause is NAS delay. Further since propagated delays grows at a greater than linear rate with its root cause delay one could expect that a 50% reduction in NAS delay would lead to a greater than 50% reduction in the associated propagated delays. Ball et al. (2010) estimate the total direct cost of delay to be $28.9 billion. The analysis we have summarized here involved models that considered passenger delay costs due to delayed flights. Not included were other cost components such as reduction in delays due to missed passenger connections or canceled flights, reduction in the costs airlines incur when delays are reduced as well as other factors. It is clear that slot control would reduce all such costs as well, although our analysis was not designed to provide a precise estimate of the degree of this reduction.

Implementing slot controls at several airports simultaneously may be a high-risk endeavor. Practical issues like equitable allocation among the airlines of the exact flights to be reduced from the congested time-slots; settlement time for the new schedules to take effect; training of various personnel in the airline industry, airports, and the FAA; adaptation of IT systems and services and so on, would need to be handled in a careful manner so as to not disrupt the passenger service. Indeed, the entire exercise may seem too daunting to undertake despite the economic benefits, although the same might be said for the alternative of capacity expansion in the case of many airports. Interestingly, the list of airports identified by our research as suitable for slot controls is diverse in many respects: it includes airports spanning the entire geography; of small, medium, and larger capacities; from mildly to highly congested regions; and has recommendations at various slot levels. This could prove useful in de-risking the entire initiative, by phasing the implementation at carefully selected lower-risk pilot airports first. The implementation may be conducted at the pilot airports, and the benefits – as well as challenges – established before taking upon the other airports.

10.4   Other Implications of Slot Controls

While our analysis primarily focuses on passenger schedule and queuing delay costs and benefits, slot control also incurs other consequences, including changes in carrier profitability, load factor, air fare, and aircraft size. Research by Vaze and Barnhart (2012), and

Le (2006) attempt to understand the impact of slot controls on these variables, focusing on New York LaGuardia (LGA). These studies, at least implicitly, consider the queuing delay – schedule delay tradeoff as well as other impacts and tradeoffs. Both studies attempt to answer the question, "What would happen if more restrictive slot controls were put in place at LGA?" They consider changes in service frequency but also impact on airline costs and profits.

In Vaze and Barnhart (2012), a game theoretical model based on an S-curve relationship between airlines' frequency share and market share is developed that explicitly characterizes frequency competition among airlines servicing LGA. They use the Nash equilibrium concept to predict the outcome of competitive situations under different slot reduction levels at LGA. Two schemes for allocating slots among airlines are considered. The first, Proportionate Allocation Scheme (PAS), distributes slots among different carriers by the same ratio as in the status quo. The PAS is likely to be considered more acceptable by major carriers, but may ignore how efficiently airlines utilize slots. In fact, airlines may differ – often substantially – in the number of passengers carried per flight or per slot. Recognizing this, the authors propose a Reward-based Allocation Scheme (RAS), under which the number of slots allocated to each airline is proportional to the total number of passengers carried by that airline. The RAS provides airlines with incentives to carry more passengers per slot, through either higher load factors or larger planes.

A wide range of slot reduction scenarios at LGA is considered in Vaze and Barnhart (2012). Assuming constant aircraft size, slot reduction reduces flight operations and forces airlines to increase load factor to accommodate as many passengers as possible. This results in a profit increase but only to a certain extent, beyond which profit starts to decline, because after achieving the maximum allowed load factor, loss of flight operations also means loss of revenue. At low slot reduction levels, airline profit gain and passenger loss bear similar changing patterns under PAS and RAS. When slot control is more substantial, RAS gives higher profit and less passenger loss than PAS.

The authors then focus on a more realistic scenario which reduces slots by 12.3% – roughly corresponding to scheduling at IMC ("bad-weather") capacity instead of VMC (currently practiced "good-weather") capacity at LGA. Consistent with our work, the authors conclude that this would result in a significant reduction in flight and passenger delays, but a very small schedule delay increase. In addition, a small reduction in total allocated capacity can considerably improve operating profit of all incumbent carriers. By allowing aircraft to partially upgauge, the loss of passengers can be reduced significantly. The authors also observe that while profit increase may be different for each individual airline under PAS and RAS, the two schemes produce similar aggregate impacts.

Two of the assumptions made in Vaze and Barnhart (2012) are that air fare and aircraft size would remain unchanged irrespective of slot control policies (although the latter is partially relaxed in their sensitivity analysis). The changes in these two factors are explicitly investigated in Le (2006) using a different approach. In her thesis, Le points out that the causes of airport congestion include (1) the HDR with grandfather rights that allocates limited slots to incumbent airlines; (2) no incentives for airlines to use larger aircraft due to weight-based landing fees; (3) slot exemptions granted to small markets served by 70-seat or less aircraft; (4) the 80% use-it-or-lose-it requirement which forces airlines to fly low load-factor flights. Instead of explicitly modeling airline competition, Le (2006) assumes a single "benevolent" airline that reacts to price elasticities of demand in competitive markets. She develops an optimization model that solves both an airline scheduling sub-problem and an airport allocation problem, and a stochastic queuing network simulation model to quantify the delay effect. Taking profit maximization by the benevolent airline and seat maximization by the government as two separate objectives, Le demonstrates the existence of profitable flight schedules at LGA that is able to accommodate passenger demand while substantially reducing flight delays.

Several metrics, including air fare, aircraft size, flight delay, and total number of seats and markets serviced, are examined by Le under multiple scenarios. She finds that, if profit maximization is the only goal, seat throughput will be significantly lower than under the throughput-maximizing scenario. Carriers tend to consolidate flights, increase aircraft size and fare, resulting in fewer passengers transported and lower airport delays given the same slot control level. Since the goal of profit seeking conflicts with that of maximizing enplanement, Le (2006) further examines two compromise scenarios by setting 80% and 90% of the unconstrained maximum profit as lower bounds on the profit of the benevolent airline, and considers such scenarios to be (1) close enough to the baseline to provide a feasible transition solution; (2) reasonably close to the optimal profit curve. Compared to the baseline, the compromise scenarios predict positive changes in total number of seats, aircraft size (assuming constant load factor), and negative changes in average fare, flight traffic, and substantial reduction in flight delay. She also looks at the number of profitable markets on a daily schedule, and finds no penalty in the number of markets with slot allocation at 8 ops/runway/15-min compared to 10 ops/runway/15-min. However, with tighter slot limits below 8 ops/runway/15-min, an increasing number of markets will become unprofitable. She concludes that having aggregate airline schedules at 8 ops/runway/15-min, which is the current IMC operation rate, would significantly reduce the congestion problem at LGA, increasing the predictability of air transportation and improve the quality of service expected by the flying public.

Despite these efforts in modeling cost and benefits from slot controls, there remain some gaps in the economic justification for slot controls that warrant further research. One is fare and competition effects. Although fare change is captured in an aggregate manner in the preceding study, modeling of the impact of slot control on fare in an explicit competitive environment is rarely seen. Conceptually, if operations are restricted in some way, then resource scarcity may result leading to higher fares. Further, to the extent that such restrictions allow one or more air carriers to increase market power, this could move fares even higher. These effects are often cited as a major detriment of market-based airport access controls. Among other challenges in addressing this question is that the degree to which there is an anti-competitive effect depends very much on how controls are implemented. For example, administrative slot controls that are based primarily on grandfather rights, would tend to preserve existing market structure and would restrict new entrants from entering the markets served by the airport. Mechanisms that allowed for some slot reallocation, for example, via auctions, would support a more vibrant competitive environment and lower fares.

An often cited concern related to the imposition of slot control is loss of service to small communities or markets. From the schedule delay perspective, loss of service represents the extreme case for increase in schedule delay (frequency goes to 0). In our work we deliberately eschew removing flights in markets that are very sparsely serviced. On the other hand, as mentioned in Le (2006), preserving small community service represents one cause for inefficient use of slots and airport congestion. Her results also reveal profitability concerns associated with small community markets. All these make small community access an important issue in the design of slot controls.

10.5   Design Issues for Slot Controls

10.5.1   Getting the Slot Level Right

Runway systems of airports have capacities and these capacities are in fact what lead to the queuing effect described previously and the need for access controls. However, the capacity of a runway system depends on many factors which can vary substantially day to day or even hour to hour. The most obvious variation is caused by changes in weather conditions. A very basic division in airport weather conditions is *VMC* versus *IMC* as discussed earlier. The degree to which capacities vary between VMC and IMC depends on airport specifics; an extreme case is San Francisco International Airport (SFO), where IMC arrival capacity is approximately one half VMC arrival capacity.

These considerations imply that careful modeling, such as what was discussed in Section 10.3, should be employed to determine the appropriate slot level; for example, it could easily be the case that the "optimal" slot level would allow some delays during poor weather days. Conversely, as shown in Section 10.3, the appropriate slot level is usually somewhat less than the airport capacity under ideal conditions. In general, the slot level should depend on the frequency of poor weather conditions. Thus, the best solution should depend on both the VMC and IMC capacity values and the relative frequency with which such conditions exist. These factors were implicitly considered in the analysis summarized in Section 10.3, since the delay arising from a given schedule is estimated over an entire month, with varying weather conditions and capacity levels. If slot levels are to be treated as essentially constant over many months or years, however, that analysis would need to be extended to cover a commensurate time period. On the other hand, there may also be benefit to varying the slot level by time of day, for example, by allowing more slots during high demand/high value periods and compensating with "cooling off" periods. (Churchill et al. 2012) describes optimization models for setting slot levels that may vary by time of day taking into account capacity scenarios and their likelihood and also variations in slot value by time of day.

It is also the case, however, that modeling challenges could be insignificant compared to political challenges related to setting appropriate slot levels. The key question is who has the authority and what criteria are used to do so. Since changes in the level of operations could have a substantial economic impact on various air carriers as well as the traveling public, the person or group who sets slot levels should be insulated from political pressures.

10.5.2   Small Community Access

Smaller communities can derive substantial value from regular service to large airports. Many of the FAA proposals for allocating slots have included specific features to protect service to small communities, and those opposing these proposals have routinely cited their adverse impact on small communities (FAA 2006 – see p. 10; Port Authority of New York and

New Jersey 2008 – see p. 4). Thus, politically acceptable airport demand management schemes may have to demonstrate their ability to insure adequate access to designated small communities. Both the review of recent public policy initiatives in Section 10.2.2 and the research reviewed in Section 10.3 highlight the importance of this issue.

10.5.3   Where Does the Money Go?

Market-based approaches may generate substantial new revenues. Obviously, how this money is spent will be of great interest to the various stake-holders. Probably the least desirable outcome from the perspective of major stake-holders such as air carriers and airport operators is that the new money goes into the general fund of the Federal Government. Legal assurances that such funds are used to enhance aviation infrastructure certainly would enhance the acceptability of any proposal. Most desirable would be that such funds are used to benefit operations at or around the airport in question. Other desirable features could be that such funds replace existing user fees or taxes such as landing fees.

In its proposal to auction a limited number of slots at LaGuardia, FAA identified two options for spending the proceeds. The first was to use them to "mitigate congestion and delay in the New York area," while the second was to give them to the carrier holding the slot that was being auctioned (FAA 2008d, see p. 1).

10.5.4   Federal versus Local Control

One might naturally assume that airport access controls would be implemented by the airport operator and that any associated revenues would go to the airport operator. However, there are many reasons this might not be the case. Airports are natural monopolies and therefore are generally highly regulated. Their revenues are almost always restricted so that only cost recovery is allowed. Thus, market based approaches that may generate revenues of arbitrary size would almost by necessity have to be implemented by another entity – most likely the Federal Government. In the US, while the airport operator controls the surface of the airport, the FAA has legal authority to control the airspace and, thus, there are strong arguments that it has right to control airport access using various means including market mechanisms (the extent of this right is subject to legal debate however). Clearly, there is a potential tension related to the question of Federal or local control and this must be managed in devising any access control solution. Taking an objective perspective, the FAA has responsibility for the efficient operation of the entire National Airspace System (NAS) and may need to control access to individual airports with a national perspective in mind. The airport operator has knowledge of its airport's characteristics and can implement and fine tune local controls. Thus, an ideal solution should take both of these perspectives into account and design an appropriate solution.

This is easier said than done, however. For example, despite consultation between the FAA and the Port Authority of New York and New Jersey in the development of the FAA's proposal to auction slots a LaGuardia, the Port Authority asserted flight operations using slots obtained through an auction "shall not be conducted" and that it would not consent to the leasing or use of terminal space for flights that used such slots (Port Authority of New York and New Jersey 2008, see p. 5).

10.5.5 Who Can Own Slots?

Assuming the use of a slot-based system one may be faced with the question of who can own slots. In fact, although market mechanisms have never been used to perform a primary allocation of slots from the Federal Government to the airports, a secondary market for slots does exist in the US. Not surprisingly slots can have substantial economic value. In fact, certain banks, as a result of airline bankruptcy proceedings, have become the owners of slots. This suggests a more general question of who should be allowed to own slots. For example, if a slot auction were held, should bidders be restricted in any way. It could be that "brokers" may wish to bid on slots and then resell them or lease them over short periods of time. There could in fact be economic value in a market for short term slot leases. Another scenario could involve a local community buying slots so as to insure access between that community and the major airport in question. Thus, there could be sound societal reasons for allowing non-air-carriers to own slots but, on the other hand, there is the potential for various unknown (and unintended) consequences, such as airport opponents purchasing the slots in order to retire them.

10.5.6 International Bilateral Agreements

Scheduled air transportation service between a pair of countries can only be conducted when authorized under a formal bilateral agreement between the countries. Such bilateral agreements typically grant access to an international city pair market to designated air carriers. Slot control regulations must be constructed in a way so as not to violate such agreements.

10.5.7 Infrastructure Investment Incentives

A well-structured air transportation system should provide an appropriate signaling mechanism to indicate when investment in new infrastructure is required. In a system devoid of slot controls or congestion pricing, the typical signal is the presence of (possibly extreme) delays. The implementation of slot controls could eliminate this signal and not replace it at all or replace it with a different (possibly better) signal, for example, high slot prices. To the extent that recurring fees, for example, from congestion pricing, depend on scarce capacity, an incentive is provided to the collectors of the revenue to not invest in new capacity. Obviously, such a situation would be undesirable and should be avoided. Appropriate incentives for infrastructure investment should be an important consideration in the design of any congestion management approach.

10.6 Conclusions

This chapter provided an overview of models that provide economic justification of slot controls and also discussed a number of design issues. We summarized our own research, which considered the tradeoff between queuing delay and schedule delay. This tradeoff is fundamental to determining the need for slot control and the optimal slot control level and our results provide strong justification for the more extensive use of slot control in the US. Other related research, specifically focused on LGA airport, supports these results and further considers other issues such as the impact on airline profits and overall passenger performance. More aggressive use of slot control at LGA is shown to substantially improve overall social welfare without disadvantaging any major constituency.

Experience and feedback from recent US Federal Government proposed rulemakings was used to identify various political and design challenges related to the implementation of slot control in the US. We provide perspectives on these challenges and, in some cases, give approaches to their resolution.

References

Ashley, K. and Savage. I. 2010. Pricing congestion for arriving flights at Chicago O'Hare Airport. *Journal of Air Transport Management* **16**(1) 36–41.

Ball, M. O., Ausubel, L. M. Berardino, F. et al. 2007. Market-based alternatives for managing congestion at New Yorks LaGuardia Airport. *Proc. of AirNeth Annual Conference*.

Ball, M. O, Barnhart, C., Dresner, M. et al. 2010. Total Delay Impact Study: A Comprehensive Assessment of the Costs and Impacts of Flight Delay in the United States. *NEXTOR Technical Report, October 2010* URL www.nextor.org (accessed February 20, 2013).

Barnhart, C. and Cohn. A. 2004. Airline schedule planning: Accomplishments and opportunities. *Manufacturing & Service Operations Management* **6**(1) 3–22.

Berardino, F. 2007. History of the slot exchange market in the US and some implications. *Proc. of NEXTOR Workshop: Allocation and Exchange of Airport Access Rights, June 6–8, 2007*. Aspen, Colorado. URL www.nextor. org (accessed February 20, 2013).

Berardino, F. 2009. New US airport slot policy in flux. *Journal of Transport Economics and Policy (JTEP)* **43**(2) 279–290.

Borenstein, S. and Netz, J. 1999. Why do all the flights leave at 8 am?: Competition and departure-time differentiation in airline markets. *International Journal of Industrial Organization* **17**(5) 611–640.

Brueckner, J. K. 2009. Price vs. quantity-based approaches to airport congestion management. *Journal of Public Economics* **93**(5–6) 681–690.

Carlin, A. and Park, R. E. 1970. Marginal cost pricing of airport runway capacity. *The American Economic Review* **60**(3) 310–319.

Churchill, A. M., Lovell, D. J. Mukherjee, A. andBall, M. O. 2012. Determining the number of airport arrival slots. *Transportation Science* at press, DOI 10.1287/trsc.1120.0438.

Czerny, A. I. 2010. Airport congestion management under uncertainty. *Transportation Research Part B: Methodological* **44**(3) 371–380.

De Wit, J. and Burghouwt, G. 2007. The impact of secondary slot trading at amsterdam airport schiphol. SEO economic research. Tech. Rep. SEO-rapport nr. 957, The Netherlands Ministry of Transport, DGTL, Amsterdam. Available online at http://www.seo.nl/uploads/media/957_The_impact_of_secondary_slot_trading_at_Amsterdam_Airport_Schiphol.pdf (accessed February 13, 2013).

Doganis, R. 1991. Interaction of airport congestion and supply. *Longer Term Issues in Transport*. Avebury, Aldershot.

Douglas, G. W. and Miller, J. C. 1974. *Economic Regulation of Domestic Air Transport: Theory and Policy*. Brookings Institution.

FAA. 2006. Congestion Management Rule for LaGuardia Airport, Notice of Proposed Rulemaking, August 29. *Federal Register* **71**(167) 51360–51380.

FAA. 2008a. Congestion Management Rule for John F. Kennedy International Airport and Newark Liberty International Airport, Final Rule, October 10. *Federal Register* **73**(198) 60544–60571.

FAA. 2008b. Congestion Management Rule for John F. Kennedy International Airport and Newark Liberty International Airport, Notice of Proposed Rulemaking, May 21. *Federal Register* **73**(99) 29626–29651.

FAA. 2008c. Congestion Management Rule for LaGuardia Airport, Final Rule, October 10. *Federal Register* **73**(198) 60574–60601.

FAA. 2008d. Congestion Management Rule for LaGuardia Airport, Supplemental Notice of Proposed Rule-Making, April 17. *Federal Register* **73**(75) 20846–20868.

FAA. 2008e. Operating Limitations at John F. Kennedy International Airport; Order Limiting Scheduled Operations at John F. Kennedy International Airport, Notice, January 18. *Federal Register* **73**(13) 3510–3542.

FAA. 2008f. Operating Limitations at Newark Liberty International Airport; Order Limiting Scheduled Operations at Newark Liberty International Airport, Notice, May 21. *Federal Register* **73**(99) 29550–29566.

Fukui, H. 2010. An empirical analysis of airport slot trading in the United States. *Transportation Research Part B: Methodological* **44**(3) 330–357.

Hansen, M. 2002. Micro-level analysis of airport delay externalities using deterministic queuing models: a case study. *Journal of Air Transport Management* **8**(2) 73–87.

Hotelling, H. 1929. Stability in competition. *The Economic Journal* **39**(153) 41–57.

Le, L. 2006. *Demand Management at Congested Airports: How Far are we from Utopia?* PhD thesis, George Mason University, Fairfax, VA.

Levine, M. E. 1969. Landing fees and the airport congestion problem. *Journal of Law and Economics* **12**(1) 79–108.

Morrison, S. A. 1983. Estimation of long-run prices and investment levels for airport runways. *Research in Transportation Economics* **1** 103–130.

Odoni, A., Morisset, T., Drotleff, W. and Zock, A. 2009. Benchmarking airport airside performance: FRA vs. EWR. *Proceedings of the 9th USA/Europe Air Traffic Management R&D Seminar.*

Port Authority of New York and New Jersey. 2008. Notice of Proposed Action, August 4.

Salvanes, K. G., Steen, F. and Sorgard, L. 2005. Hotelling in the air? Flight departures in Norway. *Regional Science and Urban Economics* **35**(2) 193–213.

Schank, J. L. 2005. Solving airside airport congestion: Why peak runway pricing is not working. *Journal of Air Transport Management* **11**(6) 417–425.

Swaroop, P., Zou, B., Ball, M. O. and Hansen. M. 2011. Do more U.S. airports need slot controls? A welfare based approach to determine slot levels Unpublished manuscript.

Vaze, V. and Barnhart, C. 2012. Modeling airline frequency competition for airport congestion mitigation. *Transportation Science* **46**(4) 512–535.

Verhoef, E.T. 2010. Congestion pricing, slot sales and slot trading in aviation. *Transportation Research Part B: Methodological* **44**(3) 320–329.

Index